SHARING GEOGRAPHIC INFORMATION

SHARING GEOGRAPHIC INFORMATION

Edited by
HARLAN J. ONSRUD
and
GERARD RUSHTON

CENTER FOR URBAN POLICY RESEARCH

New Brunswick
New Jersey

Published by the Center for Urban Policy Research
New Brunswick, New Jersey 08903

Printed in the United States of America

Cover design: Helene Berinsky

Cover graphic: Detail from "Concept for a Multi-Purpose Land Information System," by N. R. Chrisman, B. J. Niemann, Jr., J. Sullivan, and J. Sonza-Novera. © 1984 *Wisconsin Land Information Newsletter* 2, 2 (insert), Land Information and Computer Graphics Facility, College of Agricultural and Life Sciences, University of Wisconsin–Madison.

Library of Congress Cataloging-in-Publication Data

Sharing geographic information / edited by Harlan J. Onsrud and Gerard Rushton.
p. cm.
Includes bibliographical references.
ISBN 0-88285-152-7

1. Geographic information systems. I. Onsrud, Harlan Joseph. II. Rushton, Gerard.
G70.2.S54 1995
910'.285–dc20 94-19192
CIP

Contents

SECTION II
Experiences in Sharing Geographic Data

SECTION III
Economic, Legal, and Access Issues in Sharing Geographic Data

Preface

This book is based on the premise that substantial societal needs may be better addressed through increased sharing of geographic information. The ability and willingness to share geographic information are affected by the behavior and needs of individuals, organizations, and institutions and are subject to technical constraints. This volume describes past successful and unsuccessful experiences in sharing geographic information and pursues the search for formalisms and general principles that could enhance the ability and willingness to share.

The book consists primarily of papers prepared for a specialist meeting on "Institutions Sharing Geographic Information" hosted by the National Center for Geographic Information and Analysis in February 1992 in San Diego. Prior to the specialist meeting, a core planning group with representation from six universities prepared a background paper laying out the issues to be addressed at the specialist meeting and issued a public call for papers. Participants in the specialist meeting were selected through the refereeing of submitted paper proposals by the core planning group, and completed papers were subjected to a further peer review process. A few of the papers in this volume were explicitly invited or written after the specialist meeting in order to address issues not adequately covered at the workshop but necessary for comprehensive coverage of the topic.

The volume brings together contributions from the public, private, and academic sectors. Among the authors are those drawn from five different user segments who regularly share geographic data with other institutions. These user segments include federal government, state and regional government, local government, Geographic Information Systems (GIS) consultants familiar with the needs of users, and private-sector, value-added spatial data suppliers. Another group of authors consists of individuals from the academic community who specialize in the methodological questions that arise in research related to the sharing of geographic data. These include researchers in the fields of organizational theory, management information systems (MIS), and behavioral theory. A final group of authors was drawn from the GIS research and academic communities having interests in behavioral and organizational issues relating to the sharing of geographic information.

This book has two primary purposes. One is to provide insights on opportunities, options, and potential pitfalls for those organizations and individuals seeking to gain from the sharing of geographic information. The second purpose is to discuss and recommend areas of research likely to be fruitful in addressing both near- and long-term problems in the sharing of geographic information. Our hope is that improved insights drawn from experience and expanded knowledge derived from research will result in improved prescriptive strategies and prospective models for sharing geographic information.

Acknowledgments

This book represents the efforts and contributions of many people. It was an honor and a pleasure to work with each and every author. We joined with many of them for several days of presentations, writing sessions, and discussions at a specialist meeting on "Institutions Sharing Geographic Information" in February 1992 that was funded and administered by the National Center for Geographic Information and Analysis (NCGIA). We thank them for their energy and enthusiasm at that meeting and for their constructive efforts in reviewing the work of their fellow authors and for their persistence in revising their own works. In addition to their individual chapter contributions, we credit the specialist meeting participants with identifying many of the research issues discussed in the final chapter of this book.

We wish to give special thanks to the core planning group for the specialist meeting which, in addition to the editors, included Hugh Calkins, Earl Epstein, Jack Estes, Jeffrey Pinto, and Lyna Wiggins. The introduction chapter of this book is drawn from the ideas and contributions of these people. Their ideas also shaped the research topics addressed at the specialist meeting and the discussion format used at that meeting.

We also extend a hearty thanks to Kathleen Hornsby and Blane Shaw. These individuals not only carried out behind-the-scenes administrative work for the specialist meeting but worked carefully and cheerfully in the preparation of this manuscript.

Finally, we wish to thank and acknowledge the support of the NCGIA. The NCGIA is a consortium of the University of California – Santa Barbara, the State University of New York – Buffalo, and the University of Maine. The Center is supported by a grant from the National Science Foundation (SES-88-10917), which we gratefully acknowledge.

Harlan J. Onsrud and Gerard Rushton

Sharing Geographic Information: An Introduction

Geographic information is used to satisfy many needs. Members of government, business, and the general public depend on geographic information for practical decision making on a daily basis. Likewise, many scientists and educators have a frequent need to access geographic information.

The use of automated geographic information processing capabilities to aid in the direct and immediate solution of location-related problems in business and government is now almost commonplace. In addition, improved geographic information handling capabilities increasingly are being applied in addressing a wide range of more general social problems. For instance, geographic information systems are already being applied in addressing such problems as critical habitat destruction, infrastructure deterioration, soil erosion, drug use and trafficking, groundwater pollution, emergency response, hazardous materials disposal, incidence and spread of diseases, global weather changes, urban sprawl, acid rain, depletion of scarce resources, spreading of deserts, overpopulation, and many similar problems.

As local land-related problems and the broader societal problems grow and become more critical, the need for collecting, analyzing, and communicating information about the land and its resources will increase. In dealing with such problems, it becomes readily apparent that any parcel of land and the resources and cultural features tied to it are of interest to many individuals and institutions.

Technological advancements have had a substantial effect on the ability to use and share geographic information. Until recently, referencing methods for geographic information were crude and not organized to meet standards of quality and accuracy. Few location-specific questions could be asked. Locations had names and information relevant to locations was assembled by those names. So inflexible were such "geographic identifiers" that little demand existed for sharing spatial data. In the past ten years, however, with new ways of encoding geographic information and with the development of new ways of manipulating it, great interest has arisen in sharing geographic information. The concept of using custodians for widely useful thematic layers of geographically referenced data has been widely accepted. As users became accustomed to this concept of depending on others for the collection and maintenance of at least some sets of digital geographic data, a second concept extended from the first: the concept of a basic spatial data infrastructure that details the minimum geographically referenced information set to which all other thematic data may be attached. Defining this set and the institutions and mechanisms for maintaining it, and guaranteeing practical access to it, is a task that is far from complete.

The value and social utility of geographic information comes from its use. Sharing of geographic information is important because the more it is shared, the more it is used, and the greater becomes society's ability to evaluate and address the wide range of pressing problems to which such information may be applied. Failure to share geographic information is also economically inefficient and wasteful. The expertise and time it takes to collect and maintain information about land creates a need to share that information.

However, enabling the sharing of geographic information among multiple parties with varying interests on a widespread and continuing basis is far from simple. The flow of data and information may be viewed as being controlled by incentives and impediments. Some of the impediments to sharing geographic information are technical. Improvements in technical capabilities have driven much of the desire to access and share geographic information, and further technical improvements that will enhance the ability to share are underway. Among these improvements are standardizing hardware interfaces, developing spatial data transfer standards, compiling general purpose spatial databases, and building networked infrastructures capable of handling large amounts of geographic data.

Other impediments and many of the incentives for sharing are cultural, reflecting people's behavior and the behavior of the organizations to which they belong. When attempts are made to share spatial data among organizations or among the divisions of a single organization, those involved often report that the most significant impediments to sharing are institutional, organizational, or behavioral in nature rather than technical. For instance, many organizations have developed agency or corporate databases in support of their organization's primary missions but few have been willing to freely allow others outside their organization to access and copy their databases or to allow others to contribute data to them. Lack of clarification of the personal, organizational, and economic risks and rewards involved in sharing are indicated as likely substantial factors affecting this result. Technical capabilities and safeguards that could readily allow transfer and sharing may already exist but typically are not provided or facilitated due to one or more organizational, institutional, or personal impediments.

In this book, we start with the premise that an embryonic national spatial data infrastructure now exists in the United States. To facilitate sharing, the Geographic Information Systems (GIS) user and research communities must deal with both the technical and institutional aspects of collecting, structuring, analyzing, presenting, disseminating, integrating, and maintaining spatial data. Technical issues and their impact on sharing are discussed in several of the chapters in this volume. However, the primary thrust of the book is toward exploring the behavioral, organizational, and institutional considerations acting as incentives or impediments to the sharing of geographic information among and within organizations. As such, the book focuses on complex organizations and attempts to describe and discuss factors and strategies that appear to lead to successful spatial data sharing experiences as well as those that inhibit such sharing.

The results of research on behavioral and organizational issues that affect decision making in other fields promise to be useful in studying the sharing of geographic information and in developing strategies for optimizing the use of geographic information. A substantial body of literature addressing such issues already exists and must be consulted. However, our understanding of behaviors and activities needs to be considered in venues beyond the library, classroom, or study area. In particular, there is a need for observations in real settings of the process of accessing and using shared spatial data. Existing settings in which the sharing of spatial data can be observed might be categorized by institutional setting (e.g., sharing throughout a municipal government, as fostered by a state council on geographic information, among federal agencies, or between an agency and private industries), type of spatial data shared (e.g., sharing digital line graphs [DLG], Topologically Integrated Geographic Encoding and Referencing [TIGER] files, Earth Observing Systems [EOS], or local databases), or type of problem being addressed by the decision maker (e.g., sharing in order to select a suitable building site, to identify critical habitats

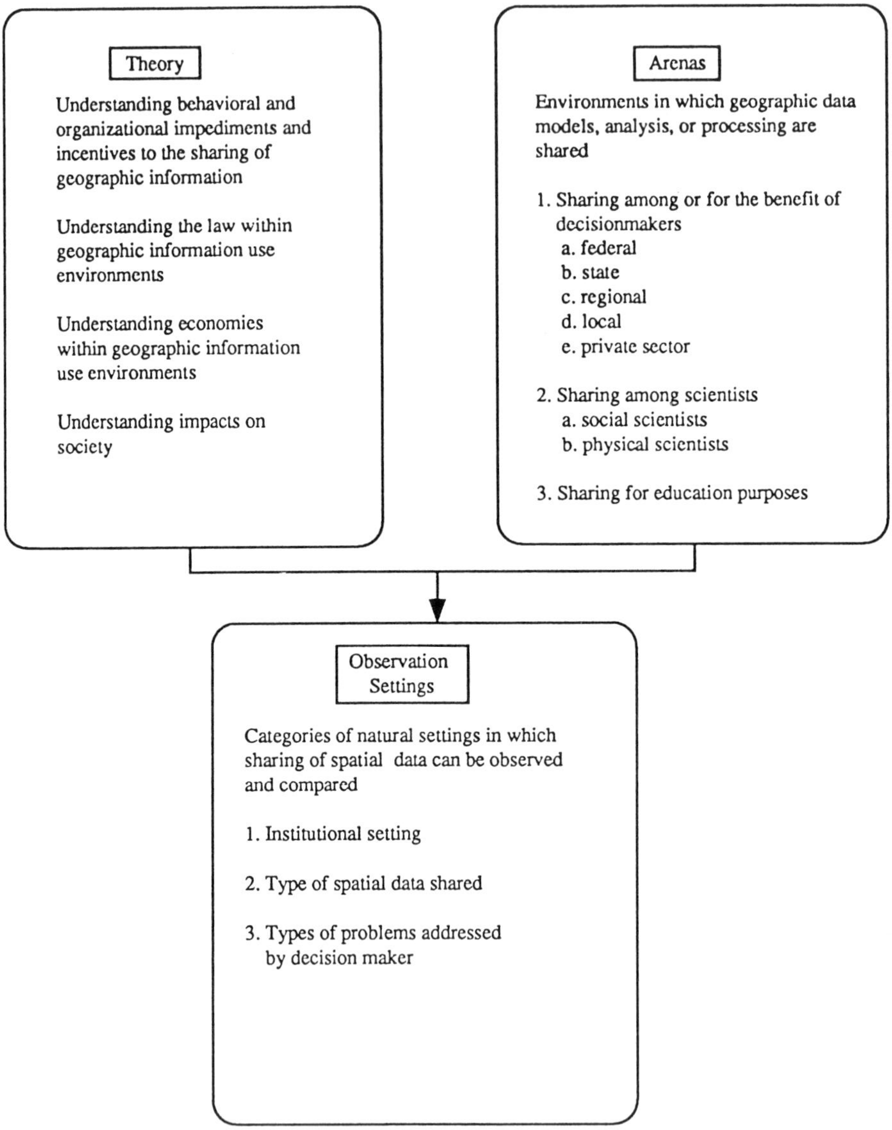

FIGURE I-1

The three primary components of a research framework: 1. the theories of individual and organizational behavior; 2. the arenas among which sharing of spatial data occurs, could occur, or could be enhanced; and 3. observations of the process of spatial data sharing in existing settings.

needing preservation, or to relocate district lines). Any studies of sharing within these or similar settings should be soundly grounded in theory and methods of observation should be selected based on their appropriateness in eliciting the forms of information desired. Solid theoretical frameworks provide a means of gauging the limits of generalizability of any principles that might be formulated from a study or series of studies.

Figure I-1 highlights three primary components of a research framework: (1) the theories of individual and organizational behavior (particularly their relevance in understanding impediments and incentives to the sharing of information); (2) the arenas among which sharing of spatial data occurs, could occur, or could be enhanced; and (3) observations of the process of spatial data sharing in existing settings. Ultimately, these three components must be brought together to produce models of behavior in regard to spatial data sharing. These results, in turn, should be useful in developing normative proposals for successful spatial data sharing in specific situations.

This book is broken into five major sections. In the first, organizational theory issues are considered in the context of geographic data-sharing environments. The chapters in this section introduce potential structures for studying sharing and provide ties to behavioral and organizational research work in other fields from which understanding of sharing issues in the geographic arena may extend. The second section provides a sampler of experiences in the sharing of geographic data. The authors relate successes and failures, and they suggest approaches and recommendations for others to consider in light of those experiences. Although the third section also addresses organizational dynamics and lessons learned from past experiences, the chapters in this section focus primarily on broader institutional issues. The authors in this section present differing perspectives on economic, legal, and access issues that they believe are important in geographic information sharing environments or will be important in the future for widespread sharing. The fourth section contains chapters that focus primarily on pragmatic considerations in the further development and improvement of a spatial data infrastructure—an infrastructure that will be useful and accessible to a multitude of users nationwide. Finally, in the fifth section, research and development activities that might facilitate increased use and sharing of geographic information are set forth and discussed.

As chapters in this volume are read, it may be useful to consider each in the context of the elements of the research framework set forth in Figure I-1. Some chapters focus on theory building, some describe experiences and lessons learned in sharing or attempting to share geographic information, and still others stress methods of observing the sharing behavior of individuals or organizations. Some authors focus on single elements or subelements within the research framework while others seek to expand the research framework and the research agenda that might be bounded by any such framework. Regardless, all the chapters in this book are concerned with gaining greater understanding of geographic information sharing environments, and all are tied to the research framework diagrammed in Figure I-1.

Exploring behavioral and organizational issues in the sharing of geographic information should be addressed through theory-focused empirical research. While it is not within the purview of this book to encourage one research approach over any other, there currently exists a broad range of quantitative and qualitative empirical methods that may be employed (e.g., mathematical modeling, controlled experimentation, field surveys, case studies, and archival and secondary research). Certainly, no one method can be deemed more appropriate than another for the study of issues in institutional sharing of GIS data. Each method or combination of methods has advantages and disadvantages as well as different assumptions, biases, and degrees of usefulness. We believe multifaceted research approaches are necessary in order to understand comprehensively the incentives and impediments to sharing geographic information among and within organizations.

The private and societal benefits to be gained from the sharing and widespread diffusion of geographic information in digital form are very substantial. The first major goal of this book is to provide insights on opportunities, options, and potential pitfalls for those organizations and individuals seeking to gain from the sharing of geographic information. The second major goal is to discuss, debate, and formulate research activities which, if accomplished by the at-large academic community, would be valuable to individuals, businesses, and governments sharing or attempting to share geographic information. Progress towards these goals constitutes the main contribution of this book.

Section I

Organizational Theory and Geographic Data Sharing

Organizational Theory and Geographic Data Sharing

Section I provides the views of people who have worked with organizations that regularly share spatial data and organizations that often depend on nationally developed geo-referenced information databases. Coming from different disciplinary backgrounds, the authors focus on different aspects of the spatial data sharing enterprise.

A common theme is that the act of sharing spatial data forges new relationships among organizations and among departments within an organization. Previously, the common relationship was for one organization to receive a spatially referenced product from another organization in the form of a map or a table. If in the form of a table, rows typically designated geographical locations, and columns designated attribute values observed at those locations. The new relationship involves one organization importing a spatially referenced data file from another and making a new product. Using its own spatial and non-spatial data as well as data from other departments and organizations, the organization produces some new computed relationship that will be of use to the organization.

Meredith begins his chapter by noting the technological developments that have allowed distributed processing environments to grow. He argues that distributed data acquisition and processing have enabled new forms of spatial data sharing to take place. Building his arguments from organization theory, he explores the principal sources of resistance to adopting distributed solutions. Meredith sees spatial data sharing as an example of the adoption of distributed processing, a mainstream issue in the field of management information systems. Working from theory in this field, he posits six normative conditions for successful data sharing and tests them against two spatial data-sharing initiatives in Louisiana.

Azad and Wiggins eschew the normative approach to defining conditions for successful data sharing and instead focus on evidentiary criteria for antecedents of inter-organizational cooperation. Cooperation, they note, is easier to advocate than to practice, particularly because geographic database sharing commonly involves a redistribution of relative power and influence. Thus, sharing changes inter-organizational relations. They argue that preserving autonomy and independence is a traditional, important goal of organizations and this objective is threatened by changes in organizational relations brought about by sharing. With the inter-organizational dynamic as their chief concern, Azad and Wiggins suggest that the probability of establishing inter-organizational relations is related to the intensity of the sharing relationship and the loss of organizational autonomy required by the relationship. Based on an extensive review of empirical studies of organization behavior, six antecedent determinants to geographic data sharing are posited and related to three case studies of spatial data sharing in Massachusetts. Concluding that none of the six determinants offers a satisfactory account of spatial data sharing and that the six together do not provide a satisfactory framework for exploring why organizations pursue cooperation, Azad and Wiggins propose a process model of the adoption of spatial data sharing. Their model involves articulation of circumstances that lead to the development of collaboration, cooperation, and coordination in data sharing. They believe that the dynamics of problem setting, direction setting, and structuring can be managed to increase the likelihood of spatial data sharing.

In the next chapter, Pinto and Onsrud describe a research framework for investigating the process by which improved cooperation and collaboration between organizations occur. In their conceptual framework, it is important to search for and find the antecedents to information sharing between organizations. For instance, the first condition they discuss involves the importance of superordinate goals. Superordinate goals are those that are highly important for all organizations involved but that any single organization will find difficult to achieve without acting in concert with others. Research, they assert, should determine whether the process of recognizing and articulating specific superordinate goals by organizations has a significant and direct effect on the sharing of geographic information. Similarly, Pinto and Onsrud set forth six additional propositions extending from their research model, each of which involves causal links between antecedents of information sharing and projected outcomes for organizations. They find support for each proposition in the existing literature, derive a testable hypothesis for each, and advocate testing the effect of the condition on sharing within geographic information environments.

The chapter by Calkins and Weatherbe stresses the importance of definitions and taxonomies in enabling better understanding of data-sharing environments. They define spatial data sharing as transferring spatially referenced information between two or more independent organizations or departments within an organization. Common distinguishing features are the different levels of government and different types of organizations that typically participate in such sharing ventures. An important aspect of the taxonomy of spatial data sharing they develop is that the relationships between entities involved are often new relationships between parties who formerly had little contact with one another. They suggest that these new relationships lead to new dependencies among departments and organizations that formerly had little or no need to coordinate their activities in time, space, and in the adoption of common standards. The taxonomy offered by Calkins and Weatherbe distinguishes different characteristics of organizations involved in spatial data sharing, different characteristics of data, different types of exchange relationships, and different impediments and constraints that effect the likelihood of successful sharing experiences.

Kevany also develops a taxonomy for spatial data sharing, but unlike Calkins and Weatherbe, the classes in his taxonomy are ones that have scales designed to measure the likelihood of success in data sharing. Implicit in his schema is that the overall likelihood of sharing success is dependent on a given situation achieving positive sharing scores on all critical factors. Any group that is planning new spatial data-sharing initiatives might score itself on the scales that Kevany provides. Though many subjective judgments are required, such an exercise might well reveal the "Achilles heel" that could cause failure in a specific data-sharing endeavor.

Taken together, the five chapters in this section amply illustrate that spatial data sharing occurs in a complex organizational environment where cooperation among organizations occurs under circumstances that are incompletely understood. Considering the large number of efforts to develop shared spatial databases that are expected in the future, it is important to better understand the cooperation process.

Paul H. Meredith

1 *Distributed GIS: If Its Time Is Now, Why Is It Resisted?*

Tremendous value may be realized by storing, sharing, merging, and manipulating large-scale databases in a distributed processing environment (Toperczer 1991 and Keen 1988). Yet there is deep-rooted and widespread resistance to distributed processing of data, even when it is limited to intra-organizational sharing of processing responsibility. When distributed strategies are advanced that transcend organizational boundaries, resistance magnifies. Why is administrative and user resistance so high, and what alternatives exist for managing or overcoming the sources of resistance? This chapter responds to that question in the following manner. First, organization theories on three types of organizational interdependence and how organizations respond to them are explored. Second, a theoretical framework for understanding various levels of interorganizational cooperation, and factors influencing their adoption, are provided. These theories are linked to a conceptual framework from the field of management information systems (MIS). The MIS material referenced identifies a set of criteria for the organizational evaluation of distributed processing proposals. Special emphasis is placed on understanding the principal sources of resistance to adopting distributed solutions. Finally, a testable framework for managing and/or overcoming those sources of resistance is proposed. Concepts found in literature on the management of technology and innovation are employed.

INTRODUCTION

Tremendous value may be realized by storing, sharing, merging, and manipulating large-scale databases in a distributed processing environment

(Toperczer 1991 and Keen 1988). Practitioners in the evolving field of Geographic Information Systems (GIS) have begun to understand and explore this potential for distributed processing of spatially related data. Some programs are well beyond the formative stage in the federal sector (Nebert 1990). The existing technological limits to realizing the benefits of distributed processing are being overcome as a result of market forces directed at vendors. To remain competitive they are being forced to design faster and interconnectable computer hardware, develop high-speed networking systems, build data-exchange-capable GIS/DBMS (database management system) software, and cooperate in the establishment of hardware and software standards for efficiently capturing, recording, and exchanging data. On a technical level, an environment may soon exist where geographically distributed users from multiple organizations will be able to deal with spatial data elements relatively independent of concerns about host devices where they are stored, the physical location of data, or the host GIS software at that location (O'Brien 1990). Achieving that independence will be costly both in terms of the initial cost of hardware and software and in the ongoing cost of learning how to coordinate and integrate operations in such an environment.

The limits to diffusion of distributed GIS technology into mainstream GIS activities are not only technical. There is deep-rooted and widespread resistance to distributed processing of data, even when limited to intra-organizational sharing of processing responsibility. When distributed strategies are advanced that transcend organizational boundaries, the resistance of key decision makers increases. Real concerns about costs, loss of control, loss of power, and loss of independence are advanced (Bozman 1989). Arguments focusing on potential but hard-to-document benefits to long-run performance and difficult-to-support claims to reduced costs in the short run have not overcome this resistance.

Administrators are doubly concerned since clear standards for justification of simple GIS investments do not exist. Acceptable criteria for evaluating and justifying GIS activities are being developed. For example, Chock (1990) outlines an excellent taxonomy for classifying costs and benefits. Yet, on balance, administrative and user resistance remains high. Why is this?

This chapter responds to that question in the following manner. First, it explores organization theories on three types of organizational interdependence. Second, it provides a theoretical framework for understanding various levels of inter-organizational cooperation and factors influencing their adoption. These theories are linked to a conceptual framework from the field of management information systems (MIS). The

MIS material referenced identifies a set of criteria for the organizational evaluation of distributed processing proposals. Special emphasis is placed on understanding the principal sources of resistance to adopting distributed solutions. Finally, a testable framework for managing and/or overcoming those sources to resistance is proposed. Concepts found in the literature on the management of technology and innovation are employed.

Informal post hoc testing of the model is carried out by subjecting three evolving GIS configurations to analysis. The first is the development of the Lafayette Parish (Louisiana) School Board's Transportation Management System (TMS), an automated school bus routing system. The second analyzes efforts to share and maintain spatial data needed for several departments of a Louisiana municipal government in conjunction with the GIS/FM (facilities management) activities at the publicly owned municipal utility system. Finally, ongoing efforts to integrate all state of Louisiana GIS activities are reviewed in light of the proposed theories and practices.

THEORETICAL FRAMEWORK

Interdependency, Complexity, and Coordination

J. D. Thompson (1967) proposed an integrated theory of organizational behavior. His basic taxonomy has been widely adopted, utilized, and extended by other organizational theorists (Galbraith 1977; March and Olsen 1976; Olsen 1981). He states that rational organizations seek to develop structure and coordinate their activities in response to technological requirements and in relation to the complexity of the task environment they face. In general, systems will seek out structures that minimize the cost of coordination. Three means of coordination are identified. From lowest to highest cost they are coordination by standardization (rules), by plan, and by mutual adjustment. The most effective means of coordination, however, depends on two factors: the type of interdependency that exists between the systems' elements and on the systems' requirements for adaptation to environmental complexity.

Thompson identifies three types of task interdependency—pooled, serial, and reciprocal—with the three forming a Guttman scale; i.e., pooled interdependency will always be found in a system if serial interdependence exists anywhere in the system. The more organizations tend toward technologies that create all three interdependencies, the more complex their structure and systems for coordination. Reciprocal interdependency, the most complex, involves relationships where units' or organizations' outputs each become the others' inputs. Thompson says that such units should be placed close together organizationally if possible and coordinated by mutual adjustment. An example is the relationship between manufacturing and maintenance in a factory. Serial interdependence takes its name from the fact that units' tasks are related to each other in a specific sequence, with each unit unable to function until the preceding unit completes its task (e.g., related as successive departments are in a production line). Units exhibiting such relationships are best located under a common administrator and coordinated by formal plans of action. Pooled or general interdependence is the simplest. Each part "renders a discrete contribution to the whole and is supported by the whole" (Thompson 1967, 54). Such relationships are best coordinated by rules and procedures, provided the situation is stable enough to establish workable rules.

Each type of interdependence can be found in one or more functional activities in GIS. Multiple interdependencies cause complexity. The task of coordinating GIS is complex, whether it is handled centrally or in a distributed fashion. Distribution, however, will increase the complexity of coordination as it attempts to relate the activities of diverse elements with several layers of administration between them. In the most extreme case, there is no common administration. This case will be discussed later when the complexities of inter-organizational coordination are explored. Thus, GIS is perceived to be administratively complex and expensive to coordinate; distributed processing of GIS is perceived as the most complex and expensive to coordinate. This is the first source of organizational resistance to adopting distributed technologies.

Second, while rules and plans may be efficient means of coordination, they both require stable task environments to be effective. GIS activities exist in environments that are on the leading edge of technology: they are changing rapidly, sometimes unpredictably, and the environmental elements are relatively heterogenous. Thompson states that under those conditions the organization will seek to protect "core" activities of the organization by forming boundary spanning units with specialized functional units to deal with the heterogenous elements in the environment. Rules and plans are generally ineffective coordinating devices in the GIS environment. However, eventually, efforts to create stable data standards by various federal agencies may simplify some of the coordination task. In fact, the agencies are bureaucratizing their relationships to both distributed data users and distributed data suppliers for that reason. They expect to reduce the complexity of administering their functions and cut the costs of coordination considerably by resorting to rules, procedures, and planning.

However, administrators considering distributed GIS correctly perceive that they will encounter greater "leading edge" complexity due to rapid changes in technology. They also see that it will be more difficult to insulate their core activities from disturbances caused by having to relate to multiple heterogenous elements which they cannot easily control. This is a second major source of resistance to sharing of responsibility for GIS activities across a distributed system.

Modes of Coordination, Ambiguity, and Resistance

March and Olsen (1976) wrote a landmark work on how organizations operate when they exist in ambiguous environments. It addresses the shortcomings of normative models, including Thompson's. They seek to account for irrational organizational behavior where irrationality is defined as decision making that appears to violate the precepts of classical decision-making models. While full explanation of this theory is beyond the scope of this discussion, certain key elements of it can be examined as they relate directly to problems in GIS adoption by organizations. Specifically, they relate to the issues emerging from discussions about shared responsibility for GIS activity. In addition, four classes of organizations are conspicuous examples of what they describe as organized anarchies: educational institutions, public organizations, R&D/technology organizations, and illegitimate (informal) organizations. These comprise many of the leading players in GIS: government, universities, vendors, professional organizations, and large corporate GIS consumers.

Irrational behavior by organizations is neither irrational nor pathological according to their "Garbage Can Theory of Management." The definition of irrationality is only flawed, as organizations are seen to behave that way when three preconditions of the rational model are not met, rendering choice situations that are ambiguous. An ambiguous environment is one in which technologies are not clear. Cause-and-effect relationships are not well understood. Most learning occurs from trial-and-error activities or from pragmatic inventions of necessity. A second dimension of ambiguity relates to preferences. In the garbage can, the preferences of organizations are problematical. Rationality assumes that a set of consistent, clearly defined goals exists for organizations. That assumption is relaxed; preferences are a loose collection of ideas and are often discovered through actions rather than being the basis of actions. Finally, participation in the garbage can organization is fluid. Who participates in decision making depends on the issue involved, what other issues are under consideration and the extent to which potential players are interested in them.

Choice situations are fundamentally ambiguous meetings, arenas for argument, interpretation of preferences, sites for problem solving, and a place for making decisions. In these meetings issues and feelings look for a place to be aired (to be legitimized as problems); solutions, espoused by persons seeking access to the decision process, look for issues that they may resolve; and participants in the process look for problems or pleasure. Choices are context dependent, depending on relative time of entry of choices and problems to the system, the access, availability, and energy level of participants and what else is going on in the organization at that time.

In the garbage can three decision styles are legitimate. If the timing is such that a choice is activated when problems are being addressed by other choices and if participants have the interest and energy to do so, choices will be made quickly with little attention to possibly important issues and problems. This is referred to as decision by oversight. In decision by flight, choices are unsuccessfully associated with a particular problem. If a better, more attractive choice presents itself, the problem may move to it, making it possible for a decision to be made even if it does not resolve any problem. In the third case decisions may resolve issues or solve problems (this is the traditional decision situation.)

Practitioners intuitively feel the theory works, most specifically in academia and government. March and Olsen (1976) sought to test and validate it in a multi-university field study. They reported that they were successful.

GIS choices are clearly being made in a garbage can environment in many cases. Advocates for GIS and distributed GIS are constantly promoting them as solutions when they talk to administrative decision makers. The GIS "fix" is being associated with a multitude of outstanding issues that exist in the public and private sector. Advocates for those issues may or may not view adoption of GIS technologies as solutions to their problems. And, with no clear means of proving the efficacy of the GIS solution, they may find it difficult to commit their support to GIS proposals. Decision makers may or may not view those same issues as the most important ones to resolve, especially if the organization has high levels of goal conflict and ambiguity. Seeking a way of moving decisions about GIS out of flight and oversight decision situations and into resolution situations would seem to be a desirable goal.

The presence of effective inter-organizational coordination mechanisms would appear to be a necessary precondition to any distributed GIS initiative. Olsen (1981) identifies four means of such coordination: by rules and procedures, by substantive rules, by ad hoc interactions, and by autonomous adjustment. Based on theoretical arguments in Thompson (1967) and March and Olsen (1976), predictions are made about conditions under which integrated participation between organizations and governments will occur. It is argued here that those predictions can be extended to the more general case of inter-organization participation in general. Olsen predicts that, based on the costs and benefits of participation, organizations will engage in integrated participation where:

1. many organizations value predictability;
2. many organizations are willing to make commitments;
3. few organizations value spontaneity and flexibility;
4. compromises can be beneficial;
5. few organizations strive to protect their ideological identities;
6. many organizations seek recognition as the legitimate representatives of causes, groups, and organizational domains;
7. organizations are willing to accept responsibility for governmental policies;
8. organizations readily delegate discretion to their representatives; and
9. few organizations value high involvement and control by members.

Ambiguity, as described previously, is predicted to impact integrated participation. Data reported suggests that is most likely when ambiguity is low about the character of the interdependence between participants and their relative power. Second, when ambiguity is low about the preferences of organizations and their preferences are stable and lie in an agreed-upon order, participation occurs. In addition, if the magnitude of differences between parties is clear and they know how much of one goal they are willing to forgo to accomplish another, participation occurs. Finally, participation is possible when ambiguity is low about the causal relationship between actions and outcomes and where it is easy to know who has competence and expertise.

It is obvious that all of those conditions are not met and are, in fact, difficult to meet in the current GIS environment. They provide major impediments to inter-organizational coordination of GIS activity. Dealing with those shortcomings is necessary if significant distributed GIS activity is to occur, no matter how desirable the outcomes for the GIS practitioner or user.

The MIS Perspective

Distributed processing is a mainstream issue in MIS. A criteria for evaluating whether a distributed system would be acceptable to an organization's management has been proposed by O'Brien. Systems with the following characteristics are desired:

- *Integrability:* Integrated delivery of telecommunications services.
- *Connectability:* Easy access and connections to telecommunications networks.
- *Shareability:* Using telecommunications to share hardware, software, and data resources among organizational information systems.
- *Phaseability:* Developing telecommunications systems and services in stages.
- *Availability:* Telecommunications hardware, software, and services that meet announced standards are available for use by the organization.
- *Reliability and security:* Providing reliable and secure telecommunications networks for vital business operations and services.

If these criteria are correct, then promoting their adoption by decision makers may be a first step in dealing with resistance to distributed GIS. Its use may also assist in dealing with reductions in ambiguity, an apparent precondition to increased interorganizational participation. Its usefulness will be tested in the analysis to be presented later.

Managing Technological Change in GIS

Burgelman (1984), Burgelman and Maidique (1988), and Walton (1989) explore how to best implement technological change in organizations. They propose specific organizational processes and structures to facilitate the strategic acceptance of technology, effective design of the actual innovation, the appropriate planning for implementation to foster user acceptance, and effective mechanisms for implementation and control of the process. Their recommended solutions will be evaluated in the analysis phase: top-level and mid-level management involvement in technology choice, use of integrated design teams with practitioner, technologist, user, and "market" feedback, and so on. Whether these approaches can be extended to multi-organizational settings is a research question to be explored later. Whether their use will deal with the ambiguity issue is yet to be resolved. Model building and testing are required.

CONCEPT INTEGRATION AND PROPOSITIONS

Based on the concepts discussed above, a set of working propositions and related hypotheses can be established. To model the total set of relationships between all of the variables identified to this point is well beyond the scope of this chaper. Neither the theoretical nor the empirical basis exists to make such a leap. Describing and testing a model, taking

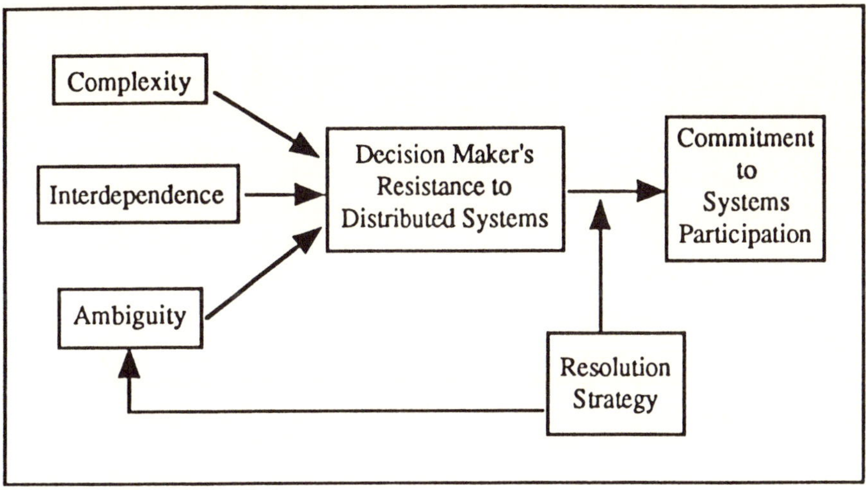

FIGURE 1.1
Resistance to distributed GIS proposed model

into consideration the relationships between the following subset of variables, will be attempted. Complexity of the distributed GIS, the degree of task interdependence between participants, and the perceived degree of ambiguity about the proposed distributed system are identified as independent variables. Dependent variables in the model will be defined as the degree of resistance to distribution exhibited by decision makers in the constituent organizations. The strategies employed for dealing with and/or resolving the sources of resistance are proposed to be mediators of the relationship between resistance and organizational commitment to interorganizational cooperation. In addition, these resolution strategies are hypothesized to influence decision makers' perceptions about the ambiguity, indirectly influencing the level of resistance. A summary of the proposed relationship between the model's variables is shown in Figure 1.1.

Operational definition of the complexity variable involves a multidimensional conceptualization. As the number of potential participants increases, complexity increases. Correspondingly, the newer and less well-accepted are the technologies required to implement the system, the higher the complexity. If the rate of change in the technologies is high, additional administrative complexity will be said to exist.

Interdependency will be operationalized using Thompson's 1967 taxonomy. Distributed systems with pooled interdependence between participants will be described as low in interdependence. The highest level will be systems with reciprocal interdependence between participants.

Ambiguity will be evaluated by considering the degree to which the three conditions for "rational" choices are met. As fewer conditions are perceived to be satisfied, decision makers will be described as increasingly ambiguous and increasingly prone to resort to irrational decision processes as outlined by March and Olsen (1976).

Resistance will be evaluated by assessing the degree to which organizations fail to consider, address, and resolve outstanding proposals for shared GIS arrangements using the systems criteria proposed by O'Brien. The greater the time delays, the more the problems are moved from one choice situation to another, and the degree to which garbage can processes are employed in dealing with distributed proposals, the greater the resistance. Note that rejection of distributed solutions is not necessarily an indication of resistance. Observation of the organizational processes leading to such a decision is necessary to make that judgment.

Resolution strategies will be evaluated based on Thompson's theory concerning least-cost solutions to coordinating interdependency, expanded to include Olsen's (1981) four strategies for coordination and integration of participants. Whether the strategies employed or proposed to be employed conform to Thompson's and March and Olsen's theories will be evaluated to test the predicted moderator effects and the predicted impact on the perceived level of ambiguity.

Commitment will be evidenced by the presence or absence of ongoing multi-unit and/or multi-organizational GIS activity. In some cases the unit of analysis will be commitment to share responsibilities across relatively autonomous departments of a large organization. In others it may be multi-departmental and multi-organizational sharing.

Propositions based on the model are as follows:

Resistance

P1: Complexity will be directly related to the level of resistance.
P2: Interdependence will be inversely related to resistance.
P3: Ambiguity will be directly related to resistance.

Ambiguity

P4: Congruent resolution strategies will be inversely related to ambiguity.

Commitment

P5: Resistance will be inversely related to commitment to participate.
P6: Use of resolution strategies congruent with satisfying Olsen's nine factors supporting integrated participation will relate positively with commitment.

POST HOC FIELD TEST OF PROPOSITIONS

The Lafayette Parish School Board Transportation Management System (TMS) is a five-year-old project to provide automated bus scheduling and routing of approximately 23,000 students to more than fifty public and parochial schools using approximately 230 buses, both public and privately owned. Two classes of clients are transported: special education and regular students. Special students are transported on buses equipped for various disabilities, both physical and behavioral. Special buses have a driver and an attendant and are required to transport students door to door. Regular students are picked up at stops and are placed on buses with a driver but no attendant. Except in rare cases, the two classes are not mixed.

To accomplish scheduling of special education students, a close working relationship must be maintained between the transportation department, which reports to the fiscal officer, and the special education department, which reports to an assistant superintendent. The only common superior in their respective chains of command is the superintendent of schools. By law, special education records are private; access to them, even within the administrative offices of the school board, is restricted. However, extensive coordination was necessary to route and schedule buses to facilitate client needs for transportation services.

Prior to the TMS project, each department kept duplicate records on students, records which in many cases were not accurate, were out of date, and were incomplete. Records notwithstanding, the only person in the system who knew where a student actually lived was the bus driver who picked the student up (parents called the drivers at home and told them where to come). Scheduling was done on an ad hoc basis with major crisis periods at the start of school and after every break. People moved during these periods and did not update their records or notify the driver. Errors were made regularly by both departments.

The active and ongoing relationship between the two departments and their respective department heads was characterized by suspicion about motives, concern over ends, and turf guarding. It was controlled through bureaucratic processes more concerned with placing blame for errors than on quality of service. Resistance to cooperation was high. However, external pressure for better service was evidenced in several ways. Direct involvement in transportation decisions by board members representing the needs of voting parents was common, confusing how to apply the few policies that did exist. Likewise, client-group parents actively lobbied for better services. In addition, bus driver groups sought to influence processes that might impact economic factors for them. And, citizen groups concerned with "cost saving" were pressuring for accountability and efficiency.

Considering the model proposed here, it is not surprising that resistance to cooperation was high. The problem being dealt with was administratively and technologically complex, the number of players was large, and their relative power to influence decisions was not clear. The relationship between transportation and special education was clearly a reciprocal relationship. Their means of adjustment were also just as clearly inappropriate—little mutual adaptation was occurring. Finally, the levels of ambiguity about ends, means, power, and players were all high.

Commitment to the development of the TMS was initially low but increased after two resolution strategies were employed. First, university personnel developing the system sought to clarify with the educators and the concerned groups the nature of their concerns about the system's impact on high priority issues that seemed to be the root causes of resistance to the system. In meetings focused on reducing the participants' ambiguity about issues that Olsen (1981) talks about, the players moved away from autonomous power positions toward some rulemaking, power sharing, and ad hoc adaptation to others' needs. Clarification of the sources of task and data interdependency and exploration and recognition of common needs for data of high integrity were priorities for the team. Finally, appeals to consider more rational criteria for evaluation of performance were made in an effort to make it possible to "trade" or "give and take" in the event that conflicts over means and ends did occur. Appeals for mutually acceptable performance measures and safeguards on issues important to the various groups were made and were generally successful. Criteria for decision making were agreed on, including factors such as length of riding time and emergency response times, and standards for protection of student record privacy were established.

Through the policy and standard-setting activities, some reduction in complexity occurred. Players working together and with the university team developed better understanding of each others' positions, needs, and priorities, reducing levels of ambiguity. The mechanisms for appropriately coordinating the remaining interdependent tasks were more effective. Resistance was reduced and decision makers became more willing to consider participation in implementing the TMS system proposal. Converting that better climate into commitment was enhanced by the other resolution strategy.

To overcome directly the remaining resistance, the university team opened up the system design and implementation process. An implementation team was established with representatives from all impacted areas using senior and mid-level administrators as members. Users were asked to evaluate and test proposed forms, reports, and procedures to increase commitment to the final design. Numerous high-profile presentations were made to the school board to generate top-level support and input. A phased implementation was made with the most popular parts being first, creating more goodwill. Every effort was made to create a success environment where administrators were willing to delegate "working out the bugs" rather than resorting to more cumbersome means of problem solving. Key personnel maintained their commitment throughout the process, and this part of the system was implemented.

The regular education transportation problem is more complex in many ways. Technically, the problem is bigger: more students, stops, and destinations. The routing problem is also more complex to solve as students must be assigned to stops based on maximum walk criteria, minimum distance from school, and other decision rules precluded when state-mandated door-to-door transportation for special education clients is considered. Administratively, the system is more complex as each school becomes a separate player in the distributed system. Schools maintain separate distributed student databases. Serial dependency exists between the TMS department and the school. Until student records began to be automated and integrated at the central office, little commitment to sharing was evident. School administrators have avoided participation overtures by raising issues external to TMS feasibility such as their increasing work load, the failure of transportation personnel to use data previously collected effectively, and their inability to obtain support from other central office departments. Efforts of the TMS team to reduce the level of ambiguity by clarifying the priorities and establishing trade-offs has had some limited success. Support for continued implementation has not been forthcoming from the board. The existence of a rule-based bureaucratic coordination mechanism and integration of command at the superintendent level only makes overcoming the remaining resistance a slow, problematical process to date. The model suggests that when ambiguity remains high, interdependence cannot move toward reciprocal relationships, and when complexity remains high, resistance will be high and system implementation will fail due to lack of commitment. In general, the predictions of the model are borne out by the regular education TMS experience. Both the experiences encountered in the earlier success and this latter failure provided the basis for developing this model.

Another distributed system proposal failed to develop. Interdependence between local agencies in a Louisiana parish, such as planning, fire, police, zoning, and the public utility, were pooled. Efforts to piggyback a GIS database access onto the utility's GIS/FM project were initially met with positive responses. As participants tried to develop a system proposal that would satisfy the users and the utility, the level of resistance increased. It became apparent that the technological factors for delivering the service were not clear and were in a state of flux. Hardware response times were seriously questioned by users who were operating real-time services. Telecommunication problems were also identified but not resolved. Users were not clear about their preferences, and almost none of the nine conditions identified by Olsen were met. Substantive discussion of the MIS issues was rare, and players were severely constrained by their constituencies so little or no compromise was possible. Adaptation often took the form of arbitrary ad hoc interactions between players and some autonomous adjustment. While some official basis for cooperation still exists, parties are currently seeking to acquire hardware and software to start separate GIS initiatives. Top management appears to have abandoned distribution even though the benefits might be large. The utilities system is still not fully operational; the users' concerns appear to have been justified. Propositions one, three, four, and six are supported in this case.

On December 1, 1989, a Memorandum of Understanding was executed among thirteen Louisiana state agencies outlining the scope of their willingness to cooperate on GIS R&D, data sharing, and technical assistance. The agreement defines a loose confederation in which parties may cooperate but none are bound to. No resource commitments are made, and no power is redistributed as a result. It is a classic garbage-can–developed document. Principles are articulated in the document. However, informal discussions with members of the agencies clearly show that none acknowledge the rights of others, are willing to negotiate or compromise, or can agree on who has technical superiority sufficient to take control of the GIS initiative. The state's GIS response has been fragmented, uncoordinated, and uneven across agencies. Little if any data sharing has been attempted. What makes the agreement interesting is that it is the result of a grass roots initiative by the agencies' staffs. It has not received executive or legislative support. And, while they all appear to agree to disagree, the ad hoc nature of the confederation permits them to address certain crucial issues such as researching how to create a statewide network that would facilitate distributed GIS activity. They carefully avoid the implications of such a network in terms of who would control it and how the responsibilities for data currency, integrity, security, and access would be handled. Those issues are literally addressed by oversight, and flight as resolution would require reassessment of the entire concept of the agreement. The agreement appears to support the propositions. It exists as a participatory statement by limiting the agencies to actions that minimize complexity, avoid interdependence, avoid ambiguity-based issues, and require only consistent resolution mechanisms.

CONCLUSIONS

A simplified model of the interactions between a subset of variables derived from several diverse streams of theory and research is proposed. Preliminary evaluation of the model in a post hoc analysis seems to support it. It is not generalizable by any means. However, it may serve as the basis for more rigorous modeling, testing, and pragmatic field experimentation. In the ideal, the model could be operationalized and tested as a path analysis and the relative significance of each construct could be assessed. To the extent that the model has validity, face or statistical, it may contribute to answering the most important question: how can the acceptance of cost-effective, efficient, distributed GIS solutions be accelerated?

REFERENCES

Bozman, Jean S. 1989. Distributed DBMS means cost, trouble. *Computerworld* 25 (February 13): 25-27.

Burgelman, Robert A. 1984. Managing the internal corporate venturing process: some recommendations for practice. *Sloan Management Review* 25, 2: 33-48.

Burgelman, Robert A., and Modesto A. Maidique. 1988. *Strategic management of technology and innovation*. Homewood, IL: Richard D. Irwin.

Chock, Margaret. 1990. The other costs of Geographic Information Systems. GIS/LIS 1990 *Proceedings*, 2: 526-531.

Galbraith, Jay R. 1977. *Organization design*. Reading, MA: Addison-Wesley.

Keen, Peter. 1988. *Computing in time: using telecommunications for competitive advantage*. New York: Ballinger.

March, James G., and Johan P. Olsen. 1976. *Ambiguity and choice in organizations*. Universitetsforlaget, Bergen, Norway.

Nebert, Douglas D. 1990. Design of a distributed spatial data library to support hydrologic investigations in the U.S. Geological Survey. GIS/LIS 1990 *Proceedings*, 1: 272.

O'Brien, James A. 1990. *Management information systems: a managerial end user perspective*. Homewood, IL: Richard D. Irwin.

Olsen, Johan P. 1981. Integrated organizational participation in government. In Nystrom, Paul C., and William H. Starbuck, eds., *Handbook of organizational design, vol. 2: Remodeling organizations and their environments*. New York: Oxford University Press.

Thompson, James D. 1967. *Organizations in action*. New York: McGraw-Hill.

Toperczer, Tom. 1991. From pyramids to peers. *BYTE* 10, 1: 101-107.

Walton, Richard A. 1989. *Up and running: integrating information technology and the organization*. Boston, MA: Harvard Business School Press.

BIJAN AZAD AND LYNA L. WIGGINS

2 *Dynamics of Inter-Organizational Geographic Data Sharing: A Conceptual Framework for Research*

This chapter addresses geographic data sharing from an organizational perspective, considering it an inter-organizational relation (IOR). The underlying principle of this perspective is that autonomy considerations dominate the decisions to engage in and maintain IORs. With organizational autonomy as its centerpiece, a research framework is proposed that covers reasons for, levels of, and stages of IORs: the basic and fundamental reasons that push public and/or private organizations to engage in IORs; how to concretely characterize different levels of IORs and the key dimension(s) along which the levels of IOR can be measured; and finally stages that usually accompany these levels of IORs. First, the problem of geographic data sharing is defined as an IOR and a simple illustrative inter-organizational typology of possible configurations for geographic data sharing is suggested. Second, relevant work from organization behavior/organization theory (OB/OT) literature is briefly reviewed and the basic and fundamental reasons for geographic data sharing are re-cited from this review of inter-organizational relations. Third, the key factor in IORs is identified as autonomy. We will argue that intensity of IORs increases as organizations move from initial inter-organizational collaboration to cooperation and finally to coordination. We will also identify a dimension that will indicate these increasing levels of IORs and therefore of geographic data sharing. Fourth, we propose – based on a review of OB/OT literature – the typical process of growth of IORs and its stages. We will then identify some key elements of these stages. Next, we illustrate our synthesis of the reasons, levels, and stages of IORs for geographic data sharing with three mini-case examples. Finally, the reasons, levels, and stages of IORs for geographic data sharing are proposed as a research framework that can be used both for theory building and theory testing activities under the National Center for Geographic Information and Analysis (NCGIA) Initiative 9.

DATA SHARING AS AN INTER-ORGANIZATIONAL RELATION

Geographic data sharing in multi-department/multi-agency settings is fundamentally an organizational affair. Building and maintaining databases is functionally much like any other organizational/corporate resource in many respects (trucks, copy machines, libraries, and so on). Therefore, if geographic data is to be shared, the sharing parties will want to have a stake in its design/specifications, building, and maintenance. As an inter-organizational resource, it is also subject to dynamics of inter-organizational relations. Looking at reasons for, levels of, and stages of inter-organization relations for geographic data sharing will help us better understand dynamics and their influence on the data-sharing outcomes. In this chapter we consider these three aspects of the dynamics: 1. the fundamental reasons that push organizations to establish inter-organizational relations (IOR) for sharing data; 2. the levels of IOR and how they manifest themselves organizationally; and 3. the typical universal stages of IOR that can be prescribed to get geographic data sharing off the ground and going.

Using Burns and McFarlan's (1987) analogy in describing the importance and the consequences of electronic data interchange in the private sector, geographic data sharing is nothing short of the organizational equivalent of thermonuclear detonation in its impact. This is because any geographic data-sharing effort, no matter how low in intensity, will involve a redefinition of existing tasks/structures and/or the redistribution of relative power and influence in the organizational environment.

If this nuclear metaphor is accurate, then geographic data sharing is a real uphill battle that has to be waged in a systematic way. This does not mean just being sensitive to organizational matters when a multi-participant geographic data-sharing effort is launched. It is, rather, a need for a full appreciation of the fact that geographic data sharing is, in essence, an introduction of information technology (IT) in organizations. The interaction of IT with the "architecture" of the organization is nothing short of drastic and therefore deserves due attention. This attention translates into viewing the geographic data sharing as changing inter-organizational relations, and not just the agreement to share a common data format (although this is often an element).

In summary, we view geographic data sharing as an activity that is being made increasingly easy by the developments in the technology of Geographic Information Systems (GIS). As such, inter-organizational dynamics will dominate the unfolding process of geographic data sharing, either helping it to deliver its potential efficiency gains or leaving unsatisfied participants behind who did not pay adequate attention to its inter-organizational dynamics.

AUTONOMY: THE CRITICAL ISSUE IN IOR

One of the valuable roles of a GIS is to integrate information across an organization so that it can be shared, interrelated, and used in common among different departments or groups.

JACK DANGERMOND, 1989

In most cases when geographic data sharing is advocated, the implicit basis for the advocacy is often gains in efficiency: 1. avoiding downstream costs by reducing the need for elaborate and complex conversions among the databases built and maintained by different organizations, and 2. realizing economies of scale by undertaking the joint conversion of analog paper-based maps to many layers of digital information that is more expensive if undertaken separately. These potential efficiencies come with some loss of autonomy, and increased interdependence, but are generally thought to be well worth the costs.

This potential loss of autonomy and increased interdependence is rarely addressed by GIS practitioners and academics in their advocacy for geographic data sharing. This is unfortunate. What at first glance may appear to be a totally technical problem in the domain of spatial and non-spatial database design, and data format interchange and communication standards, in fact requires very significant organizational changes. Jack Dangermond's comment about data sharing has been extended by other GIS actors to include data sharing across organizations, not just within one organization. In fact, organizations responsible for building and maintaining GIS are frequently scolded for not sharing data with one another. Both advocates and skeptics of GIS technology complain about the inefficiencies, redundancies, and gaps that result from independent GIS operations of various agencies with related missions. They yearn for better-coordinated GIS. With such a seal of approval, one might expect to see cooperation and geographic data sharing everywhere. But one does not. Data sharing is easier to advocate than to practice.

Therefore, we should start with the assumption that data sharing goes against the grain as far as organizations are concerned and is thus unlikely to occur. There are, of course, potential benefits from data sharing. But the benefits, we argue, can be realized only if agencies are willing to bear the perceived "costs" first. In general, it is safe to assume that agencies must be pushed or pulled into data sharing. They cannot be expected to welcome it naturally. If GIS policy makers seek to promote data sharing, they must understand how to overcome the obstacles that are in the way.

The obstacles are formidable because they stem from fundamental properties of organizational systems and, in particular, properties that are often more pronounced in the public sector (Pressman and Wildavsky 1984). In their classic study of joint public program implementation, Pressman and Wildavsky lucidly demonstrated that as decision points and interdependencies between agencies increase, the probability of successful joint implementation decreases. Each agency seeks to preserve its autonomy and independence. Organizational routines are difficult to synchronize. Goals overlap but are not identical. Actors inside and outside the organizational boundaries bring different expectations and pressures to bear on each agency. Van de Ven has succinctly articulated the uncertainty associated with cooperation, which implies a certain level of IOR (and thus data sharing), as follows:

> From an agency's point of view, to become involved in an inter-agency relationship implies (a) that it loses some of its freedom to act independently, when it would prefer to maintain control over its domain and affairs, and (b) that it must invest resources and energy to develop and maintain relationships with other organizations, when the potential returns on this investment are often unclear or intangible. (Van de Ven 1976, 28)

A SIMPLE TYPOLOGY OF INTER-ORGANIZATIONAL GEOGRAPHIC DATABASE SHARING

Up to this point we have used the terms geographic information, geographic data, and geographic database interchangeably. To narrow the scope of this discussion and to clarify its focus, the term geographic database will be used. There are many possible configurations available for organizations to enable them to cooperate in sharing a geographic database, at least hypothetically. However, the focus of this chapter is on one particular configuration—the so-called multi-participant GIS setting. It is useful to suggest a simple typology to clarify the elements of this selected configuration and distinguish it from other possibilities.

Type 1

An organization undertakes to provide geographic databases to other organizations at a nominal charge that does not reflect the production costs, or in a "one-shot" project where maintenance is not a concern. The dominant inter-organizational dynamic here is "one-way" (i.e., the organization providing the database is not affected by the user organizations wanting or not wanting to take part in the production of the database). Examples include the TIGER (Topologically Integrated Geographic Encoding and Referencing) database distributed the U.S. Bureau of the Census, and the CLUIS (Centralized Land Use Information System) project of the Boston Redevelopment Authority (one of our three mini-case examples, below).

Type 2

An organization undertakes to provide geographic databases of universal value to a variety of other organizations. However, continued maintenance and expansion of the database depends to a large degree on the using organizations picking up a portion of the tab for this effort as well as the demonstration of the usefulness of this data to the resource-providing bodies. The dominant inter-organizational dynamic is still somewhat "one-way," but it is moderated by the demand for the product by the user organizations. Examples of this type include the Vermont State GIS project and the Massachusetts Environmental Database Automation Effort (MassGIS), discussed as our second mini-case example, below.

Type 3

Several organizations undertake building and maintaining geographic databases by sharing the costs and products as a response to resource scarcity and to minimize duplication/redundancy. The dominant inter-organizational dynamic is "two-way." That is, without the mutual consent and participation of each organization the project will not take place, and its ultimate success depends largely on the inter-organizational dynamics. Examples include the Knoxville/Knox County Geographic Information System and the Boston Central Artery Project, discussed as our third mini-case example, below.

Each of these examples has an element of inter-organizational geographic database sharing, although the dominant inter-organizational dynamic in each case is quite distinct, and each case is probably characterized by unique factors that tend to inhibit or facilitate cooperation. However, what clearly stands in all cases is that from the first to the third we have an increased level of interdependence and a decreased level of autonomy. The third type is of particular interest, since we conjecture that: a. the geographic database will not be built, and therefore not shared, without at least a critical mass of participants, and b. all organizations involved will experience a certain decrease in their relative autonomy and increase in their interdependence.

To summarize, the dynamics of geographic data sharing among organizations dictate that we view it as an IOR, regardless of the form that it takes technologically or substantively. This, in turn, allows us to draw on the IOR literature from the organization behavior/organization theory (OB/OT) field to build a conceptual research framework.

FUNDAMENTAL REASONS FOR IOR

The decision dilemmas for geographic data sharing among organizations are not unique to GIS. Extensive research on the dynamics of inter-organizational relations in the private and public sector has been carried out since the late 1950s (Evan 1966). More recently, research focused on public and non-profit agencies that provide social services, health services, employment services, child-care services, urban development, rural development, mental health care, and community development has repeatedly confronted the same practical problems (Milner 1980; Mulford 1984; Whetten 1982).

Although the success of policy strategies may depend on relations among agencies that work in the same or a related policy arena or that serve the same clients or objectives, managers often have great difficulty in forging effective ways to work across agency boundaries. Structural obstacles to inter-organizational relationships appear to be quite similar across service providers (Seidman 1976). So we are left with the question of why agencies (organizations) cooperate at all.

Why Do Some Agencies Cooperate and Not Others?

The organizational behavior literature has a few theories that offer possible answers to this question. Perhaps the most popular perspective on organizational cooperation—at least in the public sector—comes from exchange theory. In exchange theory, cooperation is a voluntary transaction driven by a cost-benefit calculation; agencies engage in cooperation when two or more agencies derive mutually reinforcing benefits from doing so (Cook 1977; Sarason and Lorentz 1978). Levine and White (1961) are credited with the original formulation of this theory in the form of a resource-dependence model of organizational interaction with its environment and later adapted to account for other issues such as power (e.g., Pfeffer and Salancik 1978).

However, we believe that the exchange theory is perhaps too narrow and at times even tautological (i.e., agencies cooperate because the benefits of exchange outweigh its costs). Oliver (1990) provides a comprehensive literature review of organizational behavior in this area over the past several decades. She integrated the work of more than 160 articles related to the determinants of inter-organizational relationships and suggested six "generalizable determinants." Oliver titled her six determinants necessity, asymmetry, reciprocity, efficiency, stability, and legitimacy. We found Oliver's work to be particularly helpful to our search for reasons that drive data sharing and adopt her six determinants here. We first describe the six determinants, then turn to an analysis of their applicability to geographic data sharing.

Necessity. Most of the literature on the reasons for inter-organizational relations emphasize voluntary relationships between organizations. However, we have all observed cases, particularly of geographic data sharing, where the sharing was mandated, either by a higher authority or through legislation, regulation, or litigation. Oliver cites literature in this area that indicates that the degree of compliance to such mandated cooperation depends on the anticipated implications of non-compliance, as well as other references that deal with the perceived and actual loss of control faced by the organization under these conditions.

Asymmetry. Oliver's second determinant refers to relationships of power and control between organizations. Such power theories suggest that organizations seeking scarce resources will attempt to exert control and power over other organizations. The environments for organizations here are characterized as "negotiated arenas" with "injustice, information distortion, manipulation, exploitation, coercion, inequality, or conflict" (Oliver 1990, 244). In such environments, it is not surprising to see organizations act so as to minimize their loss of control and autonomy. In short, relationships based on asymmetry are characterized by their reliance on dominance, power, and control.

Reciprocity. The opposite to asymmetry, reciprocity assumes that the organizations cooperate in order to pursue commonly beneficial goals. Oliver states that the determinant of reciprocity is consistent with exchange theory as well as other collective strategy frameworks in the literature. Many of the studies cited here examine the health and social service organizations. Oliver concludes the discussion of reciprocity by noting several underlying assumptions for this reason: first, that resource scarcity often induces cooperation rather than competition; second, that relationships based on reciprocity often rely on a foundation of mutual support and equity; and third, that the participants expect the benefits of the IOR to exceed the costs and the loss of individual power.

Efficiency. The fourth reason identified by Oliver in the literature is efficiency. Oliver notes that since much of the inter-organizational relationship literature is based on non-market organizations, little attention has been given to efficiency motives for cooperation. Here she focuses on the input–output ratio internal to the organization as the motivating force. Also noted is that most of the literature citing efficiency determinants focuses on the negative impacts of such relationships on the market (e.g., on the restriction of competition through agreements).

Stability. Organizations facing resource scarcity in combination with high environmental uncertainty may choose to cooperate in order to create a more stable long-term environment. The pressure to establish relationships may come from the desire for predictability and reliability. Increased reliability of resource flows provides an incentive to manage relationships with other organizations.

Legitimacy. The final reason is based on the enhancement of organizational legitimacy. Oliver notes the pressures on organizations to justify their activities and products. She notes the pressures that "motivate organizations to increase their legitimacy in order to appear in agreement with the prevailing norms, rules, beliefs, or expectations of external constituents (Oliver 1990, 246). Relationships that highlight the social responsibility of an organization, or enhance its image with regulators, the public, or stockholders, may be exceedingly beneficial. Oliver notes that the literature reviews empirical cases where organizations seek to establish relationships with organizations having stronger legitimacy for the purposes of establishing reputations or increasing fund-raising capabilities.

Oliver concludes her description of the six determinants by noting that each of the six determinants, acting alone, may prove a sufficient condition to the establishment of inter-organizational relationships. However, she also notes that in many situations, multiple determinants will be at work. She discusses the implications of all of the various pairwise matching of determinants. In addition to these interactions, we also note that the determinants may change and replace each other *over time*. This is particularly true for GIS implementations requiring many years to complete and with long expected lifetimes requiring continued maintenance. We will note some of the time factors that seem to be at work in multi-participant GIS in our mini-case examples, later.

In order to see if the six determinants identified by Oliver apply to our case of multi-participant GIS, we created the matrix given in Table 2.1. In this table we have our three types of geographic database sharing as the rows and the six determinants across the columns. In the matrix cells we show examples of reasons that might account for the formation of each type of relationship. Here we note that this matrix of examples was constructed

TABLE 2.1
Fundamental Reasons for Geographic Data Sharing

TYPE	REASONS					
	Necessity	*Asymmetry*	*Reciprocity*	*Efficiency*	*Stability*	*Legitimacy*
TYPE I	Legal decision requires one-shot analysis	Organization wants to control information as sole source	Source organization has resources available; other parties' resources not relevant	Source organization sees GIS as providing efficiency gains in its own mission	Source organization has stable mandate, little environmental uncertainty	Source organization wants to improve quality of its own data collection and analysis
TYPE II	Legislative mandate encourages initiation of GIS	Lead organization wants to retain control of data quality and standards	Continuing maintenance and improvement of data over time requires more resources than lead agency has	There is a desire to limit data collection redundancy	Lead organization experiences some funding uncertainties in the future	Lead organization wishes to gain more visibility and credibility
TYPE III	State mandates a GIS effort	No single organization is able to act or to fund a GIS unilaterally	Joining together is the only way a GIS can be implemented	Each participant sees individual efficiency gains for its own organizational goals	Most of the participants are faced with funding uncertainties	Participants wish to gain higher visibility through the GIS

given our own personal acquaintance with GIS implementation efforts, and from our knowledge of the practitioner-oriented literature. We note here that Oliver, although she includes necessity (mandate) as one of the six reasons, found it difficult to include it in her summary tables. She made a decision to eliminate the factor in her summary analysis. Since we believe that mandates are frequent reasons for GIS implementation and data sharing, we have included it in Table 2.1.

As a natural progression, we also need to address another aspect of the inter-organizational relationships, that is, the question of the measurement of the *level or intensity of relationships between the organizations.* We can think of situations where the extent of inter-organizational relationship was in the form of a task force with little formal power, and of other situations where extremely formal and legalistic relationships were established. We devote the next section to a literature review of studies that have used or suggested measures of intensity.

LEVELS OF IOR: INTENSITY OF INTER-ORGANIZATIONAL RELATIONS

The degree to which inter-organizational relations reduce an organization's autonomy is a function of the type of relationship that an organization establishes. Periodic informal meetings in which information is exchanged, for example, involve less resource investment by any partner and infringe less on an organization's decision-making authority than joint ventures. Similarly, resource exchanges, such as lending equipment or personnel to another organization, have fewer implications for control over an organization's decision processes than the development of board interlocks, where external constituents acquire some formal decision-making authority over the organization's operations.

Systematic differences among types of inter-organizational relations, in terms of their required resource commitment and constraints on organizational decision-making discretion, have been tested empirically. In early research on inter-organizational relations, several studies confirmed the statistical validity and reliability of a rank ordering or scale of different types of inter-organizational relations that varied according to their degree of "intensity" (Klonglan, Warren, Winkelpleck, and Paulson 1976; Rogers 1974). In these studies, intensity of an inter-organizational relationship was defined as the magnitude of resource investment required by a relationship, where resources could include finances, personnel, clients, physical facilities, and authority. Intensity also referred to the degree of commitment required by an organization, where commitment denoted the extent of an organization's obligation to share authority and reciprocate in its relations with a linkage partner (Rogers 1974). These scales were developed to bring greater precision to the measurement of inter-organizational relations and to build on early theoretical work on inter-organizational relations in the public sector, which suggested that administrators prefer to minimize outside interference in their day-to-day operations.

Rogers (1974, 62) acknowledged "a preference among organizational administrators for maintaining autonomy" and proposed that "an organization's ability to make decisions independent of outside pressures varies with the type of contact it has with other units." He developed and tested a scale of intensity of inter-organizational relations among public agencies in which six types of relations were rank ordered from low to high intensity: director acquaintance, director interaction, information exchange, resource exchange, overlapping boards, and written agreements. The first two relations reflect minimal intensity, the third represents an increase in intensity as resources are expended in producing and exchanging information, and the fourth, resource exchange, requires further increases to the level of an organization's commitment of resources to an exchange. Board interlocks are even more intense types of relations because "one organization permits members of another organization to participate in its decision-making functions and relinquishes a certain amount of control" (Rogers 1974, 65). Finally, written agreements, according to Rogers (1974, 65), "institutionalize reciprocal obligations and costs associated with a joint activity." These types of relations are considered the most intense or committing because "formalizing such arrangements by putting them into writing represents a higher degree of resource investment" and further constrains an organization's decision-making freedom (Rogers 1974, 65).

This scale of relationship intensity was found to have high reliability and validity. In subsequent studies, other researchers (Klonglan et al. 1976) tested and expanded similar scales that rank-ordered eight types of inter-organizational relations from low to high intensity: director awareness, director acquaintance, director interaction, information exchange, resource exchange, overlapping board membership, joint programs, and written agreements. Despite the validity and potential utility of these scales, they have not been applied in other recent research on organizational relations (Oliver 1990).

In this chapter we propose a five-step scale of relationship intensity: personal meetings (which integrate director interaction and information exchange into one category and include interactions along the lines of task forces and joint committees), resource transfers, board membership, joint programs, and written contracts. These five types of relations, in the order indicated, represent progressively higher intensity of relations and lower autonomy in inter-organizational relations: organizations sacrifice the least amount of autonomy by informal personal meetings to exchange information, a greater amount by establishing resource transfers, a greater amount still by board interlocks, more still by joint programs, and the most by written contracts or agreements. The rank ordering of these relations and their possible operationalizations are summarized in Table 2.2. The general proposition advanced here is that the probability of establishing inter-organizational relations will be related to the intensity of the relationship and the loss of organizational autonomy required by the relationship.

It is difficult to label objectively the level of IOR that is signified by each of these scales that show intensity. However, it is useful to propose such a scheme for conceptual clarity. Towards that end, we agree with Whetten (1982) that coordination as a form of IOR requires the highest possible level of interdependence among organizations without them merging together (which is obviously the logical conclusion of any IOR to the maximum level). The beginning levels of IOR are perhaps harder to characterize. Again borrowing from Gricar (1981), we would suggest collaboration as the label for the lowest level of IOR among organizations. This implies that there is minimum commitment and interdependence among the organizations; nevertheless, a level of IOR exists. Finally, the mid-level IOR, whatever that level may be, we characterize as cooperation. This level of IOR requires a higher level of interdependence and loss of autonomy, than say collaboration, but lower than coordination. Now, mapping these into the corresponding IOR scale levels, it is possible to see that as any IOR increases in intensity, it should move from collaboration to cooperation to coordination. Table 2.2 graphically depicts this conceptual mapping.

STAGES OF IOR

The reasons for IOR and the levels of IOR are powerful explanatory concepts to disentangle the complex dynamics of geographic data sharing. More importantly from a normative point of view, we need to prescribe the typical stages of an IOR for geographic data sharing. This requires a process (time-dependent stage) model. Drawing on the analysis thus far and a review of the so-called "transorganization development" (Cummings 1984), "inter-organizational development" (Cummings 1984; Shermerhorn 1979) literature, we propose stages for such a model and some elements of those stages.

A process model should have the structural characteristics that Mohr (1982) specifies: it posits a process that unfolds over time—it specifies

TABLE 2.2
Intensity Scale for Inter-Organizational Relationships in Geographic Data Sharing and Mapping of Collaboration, Cooperation, Coordination

LEAST INTENSIVE	1	2	3	4	5	MOST INTENSIVE
MOST AUTONOMOUS	Personnel meetings	Resource transfers	Board membership	Joint programs	Written contracts	LEAST AUTONOMOUS
MOST AUTONOMOUS	Collaboration		Cooperation	Coordination		LEAST AUTONOMOUS

necessary preconditions for an IOR to move from collaboration to cooperation to coordination (CCC)—to occur without claiming to specify a set of conditions that inevitably lead to CCC; it describes directional forces and probabilistic processes that make CCC more or less likely to occur; and it derives its explanatory power from combinations of crucial elements rather than from magnitudes of particular variables.

This process model of CCC is a preliminary attempt to combine and reformulate the insights of the dominant theories of inter-organizational relations in the service of guiding public management of the complex and uncertain processes in the geographic data-sharing arena. Such a model can offer guidance to those who want to design and promote CCC in geographic data sharing.

A Suggested Process Model

Building on the work of McCann (1983), we propose a possible process model of transorganizational development for geographic database sharing. McCann (1983) has observed that inter-organizational relations develop through three sequential phases: problem-setting, direction-setting, and structuring. He suggests that progression through these phases is a natural process for an IOR; but, as in any sequence, internal and external forces can interrupt, enhance, or impede the cycle. Our understanding of CCC can be advanced, then, by understanding the developmental phases and by considering the conditions necessary to move through each developmental phase. Implicit in this approach is the idea that IOR dynamics can be managed to improve the likelihood that CCC relationships are achieved and sustained.

Problem-Setting or Collaboration. Problem-setting is concerned with identification of the stakeholders within a domain and mutual acknowledgment of the issue that joins them. The domain is a key term in the framework developed by McCann. It refers to the particular inter-organizational environment and the target resource to be shared. In our case, it is any geographic database jointly developed. According to McCann (1983, 18), "The primary objective of problem-setting is to give the situation an explicit form or identity that allows stakeholders to communicate about it and eventually act upon it." The importance of problem-setting should not be underestimated. Unless some consensus is reached about who has a legitimate stake in an issue and exactly what that joint issue is, further attempts at CCC will be thwarted. Problem-setting may involve focusing the attention of existing members of the domain in a direction they have previously ignored. It is also a vehicle by which previously unrecognized stakeholders achieve recognition as legitimate claimants in the issue. Through the problem-setting process, then, stakeholders negotiate issues of legitimacy and come to appreciate the interdependence that exists among themselves. CCC is initiated through appreciation of the interdependence that exists and the need for joint "appreciation" among stakeholders of the nature and substance of their interdependence.

Direction-Setting or Cooperation. In the direction-setting phase, stakeholders articulate the values that guide their individual pursuits and begin to identify and appreciate a sense of common purpose. Direction-setting gives life to the stakeholders' hopes that their desired ends can, in fact, be achieved. One vehicle for direction-setting is the "search conference" in which participants engage in joint conceptualization of the desired future of the domain (McCann 1983). This process of developing shared interpretations about the future of the domain and the articulation of commonly held values or goals serve to correlate the stakeholders' activities toward mutually desirable ends.

Structuring or Coordination. For persistent complex problems like geographic data sharing (e.g., MassGIS), the need for ongoing appreciative processes becomes essential, as does the need to manage stakeholder interactions in an increasingly systematic manner. Trist (1983) refers to this process as self-regulation of the domain. McCann (1983) refers to this third phase of CCC as structuring. CCC will be enhanced when joint appreciation of the problem dynamics is shared by stakeholders and they undertake negotiations to create a mutually acceptable regulative framework for the domain.

Through structuring, stakeholders generate a system for sustaining coincident values and establishing order within the domain. Specific goals are set, tasks are elaborated, and roles are assigned to stakeholders. Trist (1983) labels the organizational arrangements that emerge "referent structures." While the establishment of a formal structure is often thought of as occurring at a fixed point in time, we prefer to think of structuring as a developmental phase. Structuring refers to the process of institutionalizing the shared meanings and prevailing norms that emerge gradually as the domain develops. Formalization of these during the structuring phase represents an explicit ratification of a previously loosely negotiated understanding.

We recommend a research program that incorporates case study research to enable us to identify different conditions that facilitate successful

CCC during each developmental phase. Table 2.3 proposes some of the conditions we expect may be significant at each phase.

TABLE 2.3

Facilitating Conditions for Levels and Stages of IOR

Collaboration	*Cooperation*	*Coordination*
PROBLEM-SETTING	DIRECTION-SETTING	STRUCTURING
Recognition of interdependence	Coincidence of values	High degree of ongoing dependence
Identification of a requisite number of stakeholders	Dispersion of power among stakeholders	External mandates
Perceptions of legitimacy among stakeholders		Redistribution of power
Legitimate/skilled convener		Influencing the contextual environment
Positive beliefs about outcomes		
Shares access power		

MINI-CASE EXAMPLES

To illustrate our arguments in this chapter, we have selected three mini-case examples. We note, without apology, that these three cases were selected because of their geographic proximity and our familiarity with their histories. Here our purpose is limited to illustration of the frameworks we are adapting from the organizational behavior literature. Since these are not based on the full interviews of a complete case study, but rather on our interpretation of secondary literature and on brief initial conversations, we do apologize for any misinterpretations. We hope that further case-study–based research may further clarify these examples.

Mini-Case Example One: The CLUIS Project, Boston Redevelopment Authority

The Centralized Land Use Information System (CLUIS) project began in the early 1980s when an East Boston community group—Air, Inc.—filed suit against the Massachusetts Port Authority (MPA) to get the MPA to perform an impact assessment of a proposed airport expansion. The resulting settlement between MPA and Air, Inc., stipulated a comprehensive land-use database be developed to show changes in land use that had taken place over five years due to airport-related business activity.

After several false starts the final mutually agreed upon project was for the Boston Redevelopment Authority (BRA) to build a multi-departmental database of parcel-related information from various city of Boston departments. As a component of this database, thematic maps of land-use patterns were to be produced. The project manager and staff at BRA proposed to perform the database construction on a PC and subsequently link it to a parcel boundary map of East Boston so that the production of various maps and the related analyses could be more easily performed.

The database of parcel characteristics, fire locations, building permits, tax assessments, and other related information was obtained from different city departments and integrated into a PC database. Subsequently, the data was linked to a parcel-level map of East Boston. Several maps were produced and shown to the Air, Inc. group. The Air, Inc. group was very happy with the product. However, they also wanted to see continued support and update of this material. They thought this could be done jointly by MPA and BRA.

The MPA Board of Directors also saw a presentation of the system and expressed satisfaction with the product. However, in response to the community request, they indicated that MPA would prefer to end its partnership then and limit it to the funding of the project up to that point. Several reasons were given: 1. maintenance of the data created a responsibility that MPA was not prepared to take on; 2. a funding mechanism for such continuing effort beyond the suit settlement was not clear and perhaps not available; and 3. the MPA Board of Directors suggested that the BRA could take on the role of maintenance and support, as the most logical actor to do so given the nature of the agency's work.

The plea by Air, Inc. did not result in BRA adopting the maintenance and upkeep of the CLUIS database. The BRA indicated to the community that funds and staff for such efforts did not exist. Especially important was the fact that the MPA hardware, on loan to the BRA, was taken back by the MPA since they correctly claimed ownership.

The antecedent operating to initiate the CLUIS project was *necessity*, in this case by a legal mandate. The court action generated the impetus for MPA to enter into a partnership with another agency (BRA) to develop the database. However, despite strong reciprocal relationships that developed between the agency staff on both sides, MPA never saw it to be in its interest to carry the project forward. One can conjecture the reasons for the one-shot analysis. One interpretation is that MPA always saw the project as an encroachment upon its turf and was quite happy to be done with it. Another reason may be that MPA wanted to carry out the project itself, but since the community insisted on a third party becoming involved, this reduced the extent of MPA autonomy in the project and thus was not welcomed.

On the BRA side, the agency may have also refused further development and support for a number of reasons. One possible reason may be that BRA sees its critical mission as redevelopment of large commercial projects that provide employment and add to the stature of the city. It is perhaps less concerned with residential land-use planning. Furthermore, a joint program like CLUIS with MPA may have been viewed as a precursor to some loss of autonomy. In particular, the Boston-area newspapers made a big issue out of BRA proposed zoning ordinances and about whether MPA-owned property actually would be subject to those rules. This is a vague area of the law and can be defined only through legislation or litigation. BRA perhaps saw a threat to its zoning autonomy if it was seen collaborating, cooperating, and coordinating joint programs with MPA.

Concerning the intensity measure of the interaction, the CLUIS project did result in resource transfers (the second level) and in a written contract (fifth level). However, the written contract was short-term and did not establish a formal relationship over time.

Mini-Case Example Two: MassGIS

The MassGIS project was initiated by the Massachusetts Executive Office of Environmental Affairs (EOEA) in 1985. A good description of the effort to date is given in Taupier and Terner (1991). There appear to be several antecedents to the development of this statewide GIS. Initially, the system was started to help the EOEA better meet its goals—in particular, the goals of selecting suitable sites for hazardous waste in the state. The antecedent of *legitimacy*, particularly in light of public opposition to facilities, played a role here.

Taupier (1992) notes that EOEA is composed of five separate agencies and that at the time of initiation of MassGIS three of those five had expressed interest in establishing their own system. The antecedent of *efficiency* was clearly operative as the five agencies agreed to contribute to a shared system resource. Taupier notes that an analysis was completed to calculate the cost savings due to limiting redundant data collection, and that the decision to develop a shared system among the five agencies was based on this analysis (Taupier 1992, 2). He notes that there are now more than forty direct-access users of the system among the five agencies, and that the relationship is much as anticipated by the system planners at the initiation of the system.

Not anticipated at the start of the project was the demand from outside the five agencies for MassGIS data. This case provides a clear example of the change in the antecedents of inter-organizational CCC over time. Taupier describes the costing structure developed for purchase of digital and paper map products. These products are now being sold to a range of other state, regional and local agencies as well as to private-sector firms and consultants. In addition to sales, MassGIS has also been entering into data exchange agreements. Such agreements provide us with an illustration of the antecedent of *reciprocity*. The agreements provide data from MassGIS to agencies in turn for either new information (e.g., a more detailed land-use map of a town) that will enhance the MassGIS coverages, or for other updated information. Because of the long-term orientation of such statewide efforts of MassGIS, the issues of maintenance and update begin to take on an increasingly important role. Although there might be a tendency towards a relatively one-sided, *asymmetric* relationship here—with MassGIS as the sole provider, and resource-poor towns with no GIS access—this has not occurred. The reason, we suggest, is one of relative resource scarcity from all parties (including MassGIS). Here, reciprocity does, in fact, seem to be the operating factor as all the agencies seek a beneficially common cost sharing. Elements of the antecedent of *stability* are also present, as all of the parties attempt to hedge against fiscal uncertainties in a weak economy.

On the intensity scale, the second level of resource transfer appears to dominate the inter-organizational relationships to date.

Mini-Case Example Three:
The Boston Central Artery and Tunnel

The Boston Central Artery and Tunnel project is an incredibly expensive, huge, and complex endeavor to replace an existing overhead downtown freeway with a tunnel. The $4.9 billion project will take ten years to complete and will require the cooperation of a large number of multi-disciplinary teams dispersed among many private consulting firms.

Because the project will be constructed through a complex existing urban environment, detailed spatial data are required by all participants. Early in the project, the construction managers (joint venture of Bechtel Corporation and Parsons-Brinckerhoff) decided on a digital base map using GIS technology. Because the project participants had different hardware and software, the implementation had to take place in a multi-vendor environment. Rather than have each participant duplicate the work of digitizing a base map, the participants agreed to develop and maintain a single base map. Extensive agreements about standards and accuracy specifications were required to translate the geographic data between different systems. In addition to the major construction managers, up to thirty other outside consulting firms also had to agree to follow the agreed-upon standards and data formats (Van der Schaaf, Schintzel, and Petersohn 1991).

This cooperation seems to have been driven predominately by the antecedent of *efficiency*. Having large numbers of engineering consultants duplicate the base mapping effort would have been costly, as well as risky from a quality control perspective. From our interviews, we also judge there to have been an element of *legitimacy* involved as an antecedent. There was a subtle emphasis on the "showcase" aspects of this very large infrastructure project for the firms.

Over time there have been some aspects of *necessity* as well. As the subcontracting firms are selected they must show their ability and willingness to adhere to the common standards of the participants through an approved compatibility program. Operating in a multi-vendor and multi-user environment has not been trouble free for the project participants but has allowed each firm to use software and hardware that they have available. The center of the geographic data sharing in this case is the establishment and enforcement of standards.

The intensity level on our five-step scale for this case is high, due to the formal written contractual agreements between the construction managers and the subcontracting firms. However, like the CLUIS project, the written agreements are short-term and, to date, there has not been a decision about the maintenance of the data or the inter-organizational relationships in the long run.

CONCLUSIONS: A RESEARCH FRAMEWORK FOR INTER-ORGANIZATIONAL DATA SHARING

In this chapter, we have argued that the sharing of geographic data is principally an inter-organizational relation. As such, any organization's decision to engage in inter-organizational relations (data sharing) is primarily determined by the autonomy "calculations" of the outcomes; that is, each organization will attempt to minimize its loss of autonomy and therefore interdependency with other organizations.

REFERENCES

Burns, W. J. Jr., and F.W. McFarlan. 1987. Information technology puts power in control systems. *Harvard Business Review* (September–October).

Cook, K. S. 1977. Exchange and power in networks of inter-organizational relations. *Sociological Quarterly* 18 (Winter): 62-82.

Cummings, T. G. 1984. Transorganizational development. In *Research in Organizational Behavior* 6. CT: JAI Press. 367-422.

Dangermond, J. 1989. The organizational impact of GIS technology. *ARC News* 11: 3.

Evan, W. M. 1966. The organization-set: toward a theory of inter-organizational relations. In J. Thompson, ed., *Approaches to organizational design*. Pittsburgh: University of Pennsylvania Press, 172-191.

Gricar, B. G. 1981. Fostering collaboration among organizations. In H. Meltzer and W. R. Nord, eds., *Making organizations humane and productive*. New York: John Wiley and Sons, 403-420.

Klongan, G. E., R. L. Warren, J. M. Winkelpleck, and S. K. Paulson. 1976. Interorganizational measurement in the social service sector: differences by hierarchical level. *Administrative Science Quarterly* 21: 675-687.

Levine, S., and P. E. White. 1961. Exchange as a Conceptual Framework for the Study of Inter-organizational Relationships. *Administrative Science Quarterly* 5: 583-610.

McCann, J. E. 1983. Design guidelines for social problem-solving interventions. *Journal of Applied Behavioral Science* 19, 2: 177-192.

Milner, M. Jr. 1980. *Unequal care: a case study of inter-organizational relations in health care*. New York: Columbia University Press.

Mohr, L. B. 1982. *Explaining organizational behavior*. San Francisco: Jossey-Bass Publishers.

Mulford, C. L. 1984. *Inter-organizational relations: implications for community development*. New York: Human Sciences Press.

Oliver, C. 1990. Determinants of inter-organizational relationships: integration and future direction. *Academy of Management Review* 15, 2: 241-265.

Pfeffer, J., and G. R. Salancik. 1978. *The external control of organizations: a resource-dependence perspective*. New York: Harper and Row.

Pressman, J., and A. Wildavsky. 1984. *Implementation*. 3d ed. Berkeley, CA: University of California Press.

Rogers, D. L. 1974. Towards a scale of inter-organizational relations among public agencies. *Sociology and Social Research* 59: 61-70.

Sarason, S. R., and E. Lorentz. 1978. *The challenge of the resource exchange network*. San Francisco: Jossey-Bass Publishers.

Seidman, H. 1976. *Politics, position, and power*. 2d ed. New York: Oxford University Press.

Shermerhorn, J. R. 1979. Inter-organizational development. *Journal of Management* 5, 1: 21-38.

Taupier, R. 1992. An examination of incentives for the exchange of geographic information: the theory and implementation of the evolving data sharing programs in the Commonwealth of Massachusetts. NCGIA Initiative 9, San Diego, California. February.

Taupier, R., and M. Terner. 1991. MassGIS: a case study of the Massachusetts Environmental Geographic Information System. In Urban and Regional Information Systems Association 1991 *Proceedings*, pp. 87-96.

Trist, E. 1983. Referent organizations and the development of inter-organizational domains. *Human Relations* 36, 3: 269-284.

Van de Ven, A. H. 1976. On the nature, formation, and maintenance of relations among organizations. *Academy of Management Review* 1, 4: 23-44.

Van der Schaaf, F. A. Schintzel, and C. Petersohn. 1991. *GIS and the Central Artery/Tunnel Project*, Boston, Massachusetts.

Whetten, D.A. 1982. Objectives and issues: setting the stage. In D. L. Rogers, D. A. Whetten, and Associates. *Inter-organizational coordination: theory, research and implementation*. Ames: Iowa State University Press.

Jeffrey K. Pinto and Harlan J. Onsrud

3 *Sharing Geographic Information Across Organizational Boundaries: A Research Framework*

As development of geographic information systems progresses and technical problems are overcome, new problems arise. One of the frequently mentioned problems is that of the need for information and databases that may be housed in several organizations. This chapter explores the organizational and managerial roots of difficulties in sharing databases. Historically, a combination of organizational structure and operations has resulted in the fragmentation of work projects into individual tasks and growing powerlessness among manual and clerical. Such fragmentation of tasks tends to increase the power of knowledge workers and their organizations, since such workers and organizations often control information that is valuable because it is often both unique and indispensable. This chapter identifies a three-stage path model that delineates: 1. a set of facilitators of organizational information sharing, 2. the process of information sharing, and 3. consequences of enhanced information sharing.

INTRODUCTION

In recent years, Geographic Information Systems (GIS) have seen a steady rise in acceptance and use by a variety of organizations. From a technical perspective, the electronic form of data facilitates sharing. However, the rapid increase in organizations adopting GIS technologies has belied the fact that between and among organizations, there has been a general inability and often unwillingness to share data and information across organizational boundaries. The lack of information exchange among local, state, federal government, and private sectors remains a significant impediment to the more effective and efficient use of GIS throughout society.

A recent initiative sponsored by the National Center for Geographic Information and Analysis (NCGIA) has sought to call attention to the need for increased sharing of geographic data sets as a first step toward gaining better coordination among GIS-using agencies and organizations. It is widely believed that developing methodologies that provide corporate and personal incentives for internal and external sharing of geographic information in tandem with creating appropriate technical techniques and standards will help facilitate sharing of information among those who have acquired the technology. Such methods will serve also to advance the acceptance and use of geographic information within organizations that have not, to date, adopted GIS.

The success of organizations in both the public and private sectors can be greatly enhanced by the open exchange of information across organizational boundaries. One of the problems hampering the integration and widespread use of GIS has been the continued inability of various public agencies at the federal, state, and local levels to develop purposeful and productive collaborative relationships. Specifically, there is a general inability to effectively share GIS information among these organizations, resulting in the duplication of systems and services at different levels. This chapter proposes a research framework that addresses the influence of a set of antecedent constructs (superordinate goals, accessibility, communicative proximity, bureaucratization-formalized rules and procedures, the quality of exchange relationships among organizations, and resource munificence/scarcity) on the attainment of both inter-organizational cooperation and optimal use of GIS information.

ANTECEDENTS AND CONSEQUENCES OF INFORMATION SHARING

It is our goal to propose a research framework for the study of facilitators of information sharing among organizations. Within the organizational theory literature, the area of inter-organizational cooperation is one that has seen far less research to date than intra-organizational cooperation. Further, little research has examined the types of factors that can have a positive impact on influencing organizations to share information. Our goal is to offer a framework for understanding information sharing based on the expected effects of a variety of organizational factors on cooperation.

In attempting to address the issue of GIS information sharing among organizations, it is first necessary to provide a context within which such information is often needed. We propose three distinct contexts requiring inter-organizational GIS information sharing. The first information sharing context is situation specific, or project-driven. In this situation, two or more agencies come together to work on a common problem that both parties need to address. The two organizations bring their own data and expertise to the table and share information with each other in an effort to most successfully solve the problem. For example, hazardous waste disposal is an environmental problem that may require the cooperation of several federal agencies with overlapping responsibilities. In order to develop a comprehensive and effective solution to a hazardous waste problem, several agencies come together to share GIS information that they each possess so that all parties can contribute to an optimal solution. As this example illustrates, the problems associated with this context are usually non-routine and non-recurring, requiring that they each be addressed on a case-by-case basis with their own distinct solutions.

The second information sharing context is one in which different agencies may be addressing completely different problems but have a need for very similar information. Because they have needs for the same information, organizations develop procedures by which they can regularly share and exchange information with each other. For example, the Bureau of Indian Affairs and the Bureau of Land Management may determine they both need information about the same federal lands. This need for common information encourages collaborative (sharing) behavior. Before they begin to exchange information, an analysis is performed to determine the procedures (both technical and administrative) by which these two bureaus can most effectively exchange GIS information. In this context, often the first step in the sharing process is to determine some sort of protocol as to how the information of each can be accessed by the other. The focus is on the needs of the cooperating parties, and meeting those needs might result in either a case-by-case approach or a long-term data-sharing protocol. However, the protocol established to meet the needs of the participating parties may be highly inappropriate for effective sharing with additional parties or the broader community.

The third context for information sharing among organizations is based on developing standardized or generalizable patterns of exchange. In this context, information is readily accessible to all parties and can be accessed in useful forms from a central location, such as a data bank, or from networked or decentralized sources. Organizations simply engage in a routine sharing process through the central storage facility or network of decentralized facilities.

In this chapter, we focus on the first context for information sharing—

that of organizations working together in an ad hoc manner to solve a specific problem. It represents a common aspect of the information sharing problem. Many situations are readily evident and observable that require two or more agencies to exchange GIS information to solve a problem in which they have common interests. As a result, in focusing on this more common context, we are addressing concerns that are immediate and compelling to a wide range of individuals charged with the task of attempting to develop better methods for information sharing.

The chapter focuses on a three-stage model that identifies both antecedents and consequences of inter-organizational cooperation in sharing GIS information and technologies. Research in the areas of organizational power and political behavior (Pfeffer and Salancik 1978), channel relationships (Dwyer and Oh 1987), negotiation (Wall 1985), and intergroup conflict (Thomas 1976; Walton and Dutton 1969) have sought, as one of their goals, an increased understanding of the factors by which improved cooperation can occur. While this research has led to an enhanced understanding of cooperation from an organization theory perspective, an analysis of the factors that can lead to greater sharing has not been attempted in organizational environments with a focus on the dynamics of sharing geographic information. Further, the relationship between GIS information sharing and system satisfaction, effectiveness, efficiency, and decision making has not received attention to date.

Previous research on cooperation and information sharing has focused almost exclusively on intra-organizational collaboration; that is, attempting to better understand how different functional groups within a single organization can develop more cooperative relationships. This research stream, best represented by the work of Lawrence and Lorsch (1967), Ruekert and Walker (1987a; 1987b), Gupta, Raj, and Wilemon (1986), and Souder (1981; 1988) empirically examined relationships within an organization, usually between specific functional groups, and suggested factors that are important in fostering these relationships. Their findings deal with the impact of interdependence or resource dependency and coordinating mechanisms (i.e., formalized rules and procedures) on cross-functional interaction. In addition, their results suggest that the similarity of functional departments—in terms of duties and objectives—positively influences the amount and type of communication between them. Finally, this research has been very useful in offering prescriptive advice on methods to better facilitate information sharing among different functional departments.

What have been missing from the literature are systematic attempts to develop a framework of antecedent variables that can improve the likelihood of creating positive, collaborative relationships *between* organizations.

Specifically, little is known, for instance, about the reasons why governmental agencies and other GIS-using organizations will or will not share GIS-related information. There is strong evidence to suggest that considerable duplication of effort occurs across organizations because of a basic lack of will to cooperate. Although the technical ability to readily share geographic information might be accessible or achievable, the organizational and personal incentives to share are insufficient to overcome the impediments. In other instances, data and information are held closely as sources of control and power, even within government offices. In these instances, there is often outright unwillingness to share this "proprietary" information.

Little is understood about the factors that can act as facilitators, or antecedents, of information exchange among organizations. The purpose of the proposed research is to investigate the significance of a selected group of organizational and behavioral antecedents that may be necessary or beneficial in allowing the sharing or increased sharing of geographic information among various producer and user organizations. The discussion draws on literature from a number of sources, including organizational theory, inter-group dynamics, exchange theory, and political economy in order to posit a model of inter-organizational information sharing.

CONCEPTUAL FRAMEWORK

A variety of factors can act as facilitators or inhibitors of information sharing and cooperation among organizations. These factors range from individual factors such as the personalities of group members, use of information as a source of personal power, "turf" battles, interpersonal relations, and training and skills (Johnson 1975; Kelly and Stahelski 1970; Pavett and Lau 1983; Pfeffer and Salancik 1978; Schmidt and Tannenbaum 1960) to organizational factors such as political concerns, reward systems, and cultural norms (Lawrence and Lorsch 1967; Mintzberg 1979; Moch and Seashore 1981; Pfeffer 1982; Shapiro 1977). In addition to these environmental constructs, the state of computing management for each organization (Kraemer, King, Dunkle and Lane 1989) will be investigated and treated as antecedents of information sharing. We propose a research framework that will address the expected relationships between a variety of these antecedent variables and the attainment of inter-organizational cooperation, as well as investigating the "second-level" effect of inter-organizational cooperation on projected consequences or outcomes of information sharing. These consequences are assessed in terms of increased efficiency of organizational tasks, increased effectiveness of output (Gillespie 1991), and information system and partner organization satisfaction (Ives and Olsen 1983).

The antecedents shown in Figure 3.1 were chosen for several reasons. First, we believe it is necessary to identify two types of antecedents of information sharing and inter-organizational cooperation: those that were to some degree under the control of the project team members, as well as larger, organizational level constructs. Ease of communication, accessibility, and rules and procedures are proxies for organizational factors that significantly influence interactions among individuals and encompass many implicit and explicit aspects of an organization's strategy, structure, and culture (Peters 1990; Pinto, Pinto and Prescott, 1993). For example, organizations that permit their GIS departments to interact with other parties at their locations, modify their work schedules to the demands of the project, and develop their own rules and procedures to facilitate cooperation would represent a different type of culture than those organizations that do not permit the same degree of latitude. Further, the acceptance of superordinate goals by team members should transcend and mitigate the role of individual factors. Additionally, there has been little organizational-related field research utilizing superordinate goals as a method of facilitating cooperation. The quality of exchange relationships argues that positive inter-organizational relationships are marked by high levels of trust and minimal opportunism. Finally, research on resource munificence argues that organizations that operate under conditions of increasing resource scarcity are more likely to regard internally generated information as a source of power to be held over other parties. As a result, there is a strong tendency to hoard, or refuse to share, that information (Pfeffer and Salancik 1978). The establishment of means for sharing information among organizations is a nonroutine task whereas the goal is that actual sharing should become routine. Our choice of antecedents represents the need to balance past research findings that have dealt primarily with teams within an organizational unit with our objective of studying cross-functional teams working on nonroutine projects.

Figure 3.1 illustrates the research framework that identifies a three-stage path analytic model delineating the factors argued to facilitate inter-organizational GIS information sharing. The following sections discuss the importance of each of these factors, the importance of cooperation itself, and perceived outcomes, or results, of enhanced GIS information sharing.

THE NEED FOR INFORMATION SHARING

The responsibility for many project-based activities often overlaps, or is held jointly, between two or more GIS-using organizations. It has long

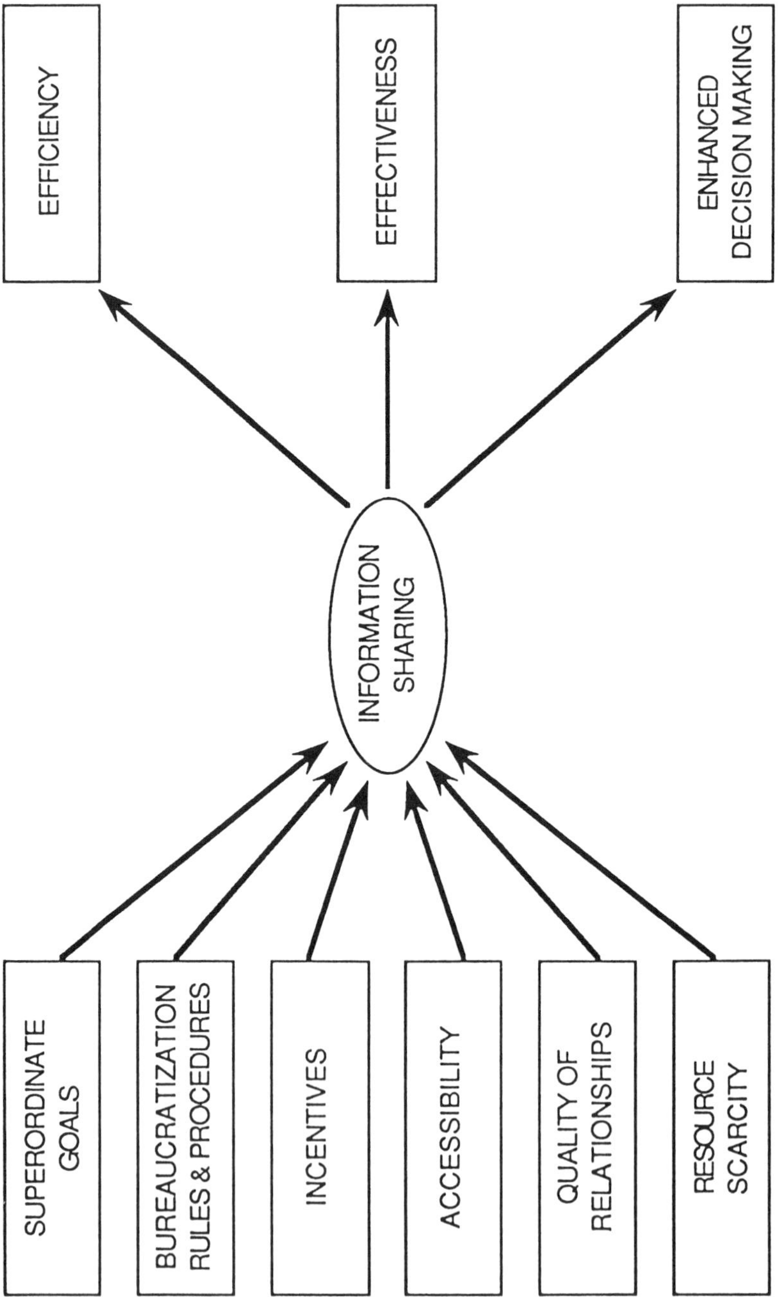

FIGURE 3.1
Antecedents and consequences of information sharing

been acknowledged, from an organizational theory perspective, that these "difficult-to-assign activities give rise to such interdepartmental issues as cooperation, coordination, conflict, and struggles for power" (McCann and Galbraith 1981, 60). To manage the development and implementation of these activities, a variety of integrating mechanisms have evolved, including task forces, liaison roles, and cross-organizational teams (Galbraith and Nathanson 1978; Lawrence and Lorsch 1967; Mintzberg 1979). These teams, or task forces, allow for lateral contact between multiple organizations (Dumaine 1990). They tend to be temporary groups that exist for the duration of the designated activity. Because of a general unwillingness for some organizations to cooperate and share information, the activities of these temporary task forces have become increasingly important for the long-term viability of the organization. Consequently, efforts must continually be made to develop policies and mechanisms that promote, rather than inhibit, cooperation across organizational boundaries.

Antecedents of Inter-Organizational Cooperation

Superordinate Goals. Classical organization theory originally established the importance of goals in organizations (Simon 1964). Since that time much has been written on the concept of an organizational goal (Kono 1984), the purposes served by goals (Quinn 1980), the multiplicity of goals in organizations (Raia 1974), and the hierarchical nature of goals (Galbraith and Nathanson 1978).

Every organization and, indeed, every manager has more than one goal that guides activities and actions. In theory, different organizations performing similar or complementary functions should possess complementary goals that are derived from a set of general, cross-organizational objectives. In practice, however, overall goals are often broken down into specific objectives that conflict with, rather than complement, each other. Consequently, in order for one organization to achieve its goals, another may be required to sacrifice, or at least compromise, its primary goals. Newman (1988) cautions that a departmental goal must be compatible with the activities of related departments so that it will not undermine the results of those activities or make them much more difficult. Consider, for example, the common conflict between state and local or state and federal agencies over GIS data sets. Because the goals of these agencies are different and may often conflict, willingness to go out of one's way to share information collected by someone else is likely to be lacking. This situation results in strong potential for duplication of data collection and maintenance efforts and may result in implementation of systems that are each underutilized by their respective organizations. Members of each organization will argue for the need for their own system and are often loath to cooperate or make available readily accessible information that may be needed by other agencies. Aware of such conflict, organizations and oversight groups are continually looking for ways of developing goals that can increase, rather than detract from, information sharing and inter-organizational cooperation.

One important, but often overlooked, type of goal in the study of cross-functional cooperation is a superordinate goal. Superordinate goals refer to "goals that are urgent and compelling for all groups involved, but whose attainment require the resources and efforts of more than one group" (Sherif 1962, 19). It is important to note, however, that superordinate goals are not a replacement for other goals the various organizations may have; rather, they are an addition to existing organizational goals. Through various controlled field and laboratory experiments, Sherif compiled impressive evidence indicating that when a series of superordinate goals are introduced into a conflict situation, intergroup conflict is reduced and cooperation is achieved. The essence of Sherif's theory is that competitive individual goals cause intergroup conflict, but superordinate goals give rise to intergroup cooperation that enhances group output. Specifically, when groups come into contact under conditions embodying shared goals or common purposes, they tend to cooperate toward common goals. Because of this cooperation, system satisfaction is enhanced; thus, evidence suggests that a two-stage process is present. Superordinate goals are assumed to lead to enhanced data sharing which, in turn, is expected to result in improved levels of organizational efficiency, effectiveness, and decision making.

The organizational research on superordinate goals has tended to be primarily conceptual in nature (Stern and Heskett 1968) or conducted in an experimental setting in which the existence of the superordinate goal(s) is manipulated (Johnson and Lewicki 1969; Stern, Sternthal and Craig 1973). From this there is strong evidence to suggest that superordinate goals can have a powerful impact on inter-organizational cooperation. What superordinate goals would cause numerous federal, state, local government, and private sector organizations to work towards geographic information sharing? Would developing the concept of a nation-wide library reference system for spatial data and providing the distributed network infrastructure allowing ready transfer of large data sets provide such a superordinate goal? What lesser superordinate goals would enhance abilities and motivations to share? The following research proposition is suggested:

P1: *Superordinate goals will have a positive and significant direct effect on geographic information exchange.*

Bureaucratization–rules and procedures. Rules and procedures refer to the degree to which activities or tasks on a project team are mandated or controlled. Rules and procedures are central to any discussion of inter-organizational cooperation because they offer a mechanism for integrating or coordinating activities, particularly those activities that cut across agency or departmental boundaries (Galbraith and Nathanson 1978). Early organizational theorists such as Taylor (1911), Fayol (1929), and Mooney (1947) relied on rules and procedures to link together the activities of organizational members. They posited that coordination could be accomplished by simply establishing rules and procedures throughout the management hierarchy. Subsequent researchers (e.g., Gouldner 1954; March and Simon 1958; McCann and Galbraith 1981) also discussed rules and procedures as a technique for coordinating activities, controlling behaviors, and maintaining organizational structure. At the departmental level, McCann and Galbraith (1981) described rules and procedures as the most common structural variable for assigning duties, evaluating performance, and minimizing the occurrence of conflicts between departments. Providing empirical support for this assertion, Ruekert and Walker (1987b) found that written or formalized rules and procedures have a significant positive relationship with the perceived effectiveness of interdepartmental relations.

Galbraith and Nathanson (1978) have noted that rules and procedures can be an effective method for achieving inter-organizational coordination. Specifically, they argue that standardization through rules and procedures is a useful integrating mechanism only when organizations are operating under conditions of relative certainty and routine tasks. As an organization's design becomes increasingly complex, however, the effectiveness of rules as a coordinating device among departments decreases.

It is important to distinguish between the concept of bureaucratic control as we use it and the effects of bureaucracy on information exchange as has been advanced by other researchers. This alternative assessment of bureaucratization argues that as bureaucracies evolve, they tend to become stifling, territorial, and will actually inhibit the flow of information across organizational borders (Obermeyer 1990). The argument points to an organization's bureaucracy as one of the principal culprits in preventing free information exchange. This position is supported in the management literature by research within a marketing context which found that bureaucratic structuring can damage exchange relationships and exacerbate opportunism between organizations (John 1984). While these arguments have merit, they do not negate but rather serve as a complement to our analysis of bureaucratization. We have chosen to examine the concept of bureaucratization from the perspective of control mechanisms (rules and procedures) rather than directly addressing an organization's bureaucracy and its potential for non-cooperativeness. In other words, when one agency sets up a series of procedural steps that ensure that their personnel will cooperate with other agencies, they are using a bureaucratic form of organizational control. This is not to gainsay the potential negative effects a large bureaucracy can have on inter-organizational cooperation but rather to argue that bureaucratization may also be viewed from another, more positive perspective.

A second conclusion relevant to our research is that rules and procedures have a means-end interrelationship. While rules and procedures are developed to ultimately facilitate the accomplishment of desired ends, they also provide a means to establish cooperation among the individuals or departments charged with a particular task. The degree of rules and procedures is related to the degree of organizational formalization. Recent studies by Deshpande and Zaltman (1987) and John and Martin (1984) found that increased formalization had a positive effect on the flow of information. Building on this research, Moenaert and Souder (1990) suggest that increased formalization between departments produces a more harmonious climate. Therefore, it appears that rules and procedures should have a direct influence on the development of inter-organizational cooperation and information sharing. Thus, if the means role of rules does not facilitate cooperation, then the end state will most probably suffer. Based on the above discussion, we offer the following proposition:

P2: *Organizational rules and procedures have a positive and significant direct effect on geographic information exchange.*

Incentives. Another logical facilitator of an organization's willingness to share information with another has to do with perceived incentives on the part of an organization to share. This argument captures the "what's in it for me?" question that individuals and organizations frequently ask before engaging in any type of personal or professional commitment. Incentives suggest that an organization or its key members must perceive a payoff to themselves arising from the act of cooperating in an information exchange relationship. Such a payoff may be in the form of creating a future bond of obligation or gaining some form of strategic or monetary advantage over rival organizations or agencies.

Under the incentive system, the willingness of one organization to participate in an information exchange lies in direct proportion to the other organization's providing the first organization with some scarce or necessary resource that it does not possess (e.g., money, access to important information, and so forth). When that other organization communicates its willingness to develop an economic exchange relationship, the first agency must determine if the incentives are valuable enough to warrant the exchange of information for that resource. If the answer is yes, the information exchange will occur. This argument suggests the basis for the following proposition:

> P3: *Incentives accruing to the sharing organization will have a positive and significant direct effect on geographic information exchange.*

Accessibility. In our model, accessibility is defined as an individual's perception of his or her liberty or ability to approach or communicate with another individual from a different organization. Factors that influence the type and amount of interaction that occur between organizational members include an individual's schedule, position in an organization, and out-of-office commitments. These factors often affect the "accessibility" among organizational members. For example, consider a setting in which an individual from a local government is physically located near an individual from a state agency. While these individuals are in close proximity to one another they may rarely interact because of different work schedules, varied duties and priorities, and commitment to their own agendas. Within a manufacturing organization, Souder (1981) has demonstrated that the lack of communication, lack of appreciation, and distrust that often exist between marketing and R&D are fostered by normal time pressures, work deadlines, and some imbalance of power and prestige. These factors often lead to a perception of "inaccessibility" among the individuals involved. Building on a set of studies, Zaltman and Moorman (1989) found that interactions between organizational members enhance trust. Interactions are easier when parties are accessible. Zaltman and Moorman (1989) suggest a causal link between interactions and trust. Communication strategies and technologies that can help overcome lack of ability to approach another caused by differences in schedules, position in the organization, and out-of-office commitments should help build trust among people and organizations with the potential for sharing geographic information. In our model, previous research suggests that accessibility should directly affect the degree of cross-functional cooperation. The fourth hypothesis pertains to this discussion of "accessibility":

P4: *Accessibility will have a positive and significant direct effect on geographic information exchange.*

Quality of exchange relationships. Research supports the view that the quality of exchange relationships with potential exchange partners represents a significant criterion for evaluating cooperative information flow (Hunt and Nevin 1974). The arguments underlying this issue suggest that member expectations of cooperation and information exchange are critical elements in maintaining quality channel relationships. If these elements are characteristics of the system, different GIS-using organizations would be expected to fulfill obligations and expect each other to desire coordination; high levels of satisfaction and morale should follow.

Following research by Dwyer and Oh (1987) that was originally compiled to address relationships among member organizations within marketing channels, we analyze the concept of exchange relationships as consisting of three key dimensions: Satisfaction, Trust, and Minimal Opportunism. Satisfaction with an exchange partner is a significant issue when evaluating information exchange relationships. Member satisfaction includes all characteristics of the relationship that the focal organization finds "rewarding, profitable, instrumental . . . or frustrating, problematic, inhibiting. . . . (Ruekert and Churchill 1984, 227). Likewise, trust refers to a party's expectations that another desires coordination, will fulfill obligations, and will pull its weight in the relationship (Anderson and Narus 1986). When one organization has built up a level of comfort and trust with another party, it is willing to engage in a more open exchange and sharing of relevant information (Deutsch 1958). This argument suggests that past history of working relationships will have an impact on future levels of trust and willingness to cooperate. Finally, opportunism is exemplified by distortion of information, failure to fulfill promises, and unwillingness to share what may be considered proprietary information. In effect, as one researcher suggested, opportunism is "self-interest seeking with guile" (Williamson 1975, 26). When the condition of opportunism is prevalent between two agencies or organizations, there is little likelihood that needed information will be shared between the parties, as each or both of the exchange members view information as a base of power to be held over the other.

The argument for the impact of quality of exchange relationships on collaborative information sharing offers the next proposition:

P5: *Exchange Relationships that are characterized by Satisfaction, Trust, and Minimal Opportunism will have a positive and significant direct effect on geographic information exchange.*

Resource scarcity. The concept of resource scarcity refers to the idea that organizations must contend with a limited pool of resources when conducting their activities. Under certain circumstances, the resource level that is available to an organization may not be constraining but may allow a wide range of activities and options. This state is referred to as munificence. More common, however, is the idea of resource scarcity in that organizations are constrained by limited budgets and the availability of technology, trained personnel, and other needed resources.

Research has long found that organizations and agencies operating under conditions of resource scarcity are often inclined to be dominated by the desire to maintain some form of control over other companion agencies. Indeed, the resource dependence model of organizational power argues that a method by which one organization can exert control over another is through creating and stockpiling some resource that is scarce and is needed by another organization (Pfeffer 1982; Pfeffer and Salancik 1978). In the case of the current discussion, the needed resource may be information of the sort provided by GIS technologies. Because information is viewed as a form of power, the agency that possesses needed information is less likely to share it with another party in fear of losing some base of power in the exchange relationship.

On the other hand, when conditions of relative munificence operate, there is less emphasis placed on hoarding resources, either material or informational. Thus, the environmental condition of munificence offers an increased likelihood of information exchange, while perceived resource scarcity is likely to create the opposite effect, in which parties possessing information are not as inclined to make it readily available to other organizations.

Based on the above discussion, the following proposition is offered:

P6: *Perceived organizational munificence will have a positive and significant direct effect on geographic information exchange.*

Information Exchange

Many concepts exist that encapsulate the meaning of organizational cooperation. Among the various terms used to describe the notion of individuals working together to accomplish a specific task are: 1. coordination (Argote 1982; Van De Ven, Delbecq and Koenig 1976); 2. collaboration (Trist 1977); 3. cooperation (Schermerhorn 1975; Sherif and Sherif 1969); and 4. integration (Gupta, Raj and Wilemon 1986; Lawrence and Lorsch 1967). The variability in terminology raises a critical question regarding what, if anything, the underlying concepts have in common. While each of the terms has a separate and distinct name, they all refer to a similar and overlapping idea as evidenced by the commonalities in their definitions. The lowest common denominator that integrates the four concepts is *joint behavior toward some goal of common interest.* For purposes of this research, organizational information sharing and cooperation are conceptualized as the degree, extent, and nature of interpersonal relationships among members from multiple organizations.

The need for information sharing and cooperation stems from the complex interdependencies among members of different organizations charged with complementary objectives. As Thompson (1967) has suggested, greater interdependence requires a greater cooperation effort. Unfortunately, problems associated with inter-organizational cooperation result not only from the interdependence of work process and technology, but from conflicts over authority and jurisdiction among team members representing different departmental units. Thus, inter-organizational information sharing is essential in the implementation of organizational decisions. It has been shown to promote productivity by helping individuals perform more effectively (Laughlin 1978). Laughlin argues that organizations that are cooperating tend to 1. understand and be influenced by each others' interests and ideas, 2. seek and give information, 3. communicate about tasks, 4. more readily assist each other, and 5. rely on division of labor.

The previous discussion has established a link between antecedents and cooperation among organizations. Research has also shown that the antecedents may not have a strong direct link to project performance. Further, GIS information exchange appears to be directly related to the outcome measures of increased efficiency and client satisfaction. Thus, in our model inter-organizational cooperation mediates between the antecedents and outcome measures. The following hypothesis reflects the mediating role of free information exchange between organizations and the need for a path-analytic framework.

P7: *GIS information exchange will mediate the relationships between the antecedents and outcome measures.*

Consequences of Cross-Functional Cooperation

As Figure 3.1 indicates, greater organizational sharing of geographic information is not simply an end to be sought for its own sake. Rather, we have posited a three-stage flow process in which information sharing represents the mediating link between the antecedent variables and organizational outcomes. In effect, our model suggests that enhanced information sharing will lead to several desired outcomes that various organizations seek: efficiency, effectiveness, and improved decision-making ability. In addressing each of these outcomes, we will be making use of previous research in the area of the measurement of GIS use and impact as developed by Gillespie (1991) and Zwart (1991).

One expected outcome of enhanced information sharing among organizations is greater internal efficiency of operations for each agency. When one organization is able to make direct contact with another party that possesses needed information, there is far less likelihood of replication of effort in creating databases. An organization's efficiency is enhanced through this sharing process. Increased efficiency from information sharing is measured, according to Gillespie (1991), through taking the difference in variable costs between the alternative methods for producing the desired output. In the case of geographic information, an estimate of the resources (labor, time, and money) needed to develop required GIS information in-house would be compared to the cost of retrieving such information from external sources with whom the organization has developed a cooperative relationship. If the difference is positive, then it could be argued that information sharing has resulted in increased organizational efficiency.

A second outcome of information sharing is that of increased organizational effectiveness. Effectiveness has been defined as the case when GIS "increases the quality of the output or produces a new output" (Gillespie 1991, A-85). Gillespie further suggests a three-step process for measuring effectiveness benefits. First, it is necessary to determine how the GIS output is different—in other words, what is being used or produced now that was not operational or available before. Second, it is necessary to determine the effect of the changes in GIS output on its users. One way of assessing the changes that can take place is to examine the activities of organizational members. Given the availability of new sources of information through organizational cooperation, do members make use of this available information? If the answer is no, then it is likely that sharing has not increased effectiveness. Finally, the third step in assessing effectiveness is to determine the value of each of the effects on the users. Obviously, this step is the most complex and difficult because it requires that some figure of value be assignable to the set of effects.

The final source of outcome to assess is the impact on decision making of the new cooperative arrangement among organizations. Zwart (1991) argues that unless we can determine that utilization of geographic information has led to enhanced or better decision-making capabilities, its impact is minimized. Nagel (1988) reinforces the importance of improved decision making by suggesting that we need to assess impact by examining the degree to which information is utilized. He argues that for information to be fully utilized by an organization, it must not simply be referred to in decision making but must actually lead to changes in organizational values or managerial decisions. As a result, if information that is shared between organizations does not lead one party to actively reassess decision priorities or value structures, the third criterion of outcome (decision making) is not fully addressed.

CONCLUSIONS

This chapter has sought to develop a research model for investigating managerial and organization-level constructs that can serve as facilitators of inter-organizational sharing of geographic information. As Figure 3.1 demonstrates, the underlying framework upon which we based our set of research propositions involved a series of hypothesized causal links between antecedents of information sharing and its projected outcomes for organizations. From the model, it is expected that enhanced information sharing will result in a number of positive outcomes for an organization. These benefits include greater efficiency, effectiveness, and improved managerial decision making.

The model and research propositions have purposely been developed in testable form in order to facilitate the second stage of our stream of research into facilitators of organizational information sharing. This second stage will consist of developing a data sample of organizations that have been required to develop cooperative, sharing relationships in order to solve mutual problems (the context for information sharing that we sought to address). Once this data sample has been identified through a series of structured interviews, a survey will be developed that will allow for large-scale sampling procedures. This instrument development and survey process should follow an accepted pattern for information system research, as was recently proposed by Kraemer and Dutton (1991). It is hoped that this first stage in the research process—that of developing a framework with testable propositions—will logically flow into the second step of conducting field research to test the proposed causal linkages.

The study of information sharing within the context of GIS environments is still in its infancy. We have attempted to create a framework for better understanding some of the causes of cross-organizational cooperation in sharing geographic information through making use of the state of the field in the areas of organization theory and marketing channel research. These streams of research have allowed us to make some supportable propositions concerning the set of causal flows between antecedents of information sharing and their likely results for an organization engaged in a cooperative venture. It is hoped that as a result of a better understanding of these linkages, managers and members of organizations charged with the responsibility of developing cooperative relationships will be in the position to facilitate geographic information sharing to the advantage of all concerned organizations.

ACKNOWLEDGMENTS

An earlier version of portions of this material was published in a chapter authored by Nancy Obermeyer and Jeffrey Pinto in *Managing Geographic Information Systems* (New York: Guilford Press, 1994). Permission of Guilford Press to reprint this material is gratefully acknowledged.

REFERENCES

Anderson, J. C., and J. A. Narus. 1986. Toward a better understanding of distribution channel working relationships. In K. Backhaus and D. Wilson, eds., *Industrial marketing: a German-American perspective.* Berlin: Springer-Verlag, 320-336.

Argote, L. 1982. Input uncertainty and organizational coordination in hospital emergency units. *Administrative Science Quarterly* 27: 420-434.

Deshpande, R., and G. Zaltman. 1987. A comparison of factors affecting use of marketing information in consumer and industrial firms. *Journal of Marketing Research* 21: 114-118.

Deutsch, M. 1958. Trust and suspicion. *Journal of Conflict Resolution* 2: 265–279.

Dumaine, B. 1990. Who needs a boss? *Fortune* (May 7): 52-60.

Dwyer, F. R., and S. Oh. 1987. Output sector munificence effects on the internal political economy of marketing channels. *Journal of Marketing Research,* XXIV: 347-358.

Fayol, H. 1929. *General and industrial management.* Translated by J.A. Conbrough. Geneva: International Management Institute.

Galbraith, J. R., and D. A. Nathanson. 1978. *Strategic implementation: the role of structure and process.* Dallas: Business Publications, Inc.

Gillespie, S. R. 1991. Measuring the benefits of GIS use. In the ACSM-ASPRS Fall Convention 1991 *Proceedings.* Atlanta. A-84-94.

Gouldner, A. W. 1954. *Patterns of industrial bureaucracy*. NY: Free Press.

Gupta, A., S. P. Raj, and D. Wilemon. 1986. A model for studying the R&D–marketing interface in the product innovation process. *Journal of Marketing* 50: 7-17.

Hunt, S., and J. R. Nevin. 1974. Power in a channel of distribution: sources and consequences. *Journal of Marketing Research* XI: 186-193.

Ives, B., and M. H. Olson. 1983. User involvement and MIS success: a review of research. *Management Science* 30: 586–603.

John, G. 1984. An empirical investigation of some antecedents of opportunism in a marketing channel. *Journal of Marketing Research* XXI: 278-289.

John, G., and J. Martin. 1984. Effects of organizational structure of marketing planning on credibility and utilization of plan output. *Journal of Marketing Research* XXI: 170-183.

Johnson, D. W. 1975. Cooperativeness and social perspective taking. *Journal of Personality and Social Psychology* 31: 241-244.

Johnson, D. W., and R. J. Lewicki. 1969. The initiation of superordinate goals. *Journal of Applied Behavioral Science* 5: 9-24.

Kelly, H. H., and A. J. Stahelski. 1970. Social interaction basis of cooperators' and competitors' beliefs about others. *Journal of Personality and Social Psychology* 16: 66-91.

Kono, T. 1984. *Strategy and structure of Japanese enterprises*. Armonk, NY: M.E. Sharpe.

Kraemer, K. L., and W. H. Dutton. 1991. Survey research in the study of management information systems. In K.L. Kraemer (ed.), *The information systems research challenge: survey research methods*. vol. 3. Boston: Harvard Business School. 3-58.

Kraemer, K. L., J. L. King, D. E. Dunkle, and J. P. Lane. 1989. *Managing information systems: change and control in organizational computing*. San Francisco: Jossey Bass.

Laughlin, P. R. 1978. Ability and group problem solving. *Journal of Research and Development in Education* 12: 114-120.

Lawrence, P. R., and J. W. Lorsch. 1967. *Organization and environment*. Homewood, IL: R. D. Irwin.

March, J. G., and H.A. Simon. 1958. *Organizations*. New York: John Wiley.

McCann, J., and J. R. Galbraith. 1981. Interdepartmental relations. In P. C. Nystrom and W. H. Starbuck (eds.), *Handbook of organizational design*. vol. 2. New York: Oxford University Press. 60-84.

Mintzberg, H. 1979. *The structure of organizations*. Englewood Cliffs, NJ: Prentice-Hall.

Moch, M., and S. E. Seashore. 1981. How norms affect behaviors in and of corporations. In P. C. Nystrom and W. H. Starbuck (eds.), *Handbook of organizational design*. vol. 1. New York: Oxford University Press. 534-565.

Moenart, R. K., and W. E. Souder. 1990. An analysis of the use of extrafunctional information by R&D and marketing personnel: review and model. *Journal of Product Innovation Management* 7: 91-107.

Mooney, J. D. 1947. *The principles of organization*. New York: Harper and Brothers.

Nagel, P. J. 1988. Customer satisfaction: a comprehensive approach. *International Journal of Physical Distribution and Logistics Management* 20, 6: 2–46.

Newman, W. H. 1988. Role of departments in multi-level strategic management. Working Paper SC65, Columbia University, Strategy Research Center, New York.

Obermermeyer, N. J. 1990. *Bureaucrats, clients, and geography: the Bailly Nuclear Power Plant battle in Northern Indiana*. Research Paper No. 216, University of Chicago, Chicago, Illinois.

Pavett, C. M., and A. W. Lau. 1983. Managerial work: the influence of hierarchical level and functional specialty. *Academy of Management Journal* 26: 170–177.

Peters, T. 1990. Get innovative or get dead! *California Management Review* 33, 1: 9–26.

Pfeffer, J. 1982. *Organizations and organization theory*. Boston: Pitman.

Pfeffer, J., and G. R. Salancik. 1978. *The external control of organizations*. New York: Harper and Row.

Pinto, M. B., J. K. Pinto, and J. E. Prescott. 1993. Antecedents and consequences of project team cross-functional cooperation. *Management Science* 39: 1281–1298.

Quinn, J. B. 1980. *Strategies for change*. Homewood, IL: Irwin.

Raia, A. P. 1974. *Managing objectives*. Glenview, IL: Scott Foresman.

Ruekert, R. W., and G. A. Churchill, Jr. 1984. Reliability and validity of alternative measures of channel member satisfaction. *Journal of Marketing Research* XXI: 226–233.

Ruekert, R. W., and O. C. Walker. 1987a. Marketing's interaction with other functional units: conceptual framework and empirical evidence. *Journal of Marketing* 51: 1–19.

Ruekert, R. W., and O. C. Walker. 1987b. Interactions between marketing and R&D departments in implementing different business strategies. *Strategic Management Journal* 8: 233–248.

Schermerhorn, J. R. 1975. Determinants of inter-organizational cooperation. *Academy of Management Journal* 18: 846-856.

Schmidt, W. H., and R. Tannenbaum. 1960. The management of scientific manpower. *Management Science* 14: B473–B489.

Shapiro, B. P. 1977. Can marketing and manufacturing coexist? *Harvard Business Review* 55: 104-114.

Sherif, M. 1962. Intergroup relations and leadership: introductory statement. In M. Sherif (ed.), *Intergroup Relations and Leadership*. New York: John Wiley. 3-21.

Sherif, M., and C. W. Sherif. 1969. *Social psychology*. New York: Harper & Row.
Simon, H. 1964. On the concept of organizational goals. *Administrative Science Quarterly* 9: 1-22.
Souder, W. E. 1981. Disharmony between R&D and marketing. *Industrial Marketing Management* 10: 67-73.
Souder, W. E. 1988. Managing relations between R&D and marketing in new product development projects. *Journal of Product Innovation Management* 5: 6-19.
Stern, L. W., and J. L. Heskett. 1968. Conflict management in inter-organizational relations: a conceptual framework. In L. W. Stern (ed.), *Distribution channels: behavioral dimensions*. Boston: Houghton Mifflin. 288-305.
Stern, L. W., B. Sternthal, and C. S. Craig. 1973. Managing conflict in distribution channels: a laboratory study. *Journal of Marketing Research* X: 169-179.
Taylor, F. W. 1911. *Principles of management.* New York: Harper and Brothers.
Thomas, K. 1976. Conflict and conflict management. In M. D. Dunnette (ed.), *Handbook of industrial and organizational psychology*. Chicago, IL: Rand McNally College Publishing. 889-935.
Thompson, J. D. 1967. *Organizations in action*. New York: McGraw-Hill.
Trist, E. 1977. Collaboration theory and organizations. *Journal of Applied Behavioral Science* 13: 268-278.
Van De Ven, A. H., A. L. Delbecq, and R. Koenig, Jr. 1976. Determinants of coordination modes within organizations. *American Sociological Review* 41: 322-338.
Wall, J.A., Jr. 1985. *Negotiation Theory and Practice.* New York: Scott Foresman.
Walton, R. E., and J. M. Dutton. 1969. The management of interdepartmental conflict: model and review. *Administrative Science Quarterly* 14: 73-84.
Williamson, O. E. 1975. *Markets and hierarchies: analysis and antitrust implications*. New York: Free Press.
Zaltman, G., and C. Moorman. 1989. The management and use of advertising research. *Journal of Advertising Research* 28: 11-18.
Zand, D. E. 1972. Trust and managerial problem solving. *Administrative Science Quarterly* 17: 229-239.
Zwart, P. 1991. Some indicators to measure the impact of land information systems in decision making. In Urban and Regional Information Systems Association 1991 Conference *Proceedings*, San Francisco, August.

Hugh W. Calkins and Richard Weatherbe

4 *Taxonomy of Spatial Data Sharing*

Modern implementations of Geographic Information Systems (GIS) function as distributed systems and can support multiple participants, each contributing data to the database. Additionally, the high cost of building and maintaining comprehensive databases comprising a variety of themes will force GIS users to share or exchange data with others. With the growth and maturation of GIS technologies and implementations it is now possible to perform surveys and case studies on spatial data sharing. These will develop technical and organizational descriptions of data-sharing activities and will assist in developing prescriptive or normative methods and structures for spatial data sharing. A taxonomy of spatial data sharing must be developed to provide a descriptive and analytical framework for such studies and to provide a common vocabulary for discussion.

INTRODUCTION

The first functioning Geographic Information Systems were typically developed for special purposes: either the analysis and solution of a specific problem or the description and management of a single resource. Standardization was irrelevant and any sharing of data consisted of distribution of printed reports and maps.

Within the last decade, the availability of commercial software for GIS creation and management, coupled with inexpensive computer hardware, has led to the use of GIS as a more general purpose spatial management tool. Commonly, spatial data sharing has been identified as an objective for Geographic Information Systems developed by governmental agencies or private firms.

There are few examples of fully implemented GISs sharing spatial data on a regular and productive basis. The relative immaturity of GIS as a management tool is a factor in this, as is a lack of technical and organizational procedures to facilitate data sharing. Another factor is the lack of a body of research into the nature of spatial data sharing, its mechanisms and characteristics, and its costs and benefits.

Technological advances, such as low-cost workstation hardware, computer networking developments, and more powerful software, are quickly removing the technical barriers to spatial data sharing. What constraints remain are mostly organizational. The first step to developing an organized body of knowledge and experience about data sharing is the development of a descriptive taxonomy in order to provide a common vocabulary.

SPATIAL DATA SHARING DEFINED

Spatial data sharing is defined as the (normally) electronic transfer of spatial data/information between two or more organizational units where there is independence between the holder of the data and the prospective user. The transfer may be in the form of periodic bulk transfers, routine daily transfers, or on-line access driven by individual transactions. The participants may be separate organizations or may be departments within the same organization. For our purposes, the distinguishing characteristic of spatial data sharing is that there be an arm's-length exchange or transfer.

GIS AND SPATIAL DATA SHARING

The goals of GISs are to serve the information needs of both operational managers and policy and decision makers. The functions supported include resource management, operations management, policy planning, and strategic planning. Broadly speaking, GISs also function as spatial decision support systems (SDSSs). These are less well-defined but are used to improve human capabilities in the area of problem analysis, including identification and evaluation of many alternative solutions to a particular problem. A fully developed spatial data-sharing activity can only improve the ability of a GIS in meeting management goals.

Spatial data, in the quantities necessary for effective use of a GIS, are expensive to collect and maintain. For an individual organization or department, one of the primary justifications for a GIS or an automated mapping/drafting system is the elimination of duplicated or redundant data gathering, organizing, and storage efforts. This elimination of effort can apply very strongly on the intraorganizational level and has inter-organizational applications as well (e.g. county assessor and city tax office). Effective use of spatial data sharing can mean that data are collected in the most logical and cost-effective way. Ideally, all data will be recorded only once, and the collecting organization or department will share it with all other units needing the data. Spatial data sharing can also ensure that decisions are made with consistent data and that decisions separated by space or time are still made with a common database.

ORGANIZING SPATIAL DATA SHARING

The overall objective for a spatial data-sharing program is to create "connections" among widely dispersed spatial databases. This implies that within an organization all data to be shared are viewed as elements of a single enterprise or corporate database, designed and managed accordingly. Data sharing among organizations implies, at the least, some common agreements on standards—which agreements may extend to data formats and database design.

Geographic data sets are commonly organized on a thematic basis within a GIS, with spatial partitioning functioning as a secondary organizational principle. Modern GIS implementations, particularly in local government, partition data-management tasks among various user or functional groups. Using local government as an example, we can find the assessor or recorder responsible for creating and maintaining parcel maps and the public works department responsible for street hardware databases.

Another outcome of the growth of commercial GIS software has been general acceptance of compatibility standards. At a minimum, most GIS software can read and write at least one data format readable by other systems.

Figure 4.1 illustrates a public-sector example of inter-organizational data sharing. Several participants are identified, with sample data that each might contribute to the others in a sharing effort.

As Geographic Information Systems have multiplied, interest has focused on assessing the use and value of both the GIS and the geographic information produced. The traditional approach has been to survey the current extent of GIS. A necessary condition of survey work is a structure to classify the results. A taxonomy defined for the purpose is required. Taxonomies in the spatial data-handling area are being proposed for the purpose of GIS survey work (Calkins and Obermeyer 1991; Ventura 1991).

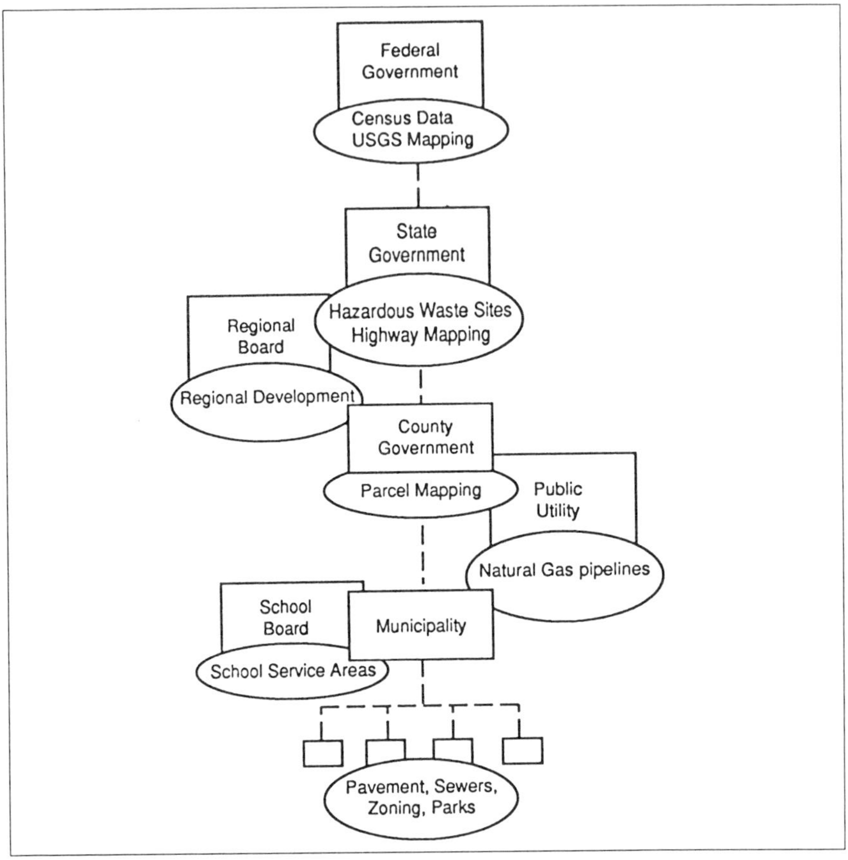

FIGURE 4.1
Spatial data-sharing example

Ventura (1991) presents a framework for collecting basic information about existing GIS, whereas Calkins and Obermeyer describe a taxonomy for investigating the use and value of geographic information in decision making. There is an implied sequence to these taxonomies. Before one can investigate either the use and value of geographic information or spatial data sharing between GISs, there is a need to identify and categorize GISs along the lines recommended by Ventura (1991, 48-70). Similarly, surveys and case studies need to be conducted to identify operational GISs where it will be productive to investigate the spatial data-sharing question. The taxonomy presented in this paper has the objective of characterizing the nontechnical aspects of the data, the data exchange, and the organizations involved in spatial data sharing. Additionally, constraints and impediments to spatial data sharing are considered.

In theory, an enterprise database can be visualized; in practice, we will still be dealing with separate spatial databases designed to meet the needs of only one or a small number of departments or agencies. Thus, spatial data sharing may be seen as only a secondary objective of some organizations operating a GIS. However, all technical and organizational issues that one would encounter in developing a corporate database will still need to be addressed. The leadership in the spatial data-sharing enterprise needs to support the development of an integrated enterprise information system by providing tools and procedures for data integration based on common definitions for locations of spatial objects and their associated data attributes.

The major difference between spatial data sharing and data sharing in other contexts is the need to have common definitions and standards for the spatial data. Most contemporary Geographic Information Systems do not directly represent spatial objects; rather, the geometry of the objects is decomposed into simple geometric primitives (point, line, and polygon) with associated attributes. Even the geometric primitives do not have standard definitions. Current national efforts are proposing definitions for the geometric primitives, but these have yet to be widely accepted (Moellering 1987).

Organizations contemplating sharing of spatial data need to carefully define spatial objects, attributes, and data quality associated with these objects (Frank 1992). Frank presents a theoretical discussion of data-related organizational issues including spatial concepts, spatial data models, entity definitions, spatial precision, and data quality. He concludes by presenting a framework of data-set acquisition (data sharing) and integration identifying four scenarios (Frank 1992, 21):

- i acquired data are used as a backdrop,
- ii acquired data are used to get started quickly,
- iii data are acquired for long-term usage,
- iv data are centrally held in a spatial database and accessed on demand.

Agreement among spatial data-sharing participants on definitions, classes, and other standards is required if spatial data sharing is to fulfill its potential of "sharing of the large expenses to keep the geographic database current and to reduce duplication of efforts of organizations for database maintenance" (Frank 1992, 20).

THE TAXONOMY

We have outlined a suggested taxonomy for research into spatial data sharing among GIS users. The classification scheme is intended to identify those characteristics that differentiate spatial data-sharing activities and will lead to useful and substantiated models of spatial data sharing for further study.

The four primary components of the taxonomy include:

1. characteristics of the organization,
2. characteristics of the data,
3. characteristics of the exchange, and
4. constraints and impediments.

Subclassifications within each of these categories are contained below.

Elements of a taxonomy

Characteristics of the organization. The first set of information is about the characteristics of the organizations sharing spatial data (Table 4.1). This information identifies who is participating and the role played in terms of several organizational characteristics. A single "data exchange" between two actors will be the basic unit of observation. For each instance of spatial data exchange, the provider and recipient can be described. If the exchange involves more than one step, it can be described as a series of exchanges between pairs of participants.

Characteristics of the data. The characteristics of the spatial data being exchanged must also be determined (Table 4.2). The taxonomy categorizes technical features such as data format, type, and nature. It also describes organizational aspects such as quality and importance. In general, categorization is from the data recipient's viewpoint.

Characteristics of the exchange. This section deals with the managerial and economic characteristics of the exchange (Table 4.3). Sharing of spatial data implies an ongoing relationship between the two participants. It may be that the success or failure of a particular sharing relationship depends more on these soft issues than on technical issues such as formats or accuracies. While the high cost of building and maintaining spatial databases is well known, evidence as to the economic value of these data in a free marketplace is still sparse.

TABLE 4.1

Characteristics of the Organization

DATA USE FUNCTION	Data collection Data management Regulation Operations management Analysis and advice Policy/decision making
TYPE OF ACTIVITY	Local government Public utility State/federal government Private sector/commercial
DEPARTMENTAL FUNCTION	Land Engineering Infrastructure management Assessment Planning Building Environmental protection Service provider Resource extraction Resource delivery
ORGANIZATIONAL IDENTITY	Self-contained Department Sub-department
DATA-SHARING ROLE	Data provider Data recipient Data broker Data vendor

Constraints and impediments. Each exchange relationship will have a different set of constraints affecting it (Table 4.4). These will depend on the nature of the data, the natures of the participants, and the national, state, and local legal environments. Impediments are more universal in nature and are time-sensitive in that most derive from the current state of the art.

TABLE 4.2

Characteristics of the Data

DATA MODE	Vector Raster Image (documents/maps) Attribute
IMPORTANCE OF DATA	Strategic Required Useful
ORGANIZATION OF DATA	Update only Selected data Complete theme
DATA TYPE	Base map Geo-referenced feature (TIGER) Thematic features
NATURE	Raw data—observations Preprocessed (classified) Sorted/organized—in source format Sorted/organized—in target format Ready to use—existing theme
QUALITY ASSURANCE	Published standard • mutual • defined by originator • defined by recipient Documented procedures None

CONCLUSIONS

Work has begun on research into the use and value of geographical information, and a taxonomy for use and value has been developed (Calkins and Obermeyer 1991). GIS technology as currently implemented will support distribution and sharing of spatial data, although such distribution and sharing are relatively rare.

TABLE 4.3

Characteristics of the Exchange

STRUCTURAL	1. *Type of Partner*	2. *Functional Relationship*
	Internal exchange	From data supplier to end user
	Same functional department	From end user to supplier
	Different functional department	Between data suppliers
	External exchange	Between end users
	Equivalent level	Between supplier and broker
	Higher	Between broker and end user
	Lower	Between data brokers
	Vendor	
OPERATIONAL	1. *Sharing Arrangement*	2. *Pricing*
	Formal contract	Free of charge
	Mandate to share or distribute	Barter
	Profit-making venture	License fee
	Voluntary/informal	Consortium membership
		Marginal cost of reproduction
		Cost recovery basis
		Market value
FUNCTIONAL	1. *Medium*	4. *Frequency*
	Paper map	
	Paper list	5. *Average Quantity*
	Tape/disk	Single observation
	On-line	Summary/aggregate observations
		Selected subset
	2. *Initiation*	Theme
	Sender	Entire database
	3. *Schedule*	
	On demand	
	On schedule	
	As needed	

TABLE 4.4

Constraints and Impediments

CONSTRAINTS	*ACCESS*	
		No published access rules
		Freedom of Information access
		Formal access policies
		Statutory access
		License restrictions
		Copyright
		Patent
	DATA CONFIDENTIALITY	
		None
		Single observation
		Summary—statistics only
		Limited number of observations
	LIABILITY	
		Minimal exposure
		Supplier liability
		User liability
IMPEDIMENTS		Communications/networking cost
		Price of data
		Format/structure incompatibility
		Conformance with standards
		Documentation

This taxonomy is intended to spark theoretical discussion of the modes and methods of spatial data sharing and will also serve as the basis for questionnaires supporting surveys and case studies into data sharing. It will evolve as surveys and case studies into spatial data sharing are carried out, and as the issues and categories are examined in more detail. In particular, the authors recognize that the second- and third-order categories presented above are not exhaustive. As it evolves, it will not only facilitate research into, and analysis of, data sharing—it will also provide a common framework for description and prescription.

ACKNOWLEDGMENTS

This paper reports research results being carried out at the National Center for Geographic Information and Analysis, supported by a grant from the National Science Foundation (SES-88-10917). Support by the NSF is gratefully acknowledged.

REFERENCES

Calkins, H. W., and N. J. Obermeyer. 1991. Taxonomy for surveying the use and value of geographical information. *International Journal of Geographical Information System* 5, 3: 341-351.

Frank, A. U. 1992. Acquiring a digital base map: a theoretical investigation into a form of sharing data. *Journal of the Urban and Regional Information Systems Association* 4, 1: 10-23.

Moellering, H. 1987. *A draft proposed standard for digital cartographic data* 48. National Committee for Digital Cartographic Standards.

Ventura, S. J. 1991. *Implementation of land information systems in local government – steps toward land record modernization in Wisconsin*. Wisconsin State Cartographer's Office.

Michael J. Kevany

5 A Proposed Structure for Observing Data Sharing

That the sharing of information or data is a desirable process appears in the Geographic Information Systems (GIS) industry to be a universally accepted "truism." Why is sharing so important? Perhaps because it offers the opportunity for great financial gain or saving, perhaps the opportunity for compatibility in information, perhaps many other benefits. Nevertheless, sharing is treated as very significant when justifying or planning a GIS. The purpose of this chapter, however, is not to justify sharing but to explore methods for analyzing the environments of sharing.

Since we generally accept the practice of information sharing as a desirable condition, it is important in a GIS project to create conditions that are conducive to sharing. Little is known, however, of what conditions create a conducive environment. The purpose of this chapter is to explore factors impacting sharing and to make a contribution toward an analytical structure for improving the level of knowledge. The chapter identifies potential factors and proposes and evaluates a structure for measuring the reality of, or potential for, information sharing in a GIS environment. The structure may be used initially to improve the knowledge of factors that affect data sharing and later to apply that knowledge to GIS situations to increase the potential for sharing.

INTRODUCTION

Information sharing is subject to numerous interpretations. Information or data sharing is a term that has no commonly accepted meaning in the GIS industry. The term "sharing" is used by persons with varying perspectives to describe a variety of practices. The sharing of geographic information may take many forms, ranging from the sale of data by one organization to another to simultaneous access of a single data set by many persons or organizations. The value of sharing to an organization is often impacted by what form it takes in a particular situation.

Since we generally accept the practice of information sharing as a desirable condition, it is important in a GIS project to create conditions that are conducive to sharing. Little is known, however, of what conditions do create a conducive environment. The purpose of this chapter is to explore factors impacting sharing and to make a contribution toward an analytical structure for improving the level of knowledge. The chapter identifies potential factors and proposes and evaluates a structure for measuring the reality of, or potential for, information sharing in a GIS environment. The structure may be used initially to improve the knowledge of factors that affect data sharing and later to apply that knowledge to GIS situations to increase the potential for sharing.

The chapter begins with a summary of the author's experiences with GIS and geographic data sharing that have provided the insight from which the proposed structure was developed. Next, the chapter identifies a set of key issues that the author believes are significant to the sharing environment. The chapter then describes a structure and its components that can be used to measure the sharing environment in existing or planned GIS projects. The structure identifies a set of factors to be measured and an initial set of measurement criteria for each. The proposed structure is evaluated with a set of empirical data for a small sample of actual cases. Finally, the results of the evaluation are reported along with recommendations for the modification of the structure and the application of the structure to future research.

BACKGROUND OF DATA-SHARING EXPERIENCES

The author has been involved in various capacities in numerous GIS design, implementation, and operation projects. He has served as a creator and user of geographic information, a system analyst for GIS design and development, a manager of GIS projects, and a consulting advisor to projects. In these capacities he has had the opportunity to observe the desire for sharing and the successful and not-so-successful achievement of that desire. As a practitioner, rather than an academic, he has observed the phenomena in an ad hoc, unstructured manner. He has therefore formed opinions concerning the conditions that support information sharing but has no empirical proof of those opinions. The need and value of information sharing in a GIS program are such that a reliable understanding of the factors that create a conducive environment should be developed. This knowledge can be very valuable to the planning and design of a GIS project. To develop the necessary knowledge and understanding, it is necessary first to analyze a set of existing GIS situations to identify the factors, their relationships, and the interactions that determine or impact the level and success of data sharing.

There are many GISs that are now operational at an adequate level to serve as the base for measurement. Most of the existing systems have achieved some level of sharing, though many have not achieved the desired or optimal level. It is possible, therefore, to analyze both the successful situations and many that have not been successful or fully successful. The identification of factors and interaction that lead to sharing, however, is a very difficult challenge. Most GIS projects are very complex endeavors. Many involve multiple organizations. Most involve multiple units within at least one organization. The sharing of data is affected by technical, institutional, and financial conditions. The identification of factors in this milieu is thus quite difficult. The interaction of these factors as they impact data sharing is even more difficult. It can be anticipated, therefore, that considerable research is required before a reliable understanding of the situation emerges.

The author is proposing a measurement structure in this chapter. The following are summary profiles of a sample of GIS projects from which the author has derived his proposed approach. It should be noted that most of these projects focus on local government although they also involve other organizations. GIS implementation in a federal government, private sector, or other environment may differ greatly from these in the factors affecting data sharing. As an example, the United States Geological Survey (USGS) National Mapping program and its GIS activities are specifically funded to create data to be shared by many organizations. This primary objective obviously affects the ease with which the resulting data may be shared.

The specific organizations observed and discussed in the following paragraphs are not named to avoid subjecting them to uninvited inquiries and to avoid focusing on some minor issue that might be unique to the specific project. The projects are profiled to indicate the sources of the various factors and measurements identified in the structure.

City/County/Utility Project

A county executive recognized the value of GIS to his organization and wished to implement the technology. An initial evaluation indicated that the cost of the system implementation would be very high and that the system and its data also might be shared by the major city and the municipally owned utility serving much of the county. All three organizations serve the same geographic area and can thus share data. The executive obtained the participation of the city and the utility in the planning and development of the system. The funding mechanism was a bond issue that would cover all development costs, followed by a use-based sharing of the costs among the three entities.

A formal agreement was executed among the three parties for the development and operation of the GIS. A specific GIS organization was established through the agreement that would operate under the direction of the three-party board. The system and database were designed to satisfy the requirements of all three organizations, and the database was developed to those specifications. The database is maintained primarily by the core GIS unit, though each organization also maintains data sets of particular interest to it.

The common components of the database have been developed and the system is now operational. The data are being updated and maintained on a continuing basis. Sharing of data is being accomplished on a regular basis in a manner that is satisfactory to each organization.

It should be noted that the forceful leadership of the county executive was significant to the establishment of the agreement and the GIS. A change in officials has brought a much less cooperative participant but, since a formal agreement was established, the GIS continues to operate effectively, and a high level of data sharing is still supported.

Multi-County and Utility Consortium

Five organizations serving two counties joined to form a GIS project. Each organization has specific geographic information needs, some of which they share with one or more other organizations and some that are unique. Considerable redundancy existed in map maintenance, in particular, that provided the impetus for the cooperative arrangement. The organizations joined to prepare a design and specifications for a shared system and database. The database has core components that are shared by all participants in addition to organization-specific components.

The five organizations executed a formal agreement providing the basis for cooperation and sharing. Funding was allocated on the basis of six shares, one each for four organizations and two for the utility serving both counties. The funding was used to prepare a detailed design and specifications, to acquire an initial system configuration, and to develop a pilot project database. A pilot project was conducted within each organization using the shared data.

The project is currently awaiting funding for full implementation. Because of this, the level of cooperation among the organizations is deteriorating, and organizations are planning to move forward independently. However, the potential for future sharing of data appears to continue to be high.

A significant issue emerged in this project: the offer of sharing, or sale, of existing digital map data by another private utility serving parts of the geographic area. The data offered by this utility have not been widely accepted due to the less than optimal specifications of the data, the fact that the data are now outdated, and the potential cost of the data offered for sale.

Single Department County Project

A tax assessment department in a large county had a windfall in revenues generated that could be applied to the development of a GIS. The director had a forceful manner and was aware of the potential complexity of cooperation among numerous organizations during the development phase. While recognizing the value of data sharing with other county departments and other organizations in the area, he decided to go forward with the design and development of a GIS with minimal involvement of other organizations. This was to ensure control and timely development of the system. This approach generated some conflict in the early stages of development, though efforts were made to inform all interested parties in the area of progress and to obtain information on the requirements of other organizations.

The system and database have been developed and implemented, and the data are now being updated. Data are now available for sharing with other organizations, though the original owner exercises considerable control over all aspects of the operation, impacting the level of sharing.

County GIS Serving Multiple Departments

This county recognized the value of GIS and decided to acquire and implement a system. The planning and design addressed the needs of virtually all county departments. A GIS department was established to develop and operate the GIS on behalf of the user departments. The GIS department manages the system, administers the database, and updates the commonly shared data components. User departments have direct access to the operation of the system for their applications.

The system has been acquired and implemented. The database development is nearing completion, and components are now being updated regularly. Applications have been developed for a few departments, and others are in the process of being developed. The data are available to all departments, and a high level of sharing has begun.

The county and GIS managers had a strong commitment to county-wide sharing throughout planning and development, and that philosophy is now guiding the operational system. The user departments, however, have not been intimately involved in the development project and must be encouraged and assisted to make use of the system.

KEY GEOGRAPHIC INFORMATION SHARING ISSUES

Several issues have been recognized as being significant to the sharing of geographic data in these and numerous other projects. These issues relate to either the providing or owning organization, the receiving or benefiting organization, or a combination of both. The issues are institutional, technical, and financial. The issues are presented in a relatively general form here but are divided into discrete components in the proposed structure that follows their discussion. The issues identified and discussed include the following:

- Sharing classes
- Project environment
- Need for shared data
- Opportunity to share data
- Willingness to share data
- Incentive to share data
- Impediments to sharing
- Technical capability for sharing
- Resources for sharing

Each of these issues provides an area for analysis to determine the factors and relationships affecting data sharing.

Sharing Classes

The sharing of geographic information can take several forms. Webster defines "sharing" as: "to divide and distribute in shares; to partake of, use, experience, or enjoy with others; to grant or give a share in."[1] Geographic information, therefore, may be divided among organizations, copies may be made, or multiple organizations may use a single data set.

Perhaps the ideal form for geographic data sharing is direct access to a single database that is shared among multiple organizations. Several other classes of sharing are also possible, however. These classes include:

- Joint compilation or acquisition of data by multiple organizations, each of which obtains a separate copy of the data
- Acquisition of a copy of one organization's data by another organization on a one-time or periodic basis
- Sale of copies of data by an organization to other organizations
- Distribution of copies of maps or geographic data reports by one organization to other organizations

Sharing may take other forms as well. For the purposes of this discussion a broad definition is used to include all forms of data access or exchange. It does, however focus on digital data sharing.

Project Environment

There are many aspects of a GIS project environment that will impact data sharing. One of the most important is whether multiple organizations have come together to develop a joint GIS or database or whether an organization wishes to share existing data of its own or data owned by another organization. Pre-planned sharing and specification appear to facilitate sharing greatly.

The level of commitment to sharing by the participating organizations is very significant. That commitment includes the attitude of the managers involved; the formality of the agreement, if any, involved; and the durability of the commitment in light of changing conditions over time.

The prior relationships and level of cooperation between and among the sharing or participating organizations are crucial to success. Prior organizational rivalries and "turf" battles are generally detrimental to the success of sharing.

Need for Shared Data

The need for sharing among organizations can take many forms. In some cases one organization needs the data of another and is willing to provide benefits to gain access to the data. In others, an organization with data wants to obtain fees or benefits of another type by sharing data that it owns. In rare cases there is no other source for necessary data than through the sharing of another organization's data. The most common case, however, is the need for the combined resources of multiple organizations to justify or afford the development and maintenance of data that is required by these organizations.

Opportunity to Share Data

The opportunity to share data is dependent on various conditions. There must be data to be shared. That is, there must be a common interest in the same data across multiple organizations. This is generally derived from a common or overlapping service area and the common interest in maps and data for that area. There must be data available to share, either already in digital format or in a form that can be converted to digital format within the resources of one or a combination of organizations. A legal and institutional structure must exist through which the sharing can be accommodated. This may be as simple as the freedom to enter into informal arrangements or as complex as enabling legislation that authorizes formal sharing agreements.

Willingness To Share Data

The owner of digital data or of the sources for data must be willing to share the data with other organizations. This is an often overlooked situation. Sharing brings impediments and responsibilities that some organizations do not wish to accommodate. The organizations that want to gain access to the data of another or to share the development of data must be willing to commit the resources necessary for sharing. That may include fees to the owner of data or source materials or the costs for compilation or conversion of data to be shared.

Incentive to Share Data

The sharing of data is encouraged where an incentive for sharing exists. That may be an incentive for an owning organization to make data available to others or an incentive for multiple organizations to join to develop a shared database. The incentive may be as simple as an owner offering to share data with others at little or no cost or as complex as a governmental program or regulation that encourages or requires sharing. Possibly the most common incentive is the spreading of development costs among multiple organizations and thus reducing the cost to each organization.

Impediments To Sharing

Numerous barriers may exist to the sharing of data between and among organizations. Technical issues, such as incompatibility between the hardware devices or software of the organizations, may restrict the ability to share. Incompatibilities in the definition or structure of data will impede sharing. Legal restrictions or confidentiality requirements may also present difficulties for sharing.

Technical Capability for Sharing

Sharing of data cannot be accomplished without the technical capability on the part of each participant to carry out the sharing. This may involve compatibility between the systems and data definitions of each organization. It may involve the technical capacity to utilize the data acquired from another organization. It may also involve the proper design of a joint system to support effective data sharing. The supplying organization must also have the technical resources to copy, deliver, or allow access to its data.

Resources for Sharing

The participating organizations must have adequate technical and financial resources to support the sharing of data. Where an initial development cost is incurred or a fee is required for access to existing data, the financial resources must be available. This may be necessary on a continuing basis where a continuous sharing is to take place. The participants must have adequate computer-processing and data-storage capacities to support the sharing as well. For initial database development and ongoing data updating and maintenance, adequate resources must be available to obtain necessary source materials and to enter the data.

MEASUREMENT CONCEPT AND STRUCTURE

The following structure is offered as a method for acquiring and analyzing information that will be useful in determining the factors and the relationships among those factors that affect geographic information sharing. This structure may be applied to existing situations initially to gain the necessary understanding of the factors that facilitate and those that impede information sharing and, later, to support the planning and design of GIS databases that can be shared successfully.

Full development of the proposed structure should happen in several steps. An initial structure is described here. That structure is also evaluated through use of readily available information for a small sample of sites. This initial evaluation focuses on the potential for the structure rather than specific findings regarding geographic information sharing. The structure should then be revised or modified based on the initial evaluation. One goal of the modification should be to reduce the number of factors involved to a smaller set that emerges as useful in the initial evaluation. The revised structure should then be applied to a large number of sites in a more rigorous manner and the results analyzed. At this stage the focus on an understanding of the factors and their relationships continues. The refined method, measures, and measuring techniques are employed to analyze the specific factors affecting information sharing. The predictive value of the factors is compared with the actual experiences of these sites. The structure will be further revised and then applied for a final confirmation of the method. At this point it is hoped that a specific set of factors can be measured and analyzed through statistical tools to derive information on the understanding of information sharing in a GIS environment. Once the factors, relationships, and interactions are understood, the structure will move to the practical design or redesign of systems that will support effective information sharing.

Measurement Factors

Thirty measurement factors have been identified for the initial structure. Initially these factors will be analyzed in a sample of sites to determine their potential value to an evaluation structure. As the factors are refined, the intent of the structure is to determine four to five measurement values that may be assigned for each factor. The structure should be applied by analyzing each of numerous GIS situations/sites and assigning values for each factor. The actual sharing experiences of the sites are also to be determined and modeled. The factors are to be analyzed to determine the probability for successful data sharing. The findings of that analysis should then be compared with the actual model of sharing for each site. The usefulness of each factor is then evaluated, and the structure is revised based on the findings of that analysis.

The objective of this analysis process is to identify specific conditions at a site that indicate the potential for information sharing. As noted above, these conditions are to be evaluated initially with the goal of identifying a set of quantifiable values for each factor that are easily and reliably measurable at each site. Each of the factors indicates conditions that raise or lower the probability for successful sharing. The combination of the factors across an organization or group of organizations in an area will present a profile of the sharing environment at the site.

In this preliminary form, the structure addresses a static condition for each factor. It is recommended that the structure be enhanced in the future to support definition of the relationships between and among factors. It appears that certain combinations of factors will magnify the potential for sharing.

The thirty geographic information sharing factors suggested are shown in Table 5.1. The following paragraphs describe each of these factors in greater detail.

Proposed Factor Definitions

Sharing Classes

1 CLASSES OF SHARING ARRANGEMENTS

Sharing of data may occur in various ways including accessing a common data set by multiple organizations, copying separate

TABLE 5.1

Geographic Information Sharing Factors

Sharing Classes	
1	Classes of sharing arrangements
Project Environment	
2	Number of organizations involved
3	Organizational goal/mission
4	Organization relationships
5	Current general relations between/among organizations
6	Historic relations between/among organizations
7	GIS relationships between/among organizations
8	Control of information
9	Growth rate for service area
10	Leadership politics/ego
Need for Shared Data	
11	Adequacy of internal data and data resources
12	Data of another organization is needed
13	Need data that can best be developed or maintained jointly
14	Level of dependence from sharer perspective
Opportunity To Share Data	
15	Organization(s) in the area have needed data
16	Organizations in the area have resources to share development/ maintenance costs
Willingness To Share Data	
17	Organization owning data willing to share
18	Organization offers/willing to share cost of data development/ maintenance
19	Level of dependence from lead organization perspective
Incentive To Share Data	
20	Governmental program or regulations that require or encourage sharing
21	Recognition of the value of sharing of data
Impediments To Sharing	
22	Real or perceived requirement for confidentiality
23	Incompatibility in the definition, specifications, or structure of available data
Technical Capability for Sharing	
24	Basis for sharing data between organizations
25	Level of planning for database or system development
26	System capacity and capability to use data
27	Data of useful technical specifications
Resources for Sharing	
28	Funding source to acquire or pay for share
29	Data or source materials for shared data
30	Service area size

sets of data for each organization, or the sale or purchase of data by an organization.

Project Environment

2 NUMBER OF ORGANIZATIONS INVOLVED

A GIS program may involve an individual department or unit within an organization, multiple departments within an organization, or multiple organizations such as city and county. The number of organizations involved will be significant to the level of sharing.

3 ORGANIZATIONAL GOAL/MISSION

The focus and breadth of the goal or mission of the organizations involved will impact sharing. Some organizations have a specific function, such as issuing permits or providing water service. Others have as their goal or mission service to other organizations, such as providing information services or mapping services. These provide sharing opportunities.

4 ORGANIZATION RELATIONSHIPS

The relationships between and among the organizations involved affect sharing. One aspect of the relationship is the leadership of one organization, equality among participants, or groupings of lead and sharing organizations.

5 CURRENT RELATIONS

Another aspect of organization relationships is the current general attitude and cooperation among organizations. This may be characterized as ranging from positive to negative.

6 HISTORIC RELATIONS

Similarly, the historic relationships among organizations may be significant to sharing. They too can be characterized along a positive to negative spectrum.

7 GIS-SPECIFIC RELATIONSHIPS

Specific GIS relationships exist also. These relationships are based on the relative requirements for geographic information. In some cases a lead organization requires the sharing of data and resources with other organizations to gain necessary development resources or to justify the development effort. In other cases organizations that require geographic information require a lead organization to create and develop the systems and database. In many cases the GIS relationship is based on a mutual need for cooperation among organizations. Finally, the lack of sharing may be because organizations do not wish to establish any relationship in regard to the GIS.

8 CONTROL OF INFORMATION
The control of information is significant to the success of sharing. The information may be controlled through a central management mechanism. The information may be controlled by individual "owning" nodes. The control may be distributed through all organizations in a distributed configuration. Control may also be established on an ad hoc basis to suit the particular needs of each situation.

9 GROWTH RATE
The rate of development growth in the area may have an impact on the sharing of information. Where development is at a high rate, the need for, and benefits of, sharing may be high while areas with stagnant development will have less need for, or incentive for, sharing. The growth rate may be divided along a continuum from high growth down through no growth, to a loss of population.

10 LEADERSHIP POLITICS/EGO
The leadership of a GIS project and/or the organizations involved can be very significant to the success of information sharing. In some cases local politics—partisan or non-partisan—will support or impede sharing. The "ego" level of the leader or leaders involved will also affect sharing. In environments where the leadership is compatible with no problems in cooperation, sharing will be successful. The compatibility of leadership may range from lack of problems and high compatibility, through minor friction, to significant problems, to a complete lack of compatibility.

Need for Shared Data

11 ADEQUACY OF INTERNAL DATA AND DATA RESOURCES
Organizations may need to share data because of an inadequacy of internal sources for acquiring the required information. In a database development effort, an organization may also have inadequate resources to acquire or develop the necessary data. Thus, the sharing of development costs and the resulting database are necessary.

12 DATA OF ANOTHER ORGANIZATION IS NEEDED
One organization may require the data of another organization to perform its functions. Sharing of information with an

organization that has or maintains the data may be the only cost-effective source for information.

13 NEED DATA THAT CAN BEST BE DEVELOPED OR MAINTAINED JOINTLY
Some data may be developed best through the cooperative efforts of multiple organizations either because separate organizations control parts of the source materials required, or joint justification and/or funding may be necessary to make the development effort cost-effective.

14 LEVEL OF DEPENDENCE FROM SHARER PERSPECTIVE
The potential organizational dependence, when viewed from the organizations that are not the leader but are sharing data, may range from high to no dependence, though the response may be to move away from the dependence of sharing toward information independence.

Opportunity To Share Data

15 ORGANIZATION(S) IN THE AREA HAVE NEEDED DATA
Sharing may be greatly facilitated if one or more organizations already maintain data that are required by other organizations.

16 ORGANIZATIONS IN THE AREA HAVE RESOURCES TO SHARE DEVELOPMENT/MAINTENANCE COSTS
The sharing of data development and/or maintenance costs is the most effective means of establishing a successful sharing environment.

Willingness To Share Data

17 ORGANIZATION OWNING DATA OFFERS OR IS WILLING TO SHARE
Where digital geographic data exist, the organization that owns or maintains that data must be willing to make the data available or share data under some circumstance including exchange, cost reimbursement, or goodwill.

18 ORGANIZATION OFFERS TO SHARE OR IS WILLING TO SHARE COST OF DATA DEVELOPMENT/MAINTENANCE
One or more organizations that are willing to share costs of data development or maintenance will create an environment for sharing.

19 LEVEL OF DEPENDENCE FROM LEAD ORGANIZATION PERSPECTIVE
The dependence of the organizations on the shared information or GIS is important to the level of information sharing. Where there is a lead organization with primary control, its perception of the level of dependence of the other organizations may

encourage it to facilitate sharing. Its perception of the level of dependence may range from a high level of dependence to no dependence at all.

Incentive to Share Data

20 Governmental Program or Regulations That Require or Encourage Sharing
Several government environmental, hazardous materials, transportation, and other programs require or encourage information sharing. The presence of such a program raises the potential for sharing other data in addition to that required by regulation.

21 Recognition of the Value of Sharing of Data
Where management or elected officials recognize the value of sharing for cost reduction or improved availability, the sharing environment will be improved. Often, sharing is mandated for program funding.

Impediments to Sharing

22 Real or Perceived Requirement for Confidentiality
When an organization believes that information must be protected for confidentiality, sharing will be hindered or rejected. Confidentiality may be a real requirement for privacy or commercial reasons, it may be only a perceived requirement, or it may be justification for refusal to share while ulterior motivations for refusing to share remain hidden.

23 Incompatibility in the Definition, Specifications, or Structure of Available Data
Sharing of data is very dependent on compatibility of the definitions, specifications, and structure of data between the organizations that want to share. If incompatibility is too great, the cost of redefinition, translation, or modification will exceed the value of sharing the data.

Technical Capability for Sharing

24 Basis for Sharing
The basis underlying sharing of information will facilitate or impede sharing. It may range from a formal agreement that addresses sharing mechanisms in detail, to a more general formal agreement, to an informal working relationship, to no agreement, and to ad hoc arrangements on a case-by-case basis.

25 Level of Planning
A program or project that involves the sharing of geographic

information among multiple organizations may require considerable planning to meet the requirements of each of the participants. In some cases all future participating organizations will join together prior to the development effort to plan and design the system cooperatively. In others, only a subset of eventual participants will join to plan the development of the project. Some sharing is not pre-planned. A single organization may implement a GIS and decide to share its information through a joint project following the development of the database. Multiple organizations may also decide to form a joint sharing arrangement following the development of individual data sets in each organization. Another possibility is the sharing of information outside of a joint project through individual arrangements with each organization.

26 SYSTEM CAPACITY AND CAPABILITY TO USE DATA
The capacities and capabilities of the participating organizations must be adequate to support the sharing of data. This may include adequate storage capacity in a central repository for the data to be shared and the necessary communications network to provide access to shared data from all participants.

27 USEFULNESS OF TECHNICAL SPECIFICATIONS
The technical specifications of data must be adequate for the users who will share the data. The scale, accuracy, and other technical specifications are very important considerations in entering into a sharing arrangement.

Resources for Sharing

28 FUNDING SOURCE
The funding source often dictates the level of information sharing. Funding sources may vary significantly with each funding approach, causing a different effect on sharing. The GIS may be funded through a single lead organization, shared in a balanced manner across multiple organizations, or shared at different levels across organizations. Information sharing may also be funded through the payment of fees for information provided. In other cases there may be no funding or payment for shared information.

29 DATA OR SOURCE MATERIALS ARE AVAILABLE FOR SHARED DATA
The data to be shared, or the source materials for creating the data to be shared, must be available to one or more organization in the sharing group.

30 SERVICE AREA SIZE
In addition to these factors that may be used to measure the conditions affecting information sharing, a set of characteristics of the service area will also be useful in evaluating sharing conditions. These characteristics include area in square miles, number of parcels, population, customers, and density of development. The type of service area development should also be determined, such as all urban, primarily urban, mixed urban, suburban and rural, primarily suburban, and primarily rural.

MEASUREMENT VALUES

For each of the thirty factors identified, a set of values should be defined through an analysis of the actual significance of the factors. The ideal is to define a set of easily and reliably defined measures. These measures can then be analyzed statistically to determine the significance of each factor and of the combinations of factors. These values will measure the factor in each site or each organization within a site. An initial set of measurement categories is illustrated in Table 5.2.

Application of the Proposed Structure

The proposed structure should be applied by collecting the necessary information for a sample of sites across the country. The candidate sites should be divided into classes to ensure that a range of conditions are considered in the evaluation of sharing factors. Five classes of sites are proposed for the sample, including the following:

1. operating multi-organization GIS program
2. in-development multi-organization GIS program
3. operating single agency GIS program
4. in-development single agency GIS program
5. site without GIS activity

For each site a profile of information should be collected. The profile should include site designation, organization(s) involved, current GIS status, date of initial GIS development activity, extent of geographic information sharing, and description of geographic information sharing. The measurement values for each factor identified in Table 5.2 should also be recorded for each site.

Information Sharing Relationships

In addition to the specific factors that impact and indicate sharing opportunities, the relationships between and among the potential information sharing factors is significant to the success of sharing. The information relationships should be diagrammed for each pair of organizations in a potential or actual sharing situation. The direction of the flow of information between each pair of organizations should be noted in the diagram. Organizations that have a common service area but are not involved in sharing should also be identified.

SAMPLE APPLICATION OF THE PROPOSED STRUCTURE

Testing of the proposed structure was performed at two levels. The first was a test of the mechanics of collecting the necessary data and assigning the proposed codes. This test was performed by collecting data for a set of actual or potential GIS sites to which the author has access. A small sample of sites was evaluated and data collected. As noted in the previous section, the specific sites are not identified in this chapter (Table 5.3).

The second test of the structure was an analysis of the data to determine how well the factors were indicators of the actual sharing experience in each site. A thorough statistical analysis should be made of the data. In this chapter, however, an abbreviated, relatively simple tabulation was made of the factors to be compared to the actual conditions of each of the sample sites.

Based on this application of the proposed information sharing measurement structure to the sample of specific sites, the following observations can be made.

Sample Measurement Values

A relatively simple tabulation and analysis of the values assigned in the sample and the comparison of the results to the known conditions of the sites resulted in several observations:

- The basic profile that emerged for each site matched the known data-sharing situation at the site.
- Several values were not assigned at any of the sites. This raises the question of whether the measurement values are germane or useful. However, a larger sample of sites might find these values assigned often.
- The attempt to define values in a manner such that low scores indicate a high potential for sharing was only partially successful. The relative differences between pairs of values appears to be not constant, and thus distortion in the values occurs. Refinement in the values and the development of statistical analysis procedures might improve this situation.

TABLE 5.2

Measurement Categories for Sharing Factors

1	*Sharing classes*	
	1	common access to a shared data set
	2	separate copies of jointly developed data set
	3	purchase/sale of copies of data
	4	no sharing
2	*Number of organizations involved*	
	1	1 unit within an organization
	2	multiple units or departments in one organization
	3	2 to 3 organizations
	4	3 to 5 organizations
	5	greater than 5 organizations
3	*Organizational goal/mission*	
	1	function specific
	2	information service
	3	mapping service
	4	other service
	5	mixed function
4	*Organization relationships*	
	1	one lead, all share
	2	all equal
	3	one group lead, one group share
	4	no sharing
5	*Current general relations*	
	1	all positive
	2	generally positive
	3	mixed/cautious
	4	generally negative
	5	all negative
6	*Historic relationships*	
	1	all positive
	2	generally positive
	3	mixed/cautious
	4	generally negative
	5	all negative
7	*GIS-specific relationships*	
	1	all positive
	2	generally positive
	3	mixed/cautious
	4	generally negative
	5	all negative

TABLE 5.2 (continued)

Measurement Categories for Sharing Factors

8	*Control of information*	
	1	central control
	2	multi-node control
	3	distributed control
	4	ad hoc control
9	*Growth rate*	
	1	high
	2	medium
	3	low
	4	no growth
	5	losing population/development
10	*Leadership politics/ego*	
	1	no problems, compatibility
	2	minor friction, incompatibility
	3	significant problems, incompatible leaders
	4	no cooperation
11	*Adequacy of internal data and data resources*	
	1	no internal data
	2	limited internal resources
	3	limited internal data
	4	adequate internal resources
	5	adequate internal data
12	*Data of another organization is needed*	
	1	cooperative organization in area with needed data
	2	organization with no existing relationship with needed data
	3	no needed data known to be available from another organization
13	*Need data that can best be developed or maintained jointly*	
	1	need data that are very costly or complex to develop
	2	need data that are moderately costly or complex to develop
	3	need data that are easily developed
	4	internal data needed only
14	*Level of dependence from sharer perspective*	
	1	high
	2	medium
	3	low
	4	no dependence
15	*Organization(s) in the area have needed data*	
	1	all data available
	2	some data available
	3	no data available

TABLE 5.2 (continued)

Measurement Categories for Sharing Factors

16 *Organizations in the area have resources to share development/maintenance costs*
- 1 resources for most costs
- 2 resources for some costs
- 3 no resources available

17 *Organization owning data offers/willing to share*
- 1 owner offers data
- 2 owner responds positively
- 3 owner refuses to share
- 4 no owner organization

18 *Organization offers/willing to share cost of data development/maintenance*
- 1 organization offers funds
- 2 organization responds positively
- 3 organization refuses to share
- 4 no sharing organization

19 *Level of dependence from lead perspective*
- 1 high
- 2 medium
- 3 low
- 4 no dependence

20 *Governmental program or regulations that require or encourage sharing*
- 1 program requires sharing
- 2 program encourages sharing
- 3 no program

21 *Recognition of the value of sharing of data*
- 1 recognize and require sharing
- 2 recognize value
- 3 no recognition

22 *Real or perceived requirement for confidentiality*
- 1 no confidentiality limits
- 2 perceived confidentiality limits
- 3 confidentiality required

23 *Incompatibility in the definition, specifications or structure of available data*
- 1 compatible data
- 2 minor incompatibility
- 3 significant incompatibility

TABLE 5.2 (continued)

Measurement Categories for Sharing Factors

24 *Basis for sharing data between organizations*
- 1 organizational responsibility
- 2 formal detailed agreement
- 3 formal general agreement
- 4 informal agreement
- 5 no agreement

25 *Level of planning*
- 1 pre-planned joint project—all organizations
- 2 pre-planned joint project—partial organizations
- 3 single organization implementation, post-project joint sharing
- 4 multiple organizations join in post-development sharing
- 5 share data outside of joint project

26 *System capacity and capability to use data*
- 1 full capability and capacity
- 2 full capability, limited capacity
- 3 lack of capability and capacity

27 *Data of useful technical specifications*
- 1 useful specifications
- 2 minimal problems
- 3 significant problems

28 *Funding source*
- 1 single lead organization
- 2 shared/balanced
- 3 shared/unbalanced
- 4 fee for sharing
- 5 no funds involved

29 *Data or source materials for shared data*
- 1 all available
- 2 most available
- 3 minimal available
- 4 none available

30 *Service area/size area in square miles*
- • NUMBER OF PARCELS
- • POPULATION/CUSTOMERS
- • DENSITY OF DEVELOPMENT
- 1 all urban
- 2 primarily urban
- 3 urban, suburban, and rural
- 4 suburban
- 5 rural

- Assigning one set of values for a site is inadequate. Since multiple organizations or departments are involved, each has a separate perspective. The method should be expanded to acquire values from each of the participants with a procedure for normalizing of the values obtained. The differences in values assigned by various participants, in fact, may offer valuable insight concerning data sharing.

Quality of Data

While the development of a measurement scheme for each factor will provide a level of objectivity, the data collected will still be subjective. It will be important to minimize the subjectivity, possibly by use of skilled investigators and training in the application of the values. The initial research into application of the proposed structure should specifically address the data collection issue. While a measurement scheme has been proposed here, it may not be practical to reduce the measurement to such specific scores. It may be necessary to develop less structured and more subjective data. Alternatively, it may be appropriate to remove factors that do not lend themselves to objective measurement and recording.

Potential Availability of Measurement Data

Based on the sample of sites used here, it appears that all data are relatively easily accessible. It will be necessary to interview one or more persons at each site who are knowledgeable about the history and environment of the site. It will probably be desirable to interview many people to identify and remove bias that a single person may hold. It may also be impractical to expect that a single person is familiar enough with the situation to provide all measures reliably.

Ease of Measurement

Most of the factors can be measured with relative ease. Most, however, involve a subjective value judgment that may vary from person to person. One person, for example, may believe that the organizations have a positive relationship in general, whereas another may be aware of frictions between organizations. The establishment of the values may require intimate knowledge of the site and its organizations. The measurement, therefore, may require extensive investigation or, more likely, access to persons who are familiar with the site, organizations and GIS project. To accomplish this it may be useful to employ a structured technique such as Delphi to obtain reliable values from multiple participants.

TABLE 5.3

Sample Geographic Information Sharing Characteristics[1]

		Site									
Factor		A	B	C	D	E	F	G	H	I	J
1	*Classes*	1	2	1	1	1	3	2	1	4	1
2	*Number*	1	5	4	1	2	1	2	2	2	3
3	*Goal*	3	5	5	1	5	3	5	3	5	5
4	*Relations*	1	2	1	1	1	4	1	1	3	1
5	*Current*	2	1	3	4	1	2	4	3	1	3
6	*Historic*	2	4	2	4	1	2	3	3	2	3
7	*GIS*	2	3	2	2	2	4	2	2	3	3
8	*Control*	1	2	1	1	1	1	1	1	2	2
9	*Growth*	2	2	3	3	4	2	3	1	2	4
10	*Leaders*	1	2	2	3	2	1	3	3	2	2
11	*Adequate*	3	3	2	4	4	4	4	5	2	4
12	*Need data*	1	1	3	3	3	3	1	3	1	2
13	*Need development of data*	1	1	1	1	1	1	2	2	2	2
14	*Depend I*	3	2	2	3	3	3	1	3	4	3
15	*Organization with data*	3	3	3	3	3	3	2	3	2	2
16	*Will share cost*	3	2	1	3	3	2	2	3	2	2
17	*Will share data*	4	1	4	1	4	1	2	4	2	2
18	*Share cost of development/maint.*	4	1	4	1	4	2	2	4	4	4
19	*Depend II*	1	2	1	2	1	4	2	1	1	2
20	*Government requirement*	3	3	3	3	2	3	1	3	2	2
21	*Recognize*	1	1	1	3	1	2	1	3	3	2
22	*Confidentiality*	1	1	1	2	1	3	2	1	1	1
23	*Incompatibility*	2	2	1	2	1	3	2	1	2	1
24	*Basis*	1	3	2	5	4	5	1	1	1	4
25	*Planning*	1	1	1	3	2	5	1	1	2	3
26	*Capacity*	1	1	1	1	1	2	2	2	3	1
27	*Specifications*	2	2	1	2	1	3	2	1	2	1
28	*Funds*	1	3	2	1	1	4	1	1	1	4
29	*Source*	2	2	2	2	3	3	2	1	3	1
30	*Size*	3	2	1	3	1	5	2	3	1	2

[1] Meanings of the numeric values shown in this table are provided in Table 5.2.

RECOMMENDATIONS

The following recommendations are offered based on the experience of developing and applying the structure and measurements and the evaluation of the data collected in the sample application.

The factors must be analyzed more rigorously to determine their significance and contribution to the sharing situation. While the pattern of actual measurement values does appear to match the level of sharing in each site and thus provides accurate information, the analysis performed in this small sample was extremely limited. Some of the factors appeared to be of marginal value in the sample but may emerge as more useful in a larger sample. The factors, therefore, require further evaluation and refinement.

The measurement values also require further refinement. Some were difficult to apply. In particular, the perspective from which the values should be assigned was difficult to determine. It is suggested that multiple persons should establish values for the factors, and those values should be combined for the site. This approach is particularly necessary where multiple organizations are involved in the sharing environment. As noted above, methods must be established to improve the objectivity of the measurements. Some values were inappropriate, and some were not assigned to any site. This indicates that further refinement is necessary.

Perhaps the most important area of research is the establishment of a base set of values for known sharing environments, both successful and unsuccessful. A standard profile should be developed against which future sites might be measured. This will require applying the structure to numerous sites for which the sharing situation is known.

An appropriate statistical analysis procedure should be developed and applied to determine values at a measured level of reliability. This was not attempted in the time limits and small sample of this chapter.

As noted above, the relationships between and among factors may be as important as the factors themselves. It is recommended, therefore, that the relationships be modeled and analyzed to gain an understanding of the dynamic interaction of the factors.

Finally, the author believes that a comparison should be made between the factors affecting data sharing in other fields with those of geographic data sharing to determine the extent to which models from other fields may be applied in the geographic information domain.

Note

1. *Webster's Seventh New Collegiate Dictionary*, G. & C. Merriam Company, Springfield, Massachusetts, 1963.

Section II

Experiences in Sharing Geographic Data

Experiences in Sharing Geographic Data

This section includes eight chapters that relate experiences, problems, and successes in sharing geographic information. The chapters are similar in that each draws upon one or more specific cases. The first six authors relate lessons they learned as involved individuals from their sharing experiences or attempts at sharing. Several hypothesize principles they believe may be useful and generalizable to other similar circumstances. The last two chapters, although still focused on experiences, are by authors who acted primarily as observers of sharing relationships.

In the first chapter, William Craig provides a personal glimpse of the numerous behavioral and institutional stumbling blocks he has encountered in the past in attempting to gain effective access to geographic information maintained by public agencies and organizations. He hypothesizes that institutional inertia is a primary culprit in limiting the ability to share geographic information and suggests several possible alternatives in overcoming that inertia.

In the second chapter, William Bamberger describes the evolution of the San Diego Regional Urban Information System (RUIS), the barriers to data sharing encountered in attempting to create the multi-participant GIS, and how those barriers have been overcome or are continuing to be addressed. He concludes with several hypothesized generalizations about data sharing derived from experiences with the RUIS project.

The chapter by Nancy Obermeyer that follows relates a similar scenario in inter-organizational information sharing. She describes the experiences of the Cincinnati Area GIS project (CAGIS) and stresses the importance of flexibility in inter-organizational agreements and the ability to renegotiate. She identifies these capabilities as critically important in keeping sharing relationships useful, efficient, and productive over time.

Next, the chapter by Kenneth Dueker and Ric Vrana examines three organizations undertaking different approaches to data and systems integration. One organization is pursuing integration by utilizing a core database concept, the second is following a database constellation strategy, and the third is attempting a data marketplace approach. The authors compare the approaches in terms of benefits, organizational changes required, and the extent and nature of organizational resistance. From their observations they suggest several implications for future research.

The fifth chapter, by Stephen Ventura, again compares three geographic data-sharing settings. In this instance, however, Ventura's work compares the development and potential effectiveness of "overarching bodies." Such bodies are charged with facilitating geographic data sharing among diverse and wide-ranging potential users. Use of this technique at three levels of government—local, state, and national—is described, comparisons in effectiveness are made, and questions in relation to organizational and behavioral research are raised.

The chapter by Nancy Tosta complements the Ventura chapter as well as contrasts with it. In her chapter, Tosta examines the progression of data-sharing activities in California state government. Whereas the Ventura chapter describes the experience of a state that has pursued use of an overall coordinating body, Tosta describes the situation in a state where there has been no overall plan for implementing GIS in state government or in creating and maintaining digital spatial data sets. She states that GIS has been implemented in California state government primarily on a "program-by-program" basis and describes how the needs for coordinating mechanisms have been met, at least partially, by inter-organizational GIS user groups, as well as by some services provided by the Teale Data Center. She goes on to suggest numerous behavioral and organizational factors derived from her experience that she believes are significant in affecting the ability and willingness to share geographic information.

The last two chapters are by academic researchers who report on their observations of data-sharing activities in local governments. These researchers are neither involved in building the sharing relationships observed nor in affecting change in them during the time of the studies. Bruce Rocheleau, a political scientist, reports on his study of the rate of sharing of information between police and other city departments in Illinois. His research suggests that such factors as hierarchical authority and accountability requirements are more important to achieving effective sharing than several other tested factors. Implications of these findings for sharing geographic information are discussed.

Finally, the chapter by Ian Masser and Heather Campbell compares information sharing in British local governments that have taken a "corporate approach" to implementing GIS versus those that have allowed each department to go its own way. The authors argue the benefits and drawbacks of taking a coordinated "corporate approach" to GIS development and then contrast them with the benefits and drawbacks of taking a "department approach." Results of their survey work divulge a nearly even split between the approaches selected by British local governments. Effects on sharing as a result of the approach selected are evaluated.

The chapters in this section present a sampling of experiences in dealing with behavioral and organizational issues in the sharing of geographic information. Although not planned as such, each chapter relates sharing experiences or attempts at sharing with, within, or among governmental bodies. The current experience base in sharing digital geographic information tends to indicate that government is involved as one of the parties in most geographic data-sharing arrangements. One may surmise that knowledge gained relative to behavioral and organizational factors from these government data-sharing experiences will also have relevancy to broader and perhaps more complex sharing environments yet to evolve.

WILLIAM J. CRAIG

6 *Why We Can't Share Data: Institutional Inertia*

Public agencies are focused on their own mission and mandates, and they have little time or incentive to worry about sharing their data with others. Individual organizations pursue their narrow goals even when a small amount of additional effort would add critically useful data and information to the corporation (e.g., state government. This is my conclusion based on a career of public policy research using data about neighborhoods, cities, counties, and state governments. This "institutional inertia" manifests itself in three general ways. The first is refusing to cooperate – either outright refusal or setting outrageous conditions, such as excessive prices. Second is the selection of technology adequate to meet limited internal needs but inadequate for others. Third is the use of narrowly defined data items and data definitions. This chapter expands on these general barriers and illustrates them with personal "war stories." I conclude that change can come only by expanding the mandates of the organization to include the sharing of data, providing incentives to do so. The imposition of mandates is within the authority of elected officials and bodies, so we must learn how best to enlighten these people. Case studies should be undertaken where sharing has been successful to measure the correctness of my hypotheses. We need to learn how critical actors were converted, whom they influenced, and how they plied that influence.

INTRODUCTION

This is a very personal paper. My twenty-five–year career has been devoted to public policy research for neighborhoods, cities, counties, and state governments. I have spent inordinate time and effort trying to pry loose data to support that research. This chapter is based on a good deal of reflection and analysis of the problems that I have encountered over the years. Given the nature of my work, all of my requests for geographic data were for ad hoc extractions or summarizations of data, not attempts to establish ongoing sharing arrangements. These are the types of requests typically made by planners, analysts, and decision makers.

Sharing data is critical to the success of GIS technology. Yet experience has shown that only a small percentage of potentially useful information actually is shared. This conclusion has been voiced both by the National Center for Geographic Information and Analysis (NCGIA) and by the Urban and Regional Information Systems Association (URISA). While concerns of confidentiality and privacy are legitimate reasons for withholding data, the primary reasons for lack of sharing are organizational and institutional. Agencies could share data, but they choose not to do so. They sometimes refuse outright but more often have created information systems (and GIS) that make data sharing difficult or of limited value. I see the major problem as "institutional inertia"; everyone is focused on the mission and mandates of the agency, and there are no incentives for activities—like sharing data—that might add to the common good. The situation will not improve until we can determine ways to change the institutional reward system.

Organizations are established to accomplish missions. It has been argued and demonstrated that formalized bureaucracies are more efficient than more open, organic systems (Daft 1989; Weber 1947). Unfortunately, the very factors that make bureaucracies efficient (formalization, centralization, standardization, and hierarchy of authority) make them unresponsive to ad hoc requests and uncaring about serving the missions of other organizations. Individual performance is rated by superiors in relationship to assigned responsibilities and to organizational mission, so there are no incentives to meet extraordinary requests (Thompson 1967, ch. 9; Evans 1986). Government bureaucracies are the potential source of most geographic information, so it is disheartening that they are even more formalized and centralized than private-sector bureaucracies (Perry and Rainey 1988). Governmental bureaucracies are more subject to the influence of external political and institutional pressure. We need to win over political and institutional leaders who have influence. Their influence needs to be exerted on the top of the bureaucratic organization and then filtered down to the technician[1] who can deliver the product.

The most common argument for sharing data is cost savings, but other reasons are equally compelling. Creating a database is the most expensive part of most GIS installations, and the ability to borrow useful data sets from others can reduce those costs. A second reason to share data is to create synergism, combining data to get better insights. For example, there might be one data set about the location of hazardous materials and another about the location of vulnerable populations; taken together, they provide a picture of potential disasters in a city and allow us to take corrective measures. A third reason is to be able to move data out of its depository system into some different system for additional analysis. A fourth reason is that data used in a shared environment becomes cleaner—more complete and correct.

URISA has long been interested in data sharing (see, for example, Ferguson 1987). In late 1987 URISA developed a research agenda that formalized this concern. The first statement of this agenda came at the 1987 International Geographic Information Systems (IGIS) Symposium, which was intended to articulate the research needs that might be addressed by the winning NCGIA proposal. "We see data issues as preeminently important. . . . Gaining intergovernmental and intragovernmental cooperation [in sharing data is a] problem in most places" (Craig 1988b, 142). The complete listing of URISA's research agenda is split into two major headings—social issues and technical issues—but concerns over data and data sharing are pervasive (Craig 1989c). URISA endorsed the NCGIA research agenda (UCSB 1988) largely because of its incorporation of issues on data sharing and on the use and value of geographic information.

BACKGROUND

This paper is based on my experiences over the last fifteen years. I looked back on ten projects that included an aspect of data sharing and tried to identify the barriers I had encountered. A listing and short description of each project follows.

1. My doctoral dissertation was based on using public records to monitor land-use change (Craig 1980). These records proved to be superior to aerial photography for identifying change, but an inordinate effort was required to convert those records into useful information.
2. I was the leading proponent of placing data issues on URISA's research agenda as it was being developed in 1987 (Craig 1988a). My arguments were articulated in a paper called "An engine without fuel: the need for research on supplying data to Geographic Information Systems." That paper opened with an analogy calling for better exploration and refining. Since much of the discussion was about sharing data, I should have continued

the analogy with calls for renewable energy and recycling. Data are not consumed when they are used.

3. In 1988 I inventoried GIS/computer-mapping systems in Minnesota. I found forty-nine different groups using thirty-three different brands of software (Craig 1989b), and the problem of sharing data was becoming more apparent. A majority said they were willing to share their data but could describe no method to do so.
4. Later that year I visited Australia and consulted with state and regional GIS/LIS (land information system) users in Victoria. One department was building the base map upon which all other agencies were to place their data (Craig 1989a). I found the lead department unwilling to communicate or negotiate with any of the dependent agencies, many of which were desperate to begin work, because the lead administrator did not want to be distracted from the job of creating the base map.
5. Back home, we conducted the Minnesota Data Exchange Project, attempting to move data among five different Geographic Information Systems using a variety of translators (Craig, Tessar, and Khan 1991). We found that a simple intermediate file standard like DXF always worked to translate data but that too often the data files had been created for a single purpose, and it was difficult to use them for a different purpose.
6. I became a user of data in an attempt to learn about lending practices in St. Paul (Craig et al. 1990). I searched the only five available data sets about this topic and found each to be purposely incomplete, restricted in organizational and geographic coverage. Furthermore, each reported results by geographic units incompatible with the others.
7. After the lending study and numerous other attempts to collect and use data about small areas, I wrote an article that summarized my experiences and frustration (Craig 1991). I found that agencies design computer systems to serve their daily needs, but they were either unwilling or unable to extract data in ways that would meet the needs of others.
8. A local foundation approached me to produce a series of maps documenting the environmental status of the Mississippi River (Craig and Anderson 1991). I was excited by the prospect but had much difficulty in obtaining useful data. Most federal agencies were willing to share their data but would not help me formulate my requests.

9. A 1991 visit to capital cities in Australasia proved to be a shock (Craig 1992). Public organizations creating the cadastral base were charging such a high price for their data that other governmental units were either dropping their GIS plans or creating their own base maps. Thus, hopes for synergism were lost.
10. I chair a committee that is monitoring an $800,000 state-funded project to create a new land-use map of Minnesota at a 1:24,000 scale using aerial photography. The previous map was created in 1969 at a scale of 1:250,000. We have put significant effort into looking at using a variety of existing data sources to keep the new map current in order to reduce the need for subsequent major (and belated) update efforts.

FINDINGS

"Institutional inertia" is the overriding reason that data sharing has been limited. Organizations have missions that justify their existence and mandates that guide their daily activity. Because there is no mandate to share data, there is no reason to cooperate. This problem manifests itself in a number of ways including refusal to cooperate, the use of limiting technology, and data capture that is inadequate for other uses. Specific examples based on my personal experience are given for each problem area.

Refusal to Cooperate

Many of my past requests for data have been met with refusal—either outright refusal or placing conditions that were tantamount to refusal. The Minneapolis City Assessor refused to produce a summary table of the count of bedrooms in a neighborhood even though we offered to pay and even though (we learned later) state law requires compliance. Likewise, the Australian administrator refused to adjust production and delivery schedules, even though he has a mandate to digitize the entire state and no reason was offered to continue digitizing maps in the order he had selected. In neither of these cases was lack of money an issue. The offices refused simply because they could not be bothered with a task outside their normal duties. In the Australian case, the administrator showed rancor toward me because I, as an outsider, had no business questioning his authority.

Outrageous prices limit sharing. The Australasian states have set a high price to gain income and make good on boasts about there being a substantial market for their base maps. The problem, of course, is that most of that market consists of other government units. The base departments are meeting their internal goals, but the corporation is worse off.

Another example of exorbitant pricing is found closer to home. The Minnesota Department of Jobs and Training has set outrageous prices to produce employment summaries for traffic zones in the Twin Cities: $90,000 for the first year and $60,000 per year thereafter. The stated reason for the high cost is the need to re-contact firms having multiple locations to determine the distribution of jobs. An additional person is hired for the year to do this work. The cost is high enough to discourage all but the most needful (and affluent) user, thus minimizing disruption of normal departmental functions.

But price is not the only prohibitive condition that can be placed on sharing data. The Mississippi River project had good cooperation in getting environmental data from federal agencies, provided the Freedom of Information Act was invoked. But we never knew what we were getting. Spill data came in gallons and pounds; locations were specified by community name rather than by river mile or geo-referenced. We could obtain neither metadata nor information about what tabulations were possible. The cycle time for a given request was usually one month, so we quickly ran out of time and stopped asking.

Protection of confidentiality is often an excuse for refusing to share data. The Federal Home Loan Bank Board (FHLBB) collects, summarizes, and distributes data on home mortgages and home loans by thrift institutions. At first they were cooperative in providing data to us but only for the Twin Cities metropolitan area as a whole. Later they refused to provide home improvement data for one six-month period. They claimed too few thrifts operated in the Twin Cities, and it might be possible to identify an individual institution. In the latest congressional amendments to the Home Mortgage Disclosure Act, similar data are *required* to be reported by institutions. Obviously, the bureaucrats had lost track of why they were asked to produce reports and needed a reminder from a higher authority.

Limiting Technology

The technology is available to do everything that is needed, but organizations each choose a more limited technology, thereby limiting themselves from extraordinary use of data. Each organization chooses the technology that is appropriate to its task. Buying or creating additional technology usually cannot be justified. Sometimes a minor addition to the technology of one organization would solve everyone's data problems. When additional technology could be added for little or no additional funds but is not acquired, no excuse exists for not doing what is needed. I see three areas where the choice of technology has restricted our ability to share data: the use of manual records or very limited computerization, the decision not to geocode records or digitize maps, and a selection of software and hardware that limits the ability to transfer data.

The most basic lack of technology is data that are not computerized. Building permits could be used to monitor land-use change, but few local authorities bother to computerize this data. In the days of inexpensive computers on everyone's desktop, it would seem reasonable to use a computer to determine fees and create summary reports.

Other data might be computerized already but are not usable by others because of technology limitations. The Minneapolis Assessor's Office, in the request mentioned above, did let us use its data file. But access was limited to one record at a time. Their computer system was designed this way to meet their operational needs. Because no one in the Assessor's Office had a need for summary reports, the system contained no option to produce one. It took my research assistant four days at a terminal to extract the required information about the number of bedrooms. No one seemed concerned about confidentiality issues that might be associated with allowing access to names, valuations, and detailed building characteristics of the 700 buildings involved.

For the GIS community, the second most basic technology problem is lack of digitization, or geocoding. There are inexpensive packages available for geocoding, i.e., determining the geographic unit of a given street address. The City of St. Paul would like summary records by census tract, but the county has no technology to accommodate this request. The county has a data field in the property record for census tract, but it has been little used in the past thirty years. A one-time geocoding effort could fill this field, but it seems impossible to mount that effort. The county doesn't need it for its purposes.

One of the most amazing stories about missed digitizing opportunities is the work done by the Agricultural Stabilization and Conservation Service (ASCS) of the U.S. Department of Agriculture in agricultural counties across the nation. The ASCS is responsible for monitoring whether farmers are in compliance with their government contracts. Photography is updated annually, and at least 15 percent of the enrolled land is checked, oftentimes 100 percent. The process involves planimetering fields and writing down results. Practically no additional effort would be required to connect the planimeter into a computer and use it as a digitizer. Much time and effort would be saved. I talked to one ASCS county official who traces the field boundaries on graph paper, because farmers are more believing if they can see a map of their field. If the fields were digitized, the computer could create these plots as a free by-product. Many other federal, county, and state offices could benefit from the digitized data for updating land-use maps and other purposes. The local ASCS offices would be happy to cooperate, but the national bureaucracy has been slow to show interest in current technology.

Finally, there are high-quality, computerized/digitized files that cannot be translated for sharing with others. Sometimes good translators simply do not exist, or the translation must be done in several stages, which degrades the quality of the final product. Other technical problems include incompatible media (e.g., tapes versus disks), and the need to compress or generalize files to make them useful to the receiving system. Some translators are simply too expensive to warrant their purchase for a one-time transfer. Many of the technical problems could have been avoided if the data-generating organization had considered the prospect of others using its data when its system was first purchased and installed. We found that a translator using simple primitives, like DXF (or the Spatial Data Transfer Standard [SDTS]), is usually available and can almost always translate files without error.

Idiosyncratic Data

Data created for one use are sometimes incompatible with the needs of others. The big question is "How much additional effort would be required to make the data more widely useful?" In many cases the answer is "none," and the only reason for not cooperating is institutional inertia. Where more effort is required, additional resources may need to be made available to the data-generating organization.

I could cite many examples of basic data files that need only minimal effort to make them more useful. Two common problems are unused fields and narrow classification systems. The Regional Multiple Listing Service data records have data fields for school district identification and finished living area that are seldom completed by realtors. The land-use classification system used by USGS shows agriculture as a single class (Campbell 1983, 28), but for an agricultural state like Minnesota it is critical to see land shifting between cultivation and pasture. When approached about the possibility of a cost-sharing program with the state paying for an initial separation of these subcategories, the USGS response was to require the state to pay for a second interpretation—after their standard product had been extracted. This option was significantly more expensive than our proposal.

The Minnesota Department of Revenue was formerly able to provide data to cities about median family income. No other reliable source is available to allow cities to monitor intercensal changes in the economic situation of their constituents at either the city or the tract levels. Since 1986 these data are no longer available. State tax forms were modified, tax credits have been lumped, and family returns can no longer be identified with certainty. Of course, copies of federal forms, which contain the necessary data, must accompany the state return. From its perspective, the Department of Revenue sees no reason to add that data item to its new computer records.

Initial digitizing standards for many operations are too focused on the limited needs of the originating organization. Generally, people cut corners where no apparent harm is done, but this laxity causes problems for subsequent users. Too many digitizers are oblivious to topological integrity and neglect to snap to closure. Too often, where the final product is to be a paper map, subdominant layers are omitted, spatial accuracy becomes relative, and features are presented without the formal identification that is necessary for a GIS. An analysis of the Minnesota Department of Transportation (MnDOT) digitizing procedures found that for as little as 5 to 10 percent additional effort all these problems could be overcome. They agreed, and their products are now used in state and local Geographic Information Systems around the state. This additional effort is small enough that MnDOT can absorb it within their existing allocation. They believe they are justified in using their resources this way and are proud of the role they now play in the corporate data system of the state.

CONCLUSIONS

I have argued that institutional inertia is the dominant cause of our inability to share data. Public organizations, in particular, have mandates that they are required to fulfill, and they have little interest in bending to accommodate the needs of others. This intolerance may result in the outright rejection of requests to share data. Or it may be the result of earlier decisions about technology or data collection procedures that were made without regard to additional uses that might be made of their data. There is a slight chance that organizations will reform themselves out of a sense of pride or an interest in cooperating.

Of course, any additional effort requires additional resources, and many public agencies are stretched to the limit now. Additional funds would need to be sought in some instances to buy new equipment, write new software, and pay more people. Clearly, these funds would be justified to the corporation if it could save money by not digitizing the same line two or more times. If the organization itself could see and appreciate these benefits, it might begin requesting additional resources. But since the data are required by a different organization, the home organization has no incentive to meet those needs, and such requests are rare. In fact, in many cases

I know of, the data-generating organization has argued against receiving these additional responsibilities even *with* additional funds. Little more evidence is needed to prove that the problem is truly institutional inertia—organizations continuing to focus on their own problems and missions.

Organizational change in a bureaucracy starts at the top with clear statements about new goals, division responsibilities, and consequences of various levels of performance (Evans 1986). The hierarchical nature of bureaucracies means that individuals and divisions are always evaluated and rewarded by superiors (Weber 1947). Because inter-agency sharing requires a vision about that of the individual department head, I believe that enlightened leadership must come from elected officials—mayors and governors, city councils, and state legislatures. Mandates and money come from these elected officials. We need to enlighten them to the benefits of decentralized data collection and extensive sharing of these resources. It is time for government to ask what its main business is (see Levitt 1960), and if the answer includes building an information infrastructure to make government and private sectors more efficient, then steps must be taken to make geographic information more accessible and more usable.

This leaves the challenge of determining how to enlighten our leadership. Pursuing this topic would be a natural follow-up to results arising from NCGIA research on the use and value of geographic information. A segment of that research has addressed the diffusion of GIS, and enlightening the leadership is part of that process. How effective are the media in promoting change? In the United Kingdom, the Association for Geographic Information (AGI) gives a Journalist of the Year Award to the GIS article that best communicates complex issues to the general reader and expert alike. Several professional organizations sponsor executive sessions; how effective are these? What about science councils? The GIS literature tells many stories of systems being purchased because of the crusading efforts of a single individual. Can the same model be used for changing the way a state or a city prepares and distributes its data?

Determining how best to effect change has been studied by sociologists, political scientists, and organizational theorists. Building on this base, the task before us should be to identify and study cases where effective data sharing is taking place. Is my hypothesis correct, and was institutional inertia the critical barrier that needed to be overcome? How much additional effort was required to make the organization's data useful to others? What net cost savings or additional productivity were gained? Most importantly, what actors were critical in forcing the data sharing, how did they become enlightened, whom did they influence, and how did they ply their influence? Systematic study will yield answers to these questions.

Those answers are critical to making more data available for sharing, to saving taxpayers money, and to using GIS technology more effectively.

Note

1. I have used this term advisedly. Daft (1989) has pointed out that "professionals" bristle at bureaucracies and perform better without bureaucratic rules and authority. Professionalization of the GIS community might solve most of the problems described in this paper.

REFERENCES

Campbell, J. B. 1983. *Mapping the land.* Resource Publications in Geography, Association of American Geographers, Washington, DC.

Craig, W. J. 1980. Monitoring Minnesota's land using public records. Unpublished Ph.D. dissertation, Department of Geography, University of Minnesota, Minneapolis.

______. 1988a. An engine without fuel: the need for research on supplying data to Geographic Information Systems. Unpublished paper presented at the 1988 URISA Conference, Los Angeles, August 7-11, 1988.

______. 1988b. The research agenda from a user perspective. International Geographic Information System (IGIS) Symposium: The Research Agenda, Crystal City, Virginia, November 15-18, 1987. *Proceedings* 1: 141-143.

______. 1989a. Land and Geographic Information Systems in Victoria. Traverse, News Bulletin of the Victoria Division of Australian Institute of Surveyors, 109, February, 8-9.

______. 1989b. A plethora of GIS use in Minnesota? *The Operational Geographer* 7, 4 (December): 3-5.

______. 1989c. URISA's research agenda and the NCGIA. *Journal of the Urban and Regional Information Systems Association* 1: 7-16.

______. 1991. Empty promises for small area data: monitoring our neighborhoods using operational records. 1991 URISA Conference, San Francisco, August 11-15. *Proceedings* 4: 220-229.

______. 1992. Why we couldn't get the data we wanted. *Journal of the Urban and Regional Information Systems Association* 4, 2: 71-78.

Craig, W. J., and W. S. Anderson. 1991. Environment and the river: maps of the Mississippi. Center for Urban and Regional Affairs, University of Minnesota, Minneapolis.

Craig, W. J., M. Goldfein, L. Heegaard, and F. W. Smith. 1990. Community reinvestment in the City of St. Paul: are residents and businesses receiving the financial services they need? Center for Urban and Regional Affairs, University of Minnesota, Minneapolis.

Craig, W. J., P. Tessar, and N. A. Khan. 1991. Sharing graphic data files in an open system environment. *Journal of the Urban and Regional Information Systems Association* 3, 1: 20-32.

Daft, R. L. 1989. *Organizational theory and design.* 3d ed. St. Paul: West Publishing.

Evans, M. G. 1986. Organizational behavior: the central role of motivation. *Journal of Management* 12: 203-222.

Ferguson, N. B., ed. 1987. Data sharing: can we afford not to? 1987 Alberta URISA Chapter Spring Conference, Edmonton, April 27-28, *Proceedings.*

Levitt, T. 1960. Marketing myopia. *Harvard Business Review* (July-August): 45-56.

Perry, J. L., and H. G. Rainey. 1988. The public-private distinction in organizational theory: a critique and research strategy. *Academy of Management Review* 13: 182-201.

Thompson, J D. 1967. *Organizations in action: social science bases of administrative theory.* New York: McGraw Hill.

University of California at Santa Barbara (UCSB). 1988. National Center for Geographic Information and Analysis: a proposal. University of California at Santa Barbara, State University of New York at Buffalo, and the University of Maine.

Weber, M. 1947. *The theory of social and economic organization.* New York: Oxford University Press.

WILLIAM J. BAMBERGER

7 *Sharing Geographic Information Among Local Government Agencies in the San Diego Region*

The high cost of developing Geographic Information Systems and the benefits of sharing geographic information are encouraging local governments and utility companies to cooperate in unprecedented ways. But there are many obstacles that make cooperation difficult. These obstacles result in redundant mapping systems and expensive duplication of geographic information. The San Diego Regional Urban Information System (RUIS), created in 1984, is one of the early attempts to create a multi-participant GIS. RUIS is now an integrated GIS, providing services to ten departments in the city and county of San Diego. RUIS has developed relationships with San Diego Gas and Electric Company and other cities to share a common base map. This chapter describes the evolution of RUIS, the barriers to data sharing that it has encountered, and how it has attempted to overcome those barriers.

INTRODUCTION

Since its inception in 1984 the Regional Urban Information System (RUIS) has been developing a Geographic Information System that ultimately will meet the needs of twenty-eight different departments in two local governments (the city of San Diego and the county of San Diego). RUIS pioneered the concept of inter-agency cooperation in the development of geographic systems and sharing geographic data in San Diego. RUIS, like other large GIS projects, has had both successes and failures.

Both the successes and failures should help to understand the nature of data sharing, the obstacles to data sharing, and how data can be shared more effectively in the future.

After a brief description and chronology of RUIS, the chapter discusses data sharing within the RUIS community—the twenty-eight departments in the city and county of San Diego for whom the system is being developed and operated. Topics in this section include the organization and management structure, organizational factors that inhibit data sharing, developing standards, sharing the base map, and several examples of sharing data for specific applications. The next major section discusses sharing data with organizations outside the RUIS community, such as other cities, special districts, and the private sector. The final section provides some generalizations from the experiences RUIS has had in sharing geographic information.

OVERVIEW OF THE REGIONAL URBAN INFORMATION SYSTEM

Background

The San Diego region, located in the southwestern corner of the United States, is a rapidly growing area with a population of 2.6 million. The region's 4,200 square miles contain a diverse landscape of coastal beaches, foothills, mountains, and deserts. The city of San Diego, the region's largest city, covers about one-tenth of the land area and is home to nearly one-half of the region's population. The Regional Urban Information System was created in 1984 by an agreement among three entities: the city of San Diego, the county of San Diego, and San Diego Data Processing Corporation (SDDPC). SDDPC is a non-profit public benefit corporation, established and owned by the city of San Diego. It replaced a traditional data-processing department and provides all data processing and telecommunication services to the city. In the RUIS agreement, SDDPC is assigned the responsibility of developing and operating the system.

Agreement by the city and county of San Diego to jointly develop RUIS was itself an unusual example of cooperation between two jurisdictions with a long history of political rivalry. The cooperation occurred primarily because of the tenacity of a city bureaucrat and encouragement by a member of the County Board of Supervisors. The bureaucrat was a city planner who had the responsibility for developing the city's Urban Information System (UIS) beginning in 1979. The UIS was a tabular (i.e., non-GIS) system to monitor growth and development in the city. The planner in charge of the UIS attempted to promote the system to other nearby cities and the county, with little success. However, in 1984, when the County

Planning Department unveiled plans to develop a similar, but independent system, the city bureaucrat persuaded a member of the Board of Supervisors (a former city councilman) to encourage the county staff to base their system on the city's UIS. The result was the RUIS agreement.

The architects of the agreement had no model upon which to base the agreement because multi-participant Geographic Information Systems were unknown in 1984. Since then, several other communities in the United States have created multi-participant GIS projects. While each multi-participant organization is unique, the goal of most is to create and share a common base map among all participants. RUIS has gone further than most in the implementation of common standards, base map maintenance, and applications. Table 7.1 shows the major events in the development of RUIS.

Costs

During the eight years of RUIS's existence, GIS development costs have totaled $7.3 million. Figure 7.1 illustrates the breakdown of costs into major categories. The RUIS cost picture is unique because only 23 percent of total costs have been spent on base map conversion. This relatively low percentage results from an arrangement, described later in this chapter, to share an existing base map with the gas and electric utility company.

Annual operating costs are averaging $0.8 million. This includes costs of hardware (file server, workstations, printers, plotters, and the like), software (Geographic Information Systems, database management systems, and so on), and staff to maintain and enhance the existing GIS applications.

DATA SHARING WITHIN THE RUIS COMMUNITY

The RUIS Organization

When RUIS was created in 1984, the city and county of San Diego were reluctant participants. There were few examples of the two jurisdictions working successfully together. Multi-participant GIS projects were unknown at that time. There were no models for an organizational structure. In spite of this the structure that was developed proved to be effective and has endured without major revision for more than eight years.

Key elements of the organizational structure are:

- RUIS management is placed in the highest administrative offices in the two jurisdictions—the County Chief Administrator's Office and the City Manager's Office.
- RUIS is *not* a separate entity but a "project" that the two jurisdictions agree to develop jointly.

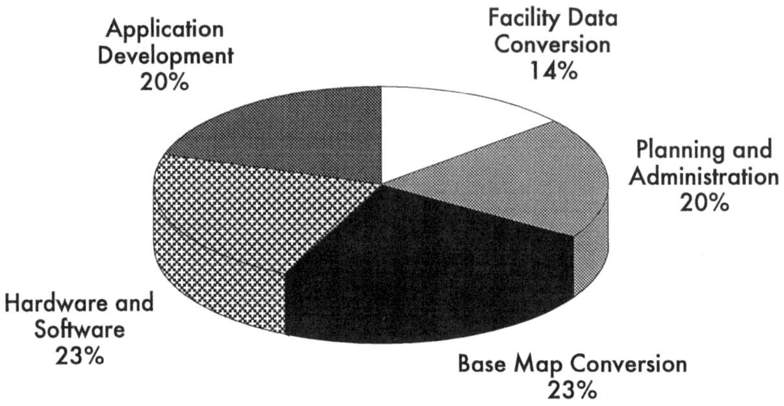

FIGURE 7.1
System development costs

- Development costs for common components of the system are shared equally; system components required by only one jurisdiction are funded by that jurisdiction. However, no specific source of funding was identified.
- Operation and maintenance costs are paid by the departments according to system usage.
- Broad-based participation by user departments is encouraged by a variety of committees. This committee structure has evolved over the years.
- A full-time administrator works for the Management Committee, consisting of the Assistant City Manager and Deputy Chief Administrative Officer.
- Both the city and county have full-time coordinators for the project.
- San Diego Data Processing Corporation has the exclusive contract to develop and operate the system.

Figure 7.2 is an organization chart for RUIS. The Policy and Budget Committee, consisting of the City Manager and County Chief Administrative Officer, is responsible for establishing policies and adopting the annual

TABLE 7.1

Major Events in the Development of RUIS

1984	Agreement establishing RUIS signed by city, county and SDDPC. RUIS task force, consisting of mid-level managers from twenty-eight city and county departments, established. Consultant hired to produce conceptual design.
1985	Tabular (non-GIS) development monitoring system completed.
1986	Conceptual design study completed. Request for Proposals for system vendor issued.
1987	Tabular building permit and road information systems completed. GIS vendor proposals evaluated and vendor selected. Consultant hired to pilot test data conversion sources and methods.
1988	Initial design of core GIS components completed.
1989	First GIS component completed (Basemap and Survey Information System—BASIS). Decision made to acquire digital base map from San Diego Gas and Electric Company (SDG&E); negotiations commence. Data conversion plan developed; DIME/ETAK file acquired from San Diego Association of Governments; initial data conversion begins. Cost-benefit analysis completed. RUIS Standards Committee established.
1990	Sublicense agreement to acquire base map signed with SDG&E. Contract signed with data conversion vendor; local staff assembled to perform quality control. Crime mapping and analysis system developed for city Police Department. Long-term technical plan developed resulting in decision to switch GIS vendors. RUIS Technical Committee established.
1991	Conversion of base map completed for entire 4,200 square miles of county. Base map maintenance system and procedures completed. RUIS provides geofile for emergency dispatch. Fire Demand Zone maintenance system completed. Water and Sewer Facility Management System initial design completed. Water and sewer facility conversion plan and pilot study completed. Water and sewer facility conversion vendors selected.
1992	GIS transferred to Unix-based, client-server technology. Census geography added to base map. Contract signed with vendor to develop digital ortho-photographs and digital terrain models for city of San Diego (approximately 400 square miles). Water meter route management system developed.

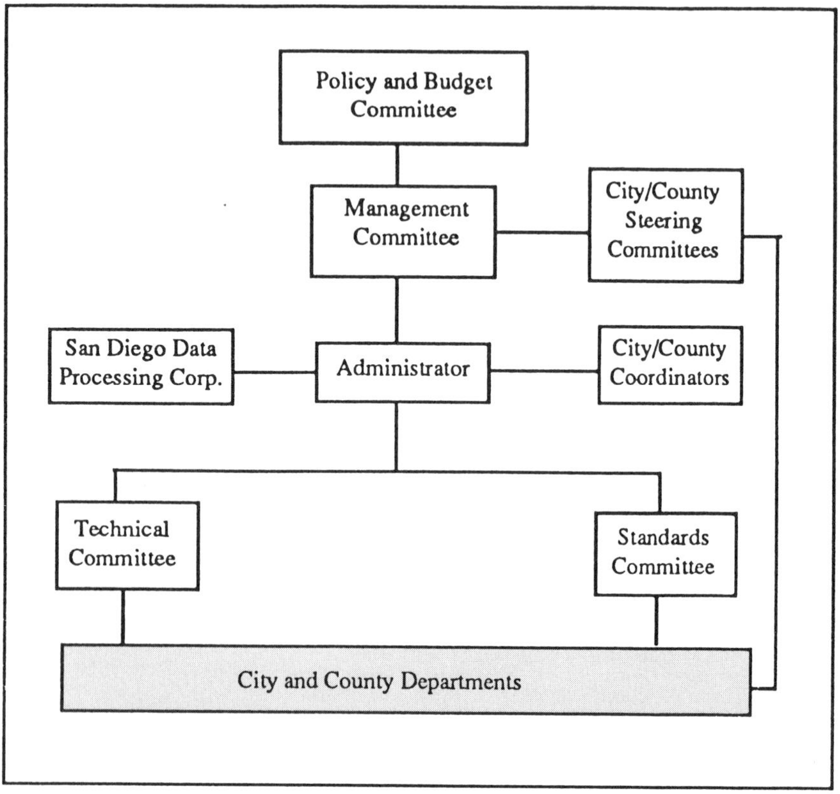

FIGURE 7.2
RUIS organization

work plan and budget. The Management Committee, consisting of deputy-level managers from the two jurisdictions, is responsible for establishing priorities and recommending a work plan and budget. The Management Committee approves all contracts and also sets policies on the sale of RUIS products and services to outside organizations.

The RUIS Administrator works directly for the Management Committee and is responsible for the day-to-day administration of the program. The Administrator provides direction to San Diego Data Processing Corporation, develops the annual budget and work plan, markets products and services, and attempts to maintain tranquillity among the various participants. The City/County Coordinators are responsible for determining the requirements for departments in their jurisdictions, determining sources of funding, and assisting with implementation issues.

The city and county Steering Committees, consisting of department heads or managers of the major departments, advise the Management Committee on the organizational, technical, and budget issues that affect their departments. The Standards and Technical Committees consist of professional staff from the key departments. The Standards Committee adopts standards regarding symbols, map products, procedures, and equipment. The Technical Committee deals with the challenges of implementing the GIS within their departments.

Organizational Conditions That Inhibit Data Sharing

From the outset, city and county management envisioned RUIS as a highly centralized system, providing all mapping and geographic needs to both jurisdictions. All city and county departments that create or use maps and other forms of geographic information must participate in RUIS, shifting control from the department to the RUIS organization. Fortunately, in 1984, all departments were using manual mapping methods. No department had a vested interest in a particular GIS or mapping system. However, department managers were leery of losing control of their functions.

The concept of a multi-department GIS, requiring departments to participate in a single system, conflicts with the basic hierarchical structure of local governments. The typical local government has a top-level manager with independent departments below. Very few functions cut across those department boundaries. Even the traditional data-processing department, which serves many departments, normally implements systems that are used by a single department. A GIS is fundamentally different. A single system is intended to meet multiple needs. Many different departments maintain the data, creating dependencies among departments that did not previously exist.

RUIS has used many approaches to overcome the organizational barriers that make reluctant departments participate.

- *Top-down Management.* Because RUIS is managed out of the City Manager's and County Chief Administrative Officer's (CAO) offices, top-level management pressure can be used to bring departments "in line." However, the county has several elected department heads that are autonomous from the CAO, including the Assessor and Sheriff. The Assessor, with a large mapping staff, has been a reluctant participant in RUIS.

- *Budget Carrots.* The core system was financed largely from the City Manager's and CAO's budgets, relieving the departments of the burden of funding the development of an expensive GIS.

- *User-directed Organization.* The RUIS committee structure promotes dissemination of information and offers many opportunities for departments to participate in decision making. This has increased their confidence in the project.
- *Communications.* RUIS has used many different forms of communication, including status reports, newsletters, slide shows, and workshops. Although a considerable effort has gone into communication, it has not been enough.

Limited funding for RUIS is another factor that has inhibited data sharing. Like many local governments in post-Proposition 13 California, the city and county of San Diego have suffered many lean years. RUIS development has depended on city and county general funds and, as a result, competes for funding with law enforcement, fire protection, health, social services, and other critical needs. Actual funding has been considerably less than estimated in the initial conceptual design, causing many expectations to be unfulfilled.

Developing Standards

A GIS forces an organization to adopt mapping and geographic standards. Prior to RUIS each department had its own standards (or none at all). The introduction of RUIS required agreement on a variety of standards, including map accuracy, GIS hardware and software, map maintenance, map graphics, and public access. Base map accuracy and maintenance are discussed separately below. This section describes the organizational structure RUIS has established to set standards.

The RUIS Standards Committee was established in 1989 as a standing committee chartered by the Management Committee. Its purpose is to develop and enforce GIS standards to be used by all departments in the city and county. The committee has developed standards for:

- Map graphics, including symbols, line types and colors
- Address and geocodes, including address formats and street name abbreviations
- Map products, which specify standard scales, map features to plot at various scales, and standard map plotting software
- Map accuracy, including coordinate geometry closure rules and procedures for merging data of varying degrees of accuracy
- Hardware, including which workstations and peripherals will be supported

- Base map maintenance, including detailed procedures for adding new subdivisions and changing features to the base map
- Application "look and feel," which ensure that all GIS applications have similar user interfaces
- Public access, which establishes the rules for distributing geographic information to the public

Broad-based department participation on the Standards Committee was encouraged by asking the department heads to select a staff member to represent their departments. The following departments are represented: Public Works, Planning, Building Inspection, Water and Sewer Utilities, Assessor, Registrar of Voters, Police, Fire, and Data Processing. The Standards Committee also has many different professions represented, including engineers, planners, land surveyors, cartographers, crime analysts, and system analysts.

The officers of the Standards Committee (chair and vice-chair) are elected for one-year terms and must be from different jurisdictions. The committee meets monthly, with detailed minutes kept for each meeting.

Selecting the GIS Software

Upon completion of the conceptual design in 1986, RUIS began the process of acquiring a hardware and software platform for its GIS. Two committees were established: an evaluation committee composed of system users and a technical committee consisting of data-processing staff. The system users favored a mini-computer or workstation-based GIS and software that provided the greatest functionality. The technical committee came to a different conclusion, favoring a mainframe-based system and software they believed provided the best data-management structure. The disagreement between these two factions was the first test of the RUIS organization to resolve differences. The conflict was initially resolved when the data-processing staff worked out an agreement with their preferred vendor to enhance their GIS product (at no cost to RUIS) to provide the functionality required by the users.

After eighteen frustrating months, and little progress on the functional enhancements, a new committee was established to evaluate the long-term technical direction for RUIS. This committee consisted of both users and data-processing staff. A three-month effort by this committee resulted in a unanimous decision to change GIS platforms to the one originally preferred by the user community. This client-server–based GIS not only provides the needed functionality but also provides tools that enable users to develop their own applications—something that was not possible with the mainframe system. Individual departments can now create and maintain their own GIS applications and data layers.

Developing the Base Map

Creating the base map is usually the most costly element of a GIS. RUIS began the process of creating its base map by hiring a consultant to make recommendations on data sources and methods. Three options were considered:

- digitizing existing manually drawn maps
- acquiring an existing digital map from the local private utility company, San Diego Gas and Electric Company (SDG&E)
- creating a new base map from scratch, using ortho-photography.

A survey of system users was conducted to determine requirements for the base map and to establish the quality of (manual) maps then in use. The survey determined that several users required a high degree of spatial accuracy (i.e., maps with a spatial accuracy of one to two feet). But the survey also revealed that the maps being used by these departments were of *unknown* accuracy. The survey also determined that the term "accuracy" has different meanings to different users. Engineers typically use the term to refer to spatial accuracy, whereas other users consider accuracy to refer to content accuracy.

The base map consultant concluded that none of the existing hard-copy map sources were worth digitizing because they were out of date, of poor quality, or for only a portion of the region. The SDG&E map, although in digital form, was based on photogrammetry with a spatial accuracy of ±10 feet. In addition, the SDG&E base map was designed to be a "background" coverage for their electric facility data, not a topologically consistent map. The third option—creating a new base map from scratch—had considerable appeal because it could meet the requirements of all RUIS participants; however, this option was estimated to cost approximately $10 to $15 million.

Because of limited financing, it became evident that developing a new base map from scratch would take many years. The SDG&E map, although imperfect, became a candidate if it could be acquired at a reasonable cost. Acquisition was complicated by the fact that SDG&E had assigned the marketing rights to a subsidiary company—Integrated Information Systems (IIS). When negotiations began, the two parties were far apart—IIS wanted several million dollars for the map, and RUIS wanted to acquire it at no cost. During the ten-month negotiation period, several events occurred:

- The president of IIS, who was representing IIS in the negotiations, was replaced.
- IIS was sold and was no longer a subsidiary of SDG&E (but retained marketing rights to the base map).
- The city of San Diego filed suit against SDG&E because of an announced merger with Southern California Edison.

Each of these events lengthened, and threatened to scuttle, the negotiations. After an agreement had been reached between RUIS and IIS, SDG&E staff decided that the agreement was not acceptable to them and insisted on changes. About the same time, the city wanted to pull out of the negotiations because of its concern that an agreement on the base map would jeopardize the lawsuit against SDG&E. The future of RUIS was at stake, so the RUIS administrator organized a broad-based effort to conclude the negotiations. This effort included enlisting high-level city and county managers as well as elected officials from the county to apply pressure on SDG&E management. Attorneys for both sides worked frantically to develop contractual language acceptable to both sides. The agreement was signed in March 1990.

The agreement specifies that RUIS acquires the SDG&E base map at no cost but is required to provide all updates and enhancements to SDG&E for a period of ten years. RUIS's ability to sell information to the private sector is limited by a minimum price provision, and RUIS must pay a royalty of 33 percent of gross sales on base map products sold to the private sector. The agreement allows RUIS a high degree of flexibility in dealing with local government agencies in San Diego County.

The base map was acquired in April 1990 and, with the assistance of a conversion vendor, was merged with several existing data files from the city and county. The final product, completed in November 1991, is an aesthetically high-quality map containing street centerlines with street names and address ranges, parcel boundaries, and other map features such as water bodies and city boundaries.

Maintaining the Base Map

Long before the base map was delivered, RUIS established a commitment to keep the base map up-to-date. San Diego is a rapidly growing area, with approximately 25,000 new parcels created each year. Because the RUIS base map is used by many different departments, including Police and Fire emergency dispatch, the base map must contain new subdivisions shortly after they are recorded.

Two options for base map maintenance were considered: centralized and decentralized. The centralized approach would require that a new base map maintenance staff be created and that the departments responsible for manual mapping relinquish control. Originally, the decentralized approach divided responsibility among three existing mapping staffs—City Engineer, County Public Works, and County Assessor—with each staff responsible for maintaining a specified geographic area.

The decentralized approach was adopted, primarily because it was the path of least resistance. It could be implemented without major disruption to the existing organization structure. The three organizations responsible for base map maintenance already had existing mapping staffs and base mapping responsibilities. The major difference was that the RUIS base map required them to maintain an automated map instead of manual maps. However, the success of the decentralized approach depends on highly complex procedures to maintain consistency, to ensure quality and timeliness, and to coordinate the different elements that make up the base map.

While still in effect, the decentralized approach to base map maintenance is not working very well. The County Assessor dropped out of the base map maintenance process because of lack of funding. The County Public Works staff has taken over the Assessor's responsibilities. Another participant—the City Engineering and Development Department—has suffered severe disruptions because of staffing problems caused by a discrimination grievance.

The decentralized approach has also resulted in complaints from user departments about the quality and timeliness of the updates. Maintaining a GIS base map is fundamentally different from maintaining a manual map. The base map maintenance personnel have had difficulty comprehending the importance of the database behind the graphic elements of the map. Significant difficulties have arisen because all of the required fields in the database are not being maintained accurately or completely. RUIS management is now considering the creation of a new, independent, centralized base map maintenance staff.

Maintaining the Computer-Aided Dispatch Geofile

In November 1989 the voters of the city of San Diego passed a bond issue to fund a new emergency dispatch system for the Police and Fire departments. The new system was to replace old, independent systems used by each department. An important element of a computer-aided dispatch (CAD) system is the "geofile" that identifies the police or fire unit that can respond the quickest to a call from any given address. The older CAD systems had tabular geofiles that were maintained manually by each department. The new system was to be "map-based" so that it could display the location of incidents on computer screens and associate a geographic coordinate with each incident.

The timing of the new CAD system corresponded with the development of the RUIS base map. Also, CAD had been previously identified in the RUIS Conceptual Design as a system that should be linked with RUIS. As a result, discussions between the CAD departments (Police and Fire) and RUIS began in an effort to determine whether RUIS could meet the geofile requirements. The CAD departments were wary of RUIS's ability to provide the geofile because at the time of the initial discussions RUIS had no existing base map. They were also skeptical of RUIS's ability to develop its base map in time to meet the CAD schedule. Finally, the CAD departments did not want to be dependent on a geofile update process they did not control.

The CAD departments had identified an alternative source for their map-based geofile—a private vendor. However this option would cost more to implement and maintain than RUIS and would result in a redundant base map. Numerous discussions were held between the CAD departments, RUIS, the City Manager's office, and the CAD vendor before the decision was made to use RUIS. The decision to use RUIS required that a multi-phase process to develop the geofile be used to ensure that a file would be available by the time the new CAD system "went live." Although this multi-phase process was somewhat more expensive, it reduced the risk of the CAD implementation being delayed because of delays in the RUIS base map development schedule.

RUIS is also responsible for providing updates to the geofile on a weekly basis. While this process is relatively new, a number of difficulties have already become obvious.

- The CAD system's data structure (where streets are organized by street hundred blocks) is very different from the RUIS data structure (where streets are modeled as intersection to intersection records). These differences require a complicated computer program to restructure data.

- The CAD system does not accept transactional updates. As a result, when updates are provided, the CAD geofile must be completely replaced. This deficiency is being corrected by the CAD vendor.

- The "real world" is much more complex than the model assumed in either the RUIS or CAD data structures. For example, addresses are not always consistent—an odd-numbered address may occasionally be assigned to the "even" side of the street, or posted addresses may differ from the address assigned by the city. The CAD system needs the address that is used, not necessarily the official address assigned by the city.
- Maintenance priorities for the CAD system differ from those of other departments. Since the CAD departments are not maintaining the geofile, their priorities are not always met.

Water and Sewer Facility Management

The city of San Diego Water Utilities Department provides water and sewer services throughout the city. Prior to RUIS, all water and sewer facility mapping was done manually. The Water Utilities Department recognized that it could achieve cost savings by automating several graphical functions. The design engineers saw benefits from a computer-aided design and drafting (CADD) system. Other divisions needed analytical capabilities that only a GIS could provide. After initial discussions it was agreed that CADD and GIS are fundamentally different technologies and that two systems should be implemented. Further, the Water Utilities staff decided that, rather than develop an independent GIS, they would use RUIS.

A significant problem for the Water Utilities Department was the slow pace of RUIS development, which was caused by limited funding. Water Utilities staff believed that the benefits of GIS would far outweigh the costs, so they agreed to provide funding to accelerate the development of the RUIS base map. The City Water Utilities Department's financial support was inconsistent with the RUIS agreement (which specified a fifty–fifty split between the city and county) but was accepted by the city because of the importance of GIS to Water Utilities. Water Utilities has also funded the city's portion of many base map enhancements, such as adding lot lines and addresses to the base map.

Water Utilities envisions many uses for the GIS, including:

- facility mapping,
- maintenance planning,
- long-range system planning,
- maintenance vehicle dispatch, and
- meter-reader routing.

Automated Regional Justice Information System (ARJIS)

The Automated Regional Justice Information System—ARJIS—is a county-wide crime reporting and analysis system. Every major law enforcement agency in the county participates in ARJIS. It is a tabular, mainframe system. Although it has no mapping capabilities, ARJIS uses a "geofile" to assign police beats and geographic coordinates to every crime incident. The geofile is based on the Census Dual Independent Map Encoding (DIME) file, which is maintained by the San Diego Association of Governments. ARJIS pays for the maintenance of the DIME file. A goal of ARJIS has been to replace the current geofile with a file provided by RUIS. This is desirable because of the inaccuracies in the DIME file and the expense of maintaining it. Also, ARJIS would like to provide a crime mapping capability and better geographic analysis tools.

Replacing the DIME file with a RUIS file is proving to be more difficult and costly than anticipated. The RUIS centerline file contains shaping records, similar to the Topologically Integrated Geographic Encoding and Referencing (TIGER) and Spatial Data Transfer Standard (SDTS) data structures. This requires program changes to ARJIS. But the existing ARJIS system is old, complex and coded in COBOL. As a result, it is expensive to modify. Also, geographic coordinates are maintained for every crime incident. The RUIS geofile uses NAD 83 coordinates (instead of NAD 27, used in the DIME file), which will require hundreds of thousands of records to be recoded with new coordinates, another expensive task.

SHARING DATA WITH EXTERNAL ORGANIZATIONS

Other Government Agencies

Although sharing data within the RUIS community has been difficult and has required overcoming many barriers, sharing data with external organizations is even more difficult. The RUIS agreement specifically states that RUIS is intended to be a "regional" system, available to all local government agencies in the county. During the initial phase of development, RUIS managers decided that other organizations should not participate in RUIS. Development of a GIS requires many complex decisions; more participants would complicate the decision-making process. In any event, no other agencies expressed a desire to participate.

As the RUIS base map neared completion, the RUIS Management Committee decided to encourage other organizations to participate. Important issues were:

- How would participation occur (on-line access, sharing the base map, and so on)?

- How would costs be shared? (The primary motivation for encouraging other agencies to participate is to reduce the costs to the city and county.)
- How would the system be "marketed"?

At the time that RUIS began actively to encourage participation of other agencies, one city had already implemented an automated mapping system and several others had established committees to study GIS. About the same time a San Diego Section of the Urban and Regional Information Systems Association (URISA) was formed. A major purpose in establishing the San Diego URISA section was to provide a forum to discuss GIS. It has been very helpful in establishing personal relationships between the professional staff from the different jurisdictions that are responsible for GIS.

The RUIS Management Committee sent letters to every city manager and the general managers of every water and sewer agency in the county. A series of workshops was conducted by RUIS to demonstrate the capabilities of the system and to discuss the needs of potential participants. The workshops were followed by telephone calls from the RUIS Administrator. The RUIS Administrator met with staff from many cities and special districts.

Initially the cities and special districts were reluctant to participate in RUIS. The reasons are not obvious, but probably include:

- lack of clarity by RUIS regarding exactly what RUIS would provide and at what cost;
- strong desire of other agencies to remain independent (specifically, the small cities were reluctant to become involved with the city of San Diego and the county because of fear of being dominated); and
- belief by the cities that developing and implementing their own independent GIS would be less costly than participating in a regional system.

However, once the economic benefits of sharing information became evident, participation in RUIS accelerated. As of June 1993, four cities and seven districts have acquired the RUIS base map to be used in their own GIS. Other districts have purchased hard-copy mylar base maps. Several other cities and districts are considering acquiring the base map. Some plan to provide their own maintenance, while others expect to purchase periodic updates from RUIS.

While organizational obstacles pose the most difficult barriers to developing relations with other organizations, data interface issues have also caused problems. Each GIS software product has its own unique data structure. The existing software available to convert data between systems is inadequate; a time-consuming and costly manual clean-up process is typically required. The high costs of data translation have made some attempts to share data with other organizations too expensive to be practical.

The Private Sector

Sharing data with the private sector involves even more complications and obstacles than sharing with government agencies. The sublicense, which provided RUIS with the SDG&E base map, places severe restrictions on RUIS's ability to provide data and services to the private sector. The agreement specifies that RUIS must pay a minimum royalty of $100 per square mile of area. Even if RUIS kept no revenue, this royalty appears to price RUIS data out of the market. (The minimum royalty for the entire base map would be $420,000.) The sublicense agreement does not account for on-line access or distribution of selected base map features. RUIS has been able to work out ad hoc deviations from the terms of the agreement but has been unable to have the agreement modified.

Another limitation in dealing with the private sector is RUIS's difficulty in defining specific products and prices. RUIS management decided early on that the sale of services and products would be used to partially recover the costs of the system. However, government information is subject to public records laws that severely limit governments' ability to charge for information. According to legal opinions, the form of the data may determine whether it can be sold or whether it must be provided for the cost of duplication. Typically, when governments provide services (such as garbage pickup or water), prices are based on the cost of the services. However, cost-based pricing for information makes less sense because the incremental cost of providing information is minimal. (The information is still there after a copy has been made and distributed.) "Value-based" pricing is more appropriate but is a concept not typically used by local governments.

Another obstacle to selling services or products to the private sector is local government's inability to act like a business. Local government is reluctant to risk taxpayers' money on marketing studies, advertising or product development. An option that RUIS is considering is developing agreements with third parties to market RUIS data. Such agreements would provide RUIS with revenue from royalties but have the marketing flexibility that is more typical of the private sector.

CONCLUSION

The "data sharing" issue is much more complicated than simply determining how data created by one organization can be used by other organizations. The ability of different organizations to cooperate will determine what data is available and what technology is used. The major benefit of GIS technology is that, unlike manually drawn maps, a *single* base map can be disseminated electronically to many users. Cities, counties, special districts, gas, electric, telephone, and cable television companies all have common needs for geographic information.

Typically, no single organization has the resources to develop a complete, automated GIS, so the benefits of GIS are lost to the community. Or limited, single-purpose Geographic Information Systems are developed that result in duplication of effort and do not provide the benefits of an integrated, community-wide database.

While there are no overt prohibitions against organizations working together to develop an integrated GIS, such systems are the exception rather than the rule. As RUIS illustrates, where such integrated systems exist, they are constantly struggling with institutional and organizational barriers. There are many reasons why local agencies have difficulty cooperating. The following are offered as generalizations based on eight years experience with the RUIS project:

- *Lack of Organizational Model.* There is no organizational structure that promotes cooperation. In cases where cooperation has occurred, it can often be attributed to the efforts of a strong-willed "project champion." Typically, organizations simply do not know how to work together.
- *Organizational Independence.* While there is often significant overlap in constituencies, organizations have independent management structures, usually based on a pyramid-type of hierarchy. Also, some of the potential players are elected officials (e.g., assessors) who answer directly to voters.
- *Distrust.* Individuals in organizations often do not understand or trust the motives of individuals in other organizations. Because of this there is a reluctance to relinquish control to another organization.
- *Different Requirements.* While many organizations share the need for certain geographic information, exact specifications and priorities differ. For example, a city engineer may need highly

detailed and accurate information about subdivision lots, while a telephone company could get by with a rough schematic of the lot lines.

- *Lack of a Mandate.* While there is a financial incentive to cooperate in the development of a GIS, there is no mandate to do so. Without a mandate, the management of the various organizations considers the loss of control associated with a cooperative GIS to be risky.
- *Financial Structure.* Utility companies can often include the costs of their GIS in their rate base and, therefore, receive a return on their investment. As a result, their financial incentive to cooperate is less.

Because of these types of barriers to cooperation, communities are paying for the development of redundant systems or are completely missing the benefits of modern land records and geographic information management.

Nancy J. Obermeyer

8 *Reducing Inter-Organizational Conflict To Facilitate Sharing Geographic Information*

The implementation of large-scale Geographic Information Systems requires an openness to both a new technology and new procedures, especially procedures that facilitate the sharing of geographic information across organizational boundaries. In an earlier paper, I suggested that negotiation is an important strategy in establishing inter-organizational information-sharing strategies. The experience of the Cincinnati Area GIS project (CAGIS) supports this concept and takes it further, suggesting that ongoing readjustments are often required to enable the project to continue to function smoothly. These readjustments are made possible by flexibility within the overall structure of the inter-agency agreement. This chapter examines the experience of the Cincinnati Area GIS project, paying special attention to the role of flexibility in the ability of the member organizations to share geographic information across organizational boundaries in a long-term, open-ended project. The case study also reinforces the importance of a project champion.

INTRODUCTION

One of the hallmarks of a bureaucracy (into which category most organizations fall) is the routinization of tasks that is designed, on the one hand, to streamline and expedite frequently repeated organizational duties and, on the other, to assure that the performance of such duties occurs "without regard to person," that is, that all who interact with a given bureaucracy receive equal treatment (Weber, in Gerth and Mills 1946). All too often, the result is an organization that is unable to shift gears and respond to new and changing situations in a timely manner.

As broad-based, integrated Geographic Information Systems gain greater acceptance, there has been a growing need for organizations to share information across organizational boundaries. This chapter assumes that inter-organizational information sharing is an activity that occurs within the framework of an inter-organizational alliance.

Such alliances can result from one or more of a variety of pre-conditions. In some instances, inter-organizational alliances form because one organization holds power over another and can therefore force the subordinate organization to cooperate (Keller 1984; Lindahl 1919; Milward 1982; Stern 1979; Weber, in Roth and Wittich 1978); in other instances, appeals to professionalism and common goals motivate organizations to ally voluntarily (Keller 1984; Milward 1982; Olson and Zeckhauser 1966); in many instances, organizations of more or less equal power status negotiate with one another to achieve an acceptable arrangement for inter-organizational cooperation.

This chapter examines the alliance growing from the negotiation among relative equals. While inter-organizational appeals to professionalism and negotiations may succeed in achieving an initial inter-organizational agreement, such an agreement among relative equals may also, in fact, lay the groundwork for ongoing readjustments to maintain an equilibrium between or among the participating organizations (Keller 1984; Milward 1982). Such readjustments require flexibility, both in the agreement itself, and in those charged with implementing the agreement.

The present chapter explores the role of flexibility in the formation and maintenance of inter-organizational alliances for voluntarily sharing geographic information, taking as an example the specific case of the Cincinnati Area Geographic Information System project (CAGIS), which thus far appears to be a successful inter-organizational alliance among equals for the purpose of sharing geographic information.

The chapter begins with a discussion of the conditions that may precipitate a need for negotiation or renegotiation of information sharing agreements within the context of an inter-organizational alliance. The following section examines the types of flexibility required within the alliance to facilitate such a renegotiation. The final section describes the CAGIS experience as an ongoing success story in this respect and offers suggestions that may be pertinent to other GIS projects that are dependent on sharing geographic information across organizational boundaries.

Inter-organizational Alliances and the Need for Adjustments

The reasons why a particular inter-organizational alliance comes into existence and may later require adjustment of its agreement, and the frequency with which it will require such an adjustment will, of course, vary in specific details from one alliance to another. However, we can identify several general conditions that may precipitate these events.

There are several bases for the formation of inter-organizational alliances, as noted above. Among them, two are relevant to this discussion: *appeals to professionalism* and *negotiation*. Olson and Zeckhauser (1966) and McGuire (1974) suggest that even when organizations are devoid of feelings toward one another (either positive or negative), they may find that it is in their interest to organize for the purpose of providing collective goods within a specific functional area.

Appeals to professionalism may sometimes be a motivating factor in the formation of inter-organizational alliances that can facilitate inter-agency information sharing. We can characterize appeals to professionalism as a low-cost (and frequently a no-cost) strategy that relies on persuading another organization to join a common effort designed to promote common organizational goals or the public interest. Keller (1984) and Milward (1982) note the importance of functional interests and professionalism in laying the foundation for inter-organizational alliances, while Gage (1984, 136) argues for the importance of understanding such networks as " . . . instruments for establishing and maintaining political networks to accomplish policy objectives." The emphasis on common objectives and the ability to enhance the performance of inter-related functions appear to be directly relevant to inter-organizational sharing of geographic information across organizational boundaries.

In some cases, professionals may respond to appeals to a shared sense of professionalism, forming alliances with new partners or even putting aside inter-agency rivalries to pursue a common goal. These alliances may require the short-term sublimation of the organization's primary objectives or its independence in order to achieve a more important long-term common goal. Sometimes, the benefits may be so great and the sublimation so minor that the participants agree rather easily; in other instances, the alliance will be hard-won and require much negotiation and compromise. In addition to the achievement of the specific objectives of the alliance, other indirect benefits of such alliances may accrue to the participants. For example, such ventures may stimulate employment opportunities in the field and foster a higher level of public support for the organizations in their specific policy or functional areas by giving the impression that the participants are acting as good public citizens.

In some instances, as described above, the development and maintenance of alliances will be more difficult, depending on the negotiation of an acceptable exchange among the parties involved. As Weber asserts, "Rational exchange is only possible when both parties expect to profit from it or when one is under compulsion because of his own need or the other's economic power" (Weber, in Roth and Wittich 1978, 73). Other authors note that, even in cases where inter-organizational alliances develop as the result of negotiations, an uneven distribution of power is the norm rather than the exception (Keller 1984; Lindahl 1919; Milward 1982; Stern 1979). Within this environment of uneven power distribution, we can expect factions to compete against their alliance partners for position and to assure that their individual organizational goals are not given short shrift as the alliance continues (Milward 1982, 470). The result is an ongoing search for equilibrium among the members of the alliance (Keller 1984; Milward 1982). Equilibrium is not a static condition but rather a process outcome that the members of the alliance achieve through their continuing efforts to assure their positions or to gain power. Inasmuch as the quest for position is ongoing, the equilibrium point for the system as a whole changes frequently. These authors and others (Buntz and Radin 1983) argue that inter-organizational conflict in alliances is inevitable.

Whether or not inter-organizational conflict is inevitable is debatable; however, we can identify the potential sources of conflict where it does occur. The first potential source of inter-organizational conflict within an alliance is the uneven distribution of power described above. For example, where one organization holds statutory authority over another (or other) member(s) of the alliance, conflict may arise from an incompatibility between the goals and objectives set by the organization with statutory authority (usually a public agency) and the desire for organizational or local autonomy (Guess 1985; Stenberg 1985).

Another important potential source of inter-organizational conflict is conditions in the external environment that may affect the ability of participants to continue as members of the alliance. For example, any discrepancy between the demands placed on members of the alliance and their continuing fiscal abilities to meet those demands may create a negative situation (Guess 1985; Stenberg 1985). In the current environment of a weak economy and the resulting fiscal austerity, such externally driven problems seem particularly plausible as some organizations may face budgetary cutbacks that may limit or curtail their participation in any activity not viewed as central to their organizational mission.

Geographical and cultural differences may also generate conflict (Agranoff and Lindsay 1983; Stenberg 1985); these differences may become a growing source of conflict within the context of alliances formed for the express purpose of sharing geographic information as Geographic Information Systems approach the global scale.

Internal conditions within an organization may also generate conflict that may in turn precipitate renegotiation of specific aspects of the original inter-organizational agreement. Personnel changes or internal reorganizations, for example, may alter the effort that an organization puts into its participation. Many authors (e.g., Caudle 1987; Epstein and Leidner 1987) have noted the importance of "champions" within organizations as important agents of organizational change in general; other authors writing specifically in the area of GIS agree that champions are critical to the successful implementation of the technology (e.g., Anderson and Preecs 1989; Huxhold 1991; Levinsohn 1989; Somers 1989; Wentworth 1989). Internal changes within an organization have the potential to replace a GIS champion (or, to use Caudle's more colorful term, "changemaster") with someone who may be indifferent or—worse—antagonistic towards GIS technology. Even when written inter-organizational agreements are in place, we cannot assume that compliance with the letter of the law is the same in terms of either practice or outcome as compliance with the spirit of the law.

While it seems to be inevitable, conflict should not be viewed automatically as a negative condition (Buntz and Radin 1985; Pondy 1967). Rather, conflict is the means by which the intergovernmental network maintains an equilibrium of power. In some instances, conflict can produce negative effects that may even shatter an alliance; in others, resolving the conflict serves an important integrative function within the network (North, Koch and Zinnes 1960; Pondy 1967).

ENHANCING FLEXIBILITY

As noted above, it is not impossible to overcome the difficulties that arise when organizations form alliances to share geographic information. To do so, however, requires flexibility within the organizations. Obviously, inter-organizational agreements will inherently bind the alliance together for specified periods of time and provide a framework for adjustments among the alliance members. Frequently (and fortunately), such agreements leave many details to be worked out later, opening the door for adjustments as needed.

The continued presence of at least one champion with the commitment to explore a variety of adjustments to the original agreement appears to be critical to maintaining the alliance. The champion(s) must act not only as innovator but also as caretaker, working over the long haul to achieve the objectives of the project. The champion(s) must take professional and personal responsibility for the success (or failure) of the project and work to see it through for maximum effect.

It is essential that the project champion has a high level of authority that enables him/her to make and implement decisions that will improve the chances that the project will succeed. Without a high level of authority, the champion can do little more than rely on appeals to professionalism to rally the troops. Indeed, if the champion holds a relatively low-level position, his/her abilities to persuade supervisors and equals are likely to be limited. This high level of authority recommended for the project champion should include the capacity to provide tangible incentives and assistance to alliance partners on an as-needed basis.

Finally, the project must be a priority of all the members of the alliance. This will, to some extent, be dependent on the original stimulus for the project remaining in place to provide a consistent common objective among the project participants. In the case of the CAGIS project, all of these conditions were present.

THE CAGIS EXPERIENCE

The CAGIS had its beginnings with the recognition that the infrastructure in the city of Cincinnati and the surrounding county (Hamilton) was in a state of serious disrepair. This initial recognition resulted in the formation of a task force headed by the CEO of Proctor and Gamble, a major manufacturer in the region (and in the United States), to discuss the problem and develop strategies for improving the infrastructure of the Greater Cincinnati area. Members of business, industry, and government comprised the task force. This broad-based coalition laid the foundation of strong general support for the task force recommendations. Among the recommendations of the task force was the suggestion that the public and private sectors in the Cincinnati area employ a broad-based GIS to provide a means of tracking the state of the infrastructure in the area.

The recommendation to employ a GIS was accepted and, on January 25, 1990, the "CAGIS Agreement" united the city of Cincinnati, the Board of Commissioners of Hamilton County, the Cincinnati Gas and Electric Company (CG&E), and the Cincinnati Bell Telephone Company (CBT) in an inter-organizational alliance for the purpose of developing and implementing a broad-based integrated Geographic Information System (CAGIS Agreement 1990). The CAGIS Agreement specifically identifies the development and management of " . . . a shared database of geographic and infrastructure data and a comprehensive land related digital map . . . " as a key reason for the creation of the alliance (CAGIS Agreement, Sec. 1).

Terms of the Agreement

Not surprisingly, and consistent with the notion of appealing to professionalism, the agreement emphasizes the "public purpose" that is served by the collaboration of the four parties to the agreement (CAGIS Agreement, Sec. 1). The relationship among the parties is specifically defined as a contractual relationship, not a joint venture, and the Agreement specifies both the rights and the responsibilities of each of the parties involved within a broad legal framework.

The specifics of the agreement call for the participants to share in the costs of all "hardware, software, goods, services, or shared data elements" needed for the project (CAGIS Agreement, Sec. 3). In this sense, the agreement surpasses simple geographic information sharing, including all the technology that makes such sharing possible. Specifically, the CAGIS core cost allocation model for the conversion/implementation period calls for the city of Cincinnati and the county of Hamilton each to cover 42.5 percent of the budget, while Cincinnati Gas and Electric covers 10 percent and Cincinnati Bell, 5 percent (CAGIS Agreement, 3, Ex. 1).

Not only does the agreement allocate fiscal responsibility among the participants, it also charges each of the members with responsibility for acquiring, entering, and maintaining specific data sets with specific attributes (CAGIS Master Agreement, Ex. 2). This requirement means that each of the parties also contributes staff to implement the project. In this way, each of the participants plays an active, ongoing role in CAGIS.

The CAGIS Agreement delegates specific responsibility for organizing and implementing the project to the city of Cincinnati, which developed a specialized department within the Regional Computer Center of the city of Cincinnati to handle the responsibility (CAGIS Agreement, Sec. 1). The CAGIS Administrator reports to the City Manager (CAGIS Agreement, Sec. 10). This position is critical and is vested with a high level of authority, as it should be. The CAGIS Administrator has a small staff to assist in the performance of core duties. If the City fails to perform its responsibilities satisfactorily, the project participants can terminate the City's role in the project (CAGIS Agreement, 39).

The project is envisioned as an ongoing effort, and the Agreement calls for the participants to "exercise their best efforts to negotiate a new agreement for the operation and management of CAGIS to take effect upon the termination of this Agreement" (CAGIS Agreement, 18).

The participants to the Agreement are required to continue the current practice of providing specified information to other participants upon request (CAGIS Agreement, 28). However, the participants may not be "compelled to provide any non-core data or other data that is proprietary in nature or that may jeopardize the security or safety" of their own property (CAGIS Agreement, 28). Hence, while the agreement calls for the participants to share information relatively freely, they are not required to give up the store.

The CAGIS Case

According to discussions in October 1991 with CAGIS Administrator Ralph Goldsmith, the CAGIS project appeared to be on time and within budget two years into its implementation. Achieving this level of accomplishment has not been without its difficulties, however. In particular, maintaining the cooperation of the Assessor has been problematic.

Not surprisingly with a project of this size, the technical tasks of data collection and input have been challenging. In order to assure reasonably accurate base maps, CAGIS chose to produce aerial photographs as the basis for the entire project area. Fitting known property lines and other information from existing hard-copy maps into the framework provided by the aerial photographs was both a technical as well as an administrative challenge. Working with the county Assessor's office and the project consultant, Woolpert Associates, the CAGIS Administrator was able to overcome the technical aspects of this problem.

However, the problem was more than just technical in nature. Personnel changes brought a new Assessor into the picture shortly after the project began formally, and the new Assessor did not have a strong interest in the success of CAGIS. Work within the Assessor's office was not dependent on the success of CAGIS; however, the success of CAGIS requires the participation of the Assessor's office.

Using the authority and resources of his position in concert with his skills of persuasion, the CAGIS Administrator renegotiated the details of the existing agreement with the Assessor to assure his participation. One of the difficulties that the Assessor faced was the technical matter of trying to fit property lines into the framework of the aerial photographs; the difficulty of the task made him hesitant to cooperate. As described above, the Administrator was able to gain the assistance of the project consultant to help resolve the technical details of the problem. At the same time, and perhaps more important, the Administrator was able to supply members of his own CAGIS staff to the Assessor to assist in the digital entry of the Assessor's data.

As described earlier, there are several important features that facilitated the resolution of this problem. In the first instance, the CAGIS

Administrator's high level of authority meant that he had the ability to mold and modify specific details of the inter-organizational relationship within the basic framework of the CAGIS Agreement to assure the achievement of the project objectives. Moreover, he had at his disposal *specific tangible resources* that he could use as incentives to participation.

The CAGIS Administrator's personal characteristics also appear to have been important to his ability to overcome this specific problem. In talking with him, it is clear that he acts as a champion for CAGIS. He appears to be dedicated to the success of the project and acts accordingly.

One aspect of the CAGIS Agreement itself that may encourage the kind of flexibility demonstrated in resolving the problem of gaining the Assessor's participation is the required contribution of both money and personnel by the participants, which started on day one of the project. This requirement assures a high level of commitment from each of the parties from the start which, in turn, encourages ongoing commitment, lest the parties risk wasting their initial contribution by failing to take action to ensure the long-term success of the project.

CONCLUSION

The CAGIS project is just one example of organizations sharing geographic information across organizational boundaries. Although we are a long way from the end of the story, we learn some important lessons from the success to date of the project. For example, an inter-organizational agreement that provides a broad framework while leaving the details vague provides flexibility in fine-tuning those details and modifying them as conditions change. In addition, an agreement that gives the project administrator a high level of authority allows the administrator to be creative in his/her approach to problem resolution. This also has the advantage of facilitating prompt resolution, as opposed to situations where the administrator must seek approval from higher authorities for most administrative actions.

An inter-organizational agreement that requires the contribution of both money and actual ongoing effort also seems to work to advantage by increasing the stakes of each of the participants from the start. This arrangement seems to encourage each participant to work diligently to hold up its end of the bargain by requiring responsibility and accountability of each organization throughout the course of the alliance.

The personal characteristics of key staff can be critical. The lead person must perceive himself/herself as a champion of the project—and act as one—in order to maintain a high level of support for the project. If that individual is persuasive and creative, so much the better.

As time goes on, more broad-based integrated GISs will be implemented. Each will have its own specific purposes and negotiate agreements appropriate to those purposes. In this respect, each particular inter-organizational information-sharing alliance will be unique. However, we can learn something from those that have already occurred, such as CAGIS, in order to improve our general understanding of the role of bureaucratic flexibility in sharing geographic information.

REFERENCES

Agranoff, R., and B. Lindsay. 1983. Intergovernmental management: perspectives from human services problem solving at the local level. *Public Administration Review* 43: 227-37.

Anderson, C. S., and B. Preecs. 1989. Beyond the ad hoc approach to staffing a GIS. In William F. Gayk (ed.), *Proceedings of the Urban and Regional Information Systems Association,* vol. 4. Washington, D.C.: Urban and Regional Information Systems Association. pp. 395-405.

Buntz, C. G., B. A. Radin. 1983. Managing intergovernmental conflict: the case of human services. *Public Administration Review* 43: 403-12.

CAGIS Agreement. 1990. Cincinnati Area GIS (unpublished).

Caudle, S. L. 1987. High tech to better effect. *The Bureaucrat* 16: 47-52.

Epstein, P. D., and A. Leidner. 1987. Technology transfer in the trenches. *National Civic Review* 76: 130-136.

Gage, R. W. 1984. Federal regional councils: networking organizations for policy management in the intergovernmental system. *Public Administration Review* 44: 134-44.

Gerth, H. H., and C. W. Mills. 1946. *From Max Weber: essays in sociology.* New York: Oxford University Press.

Guess, G. M. 1985. Role conflict in capital project implementation: the case of Dade County Metrorail. *Public Administration Review* 45: 576-85.

Huxhold, W. E. 1991. *An introduction to urban Geographic Information Systems.* New York: Oxford University Press.

Keller, L. F. 1984. The political economy of public management. *Administration and Society* 15: 455-74.

Lindahl, E. 1919 (1958). Just taxation—a positive solution. In Musgrave, R., and J. Peacock (eds.), *Classics in the economics of public finance.* London: MacMillan.

Levinsohn, A. G. 1989. A strategic planning based approach to the design of land information systems. In *Proceedings of the Urban and Regional Information Systems Association,* vol. 2. Washington, D.C.: Urban and Regional Information Systems Association. pp. 30-38.

McGuire, M. 1974. Group segregation and optimal jurisdictions. *Journal of Political Economy* 80: 112-32.

Milward, H. B. 1982. Interorganizational policy systems and research on public organizations. *Administration and Society* 13: 457-78.

North, R. C.; H. E. Koch, Jr.; and D. Zinnes. 1960. The integrative functions of conflict. *Journal of Conflict Resolution* 4, 3: 355-374.

Olson, M., and R. Zeckhauser. 1966. An economic theory of alliances. *Review of Economics and Statistics* 48: 266-79.

Pondy, L.R. 1967. Organizational conflict: concepts and models. *Administrative Science Quarterly* 12: 296-320.

Somers, R. 1989. Organizational change for successful GIS implementation. In *Proceedings of the Urban and Regional Information Systems Association*. vol. 2. Washington, D.C.: Urban and Regional Information Systems Association. pp. 39-51.

Stenberg, C. W. 1985. States under the spotlight. *Public Administration Review* 45: 319-26.

Stern, R. N. 1979. The development of an interorganizational control network: the case of intercollegiate athletics. *Administrative Science Quarterly* 24: 242-46.

Weber, M. 1978. In Roth, G., and C. Wittich (eds.), *Economy and society*. Berkeley: University of California at Berkeley.

Wentworth, M. J. 1989. Implementation of a GIS project in a local government environment: the long and winding road. In *Proceedings of the Urban and Regional Information Systems Association*. vol. 2. Washington, D.C.: Urban and Regional Information Systems Association. pp. 198-209.

Kenneth J. Dueker and Ric Vrana

9 *Systems Integration: A Reason and a Means for Data Sharing*

Data sharing takes place within and among organizations when they operate in common geographic areas and when the costs of original compilation and data capture are significant. When geographic data can be shared, efficiency, effectiveness, and enterprise benefits are made possible. Implementing systems integration to facilitate this data sharing may require some organizational restructuring to take full advantage of new technologies. Some resistance to organizational restructuring, however, can be anticipated and, in fact, may be directly proportional to the level of expected benefits. This chapter examines integration issues and impediments to change in three case studies offering different approaches to data and systems integration. It is found that organizational structure, type of analytical system(s) to be integrated, and the degree to which users access a common database are important considerations for anticipating resistance as well as evaluating the benefits to be derived from better integration.

INTRODUCTION

The beneficial sharing of geographic information within and among organizations results from, and is a reason for, systems integration. On the one hand, the benefits of systems integration occur in response to the mission-driven demand for sharing data resources and analytical systems. On the other hand, successful integration produces further sharing of data and resources. The end objective of systems integration is common, consistent, and relatable geographic data that increase in use and value over time and for many applications.

The focus of the National Center for Geographic Information and Analysis (NCGIA) Initiative 9, "Institutions Sharing Geographic Data," is to aid inter-organizational sharing of geographic information and examine issues that facilitate or impede that process (NCGIA 1991). The initial proposed research agenda has been defined largely in terms of identifying behavioral and organizational impediments to GIS use. This can be investigated by testing hypotheses concerning various forms of organizational resistance to change associated with the introduction of new technology (Onsrud, Pinto, and Azad 1992). This study accepts these perspectives as worthwhile and will identify trends and resistance in terms of the formulated hypotheses where possible. We intentionally broaden the research question, however, in order to examine how the benefits and costs of sharing geographic information are related to various forms of systems integration.

We begin by considering the nature of the benefits of better geographic information technology. Of particular interest are benefits derived from better integration. A working typology of systems integration is proposed and applied to define at least three broad strategies for integration.

Next, we examine three case studies of organizations undertaking these integration strategies and discuss the controlling factors that influence their decisions. With each strategy comes some degree of organizational change and resistance. The technical issues of GIS implementation are well recognized, but organizational change and institutional resistance to that change is not. We examine how the type of integration objective influences the need for organizational change in these three cases and point out how these can be evaluated against the nature of the expected benefits.

BENEFITS OF GEOGRAPHIC INFORMATION TECHNOLOGY IN ORGANIZATIONS

Institutional incentives for adopting GIS technology are frequently discussed in cost/benefit terminology since this is usually an explicit requirement of the procurement decision process. This procedure varies among organizations, but it usually consists of assessing the needs of an organization through a series of interviews and iterations in formulating a user requirements analysis (Guptill 1988).

At this stage in the process, the difficulty of adopting a cost/benefit model to GIS implementation has been widely acknowledged and in particular revolves around assessing GIS benefits. This is compounded by problems defining the nature of the resulting information products and their market demand curves (see Dickinson and Calkins 1988, 1990; Wilcox 1990). An attempt to categorize these benefits appears in Prisely and Mead (1987), which defines certain efficiency benefits and effectiveness benefits for implementing GIS technology. Gillespie (1991) identifies benefits and formulates an approach to measuring them that accounts for the mix of fixed and variable costs and separates subjective and objective measures of expected benefits. He further distinguishes between efficiency benefits and effectiveness benefits. To this breakdown, we will add a class of benefits pertaining to enterprise goals.

Efficiency Benefits

Efficiency benefits are identified in a cost/benefit analysis in terms of producing the existing line of products and services at a lower unit cost. A realistic appraisal, however, often indicates that efficiency benefits do not accrue right away. Start-up equipment and training costs may delay the onset of unit efficiencies for several years in GIS installations. Efficiency benefits are most easily identified where GIS implementation automates an existing manual analytical or mapping operation reducing production time and personnel hours. Therefore, organizations that derive the largest efficiency benefits are those agencies that have large mapping or exhaustive analysis and inventory programs to complete. These tend to be large agencies, especially in the public sector, with a mandate to provide data services and products in an era of declining fiscal inputs.

Effectiveness Benefits

Effectiveness benefits are derived from providing new products, services, and analyses to serve existing organizational mandates. Effectiveness can be difficult to measure. In much of the cost/benefit literature, effectiveness is characterized as an "intangible benefit," and various strategies for estimating a quantitative monetary value, or justifying procurement on the basis of efficiency benefits alone, are suggested (Emery 1971; Gramlich 1990). A complicating problem is that the cost of database and applications development that enables organizations to reap effectiveness gains is quite tangible and begs to be justified in cost-recoverable terms.

Generally, effectiveness benefits are equal to the value of the changes that result from the use of a GIS in a decision-making or information communication context. Estimating that value can involve a mix of objective and more subjective procedures (Gillespie 1991). However, the effects of changes often allow an organization to improve the spatial resolution of analytical procedures, to expand services to clients and customers, and to better monitor and enforce mandated programs. Increased resources are brought to bear on problems, and better insight is gained into difficult decision-making contexts. Whether the outcome is quantifiable or remains an "intangible" benefit, the focus is on providing a new capability to carry on an existing mandate of the organization's mission.

Enterprise Benefits

There is a third category of benefits to be evaluated when a unit within an organization alters its technology for working with geographic information. From the point of view of the unit making the technological investment, this category of benefits is not easily subsumed under either efficiency or effectiveness gains because its primary effect is on the organization as a whole; that is, as an enterprise (Seaborn 1990). Enterprise benefits consist of better information and communication throughout the organization and therefore result from, and provide greater opportunities for, the integration of data and analytical systems.

Efficiency benefits from data integration accrue to individual units in the form of the lower costs of shared data, but at the same time, a benefit accrues to the organization as a whole if data consistency resolves ambiguities and errors. However, there may be additional costs. More data sharing requires transfer procedures and data standards, but better standards for data and analysis assure higher quality products. Improved data access from disparate units provides users with an appreciation for how their activity affects the entire agency. Managerial support is enhanced by additional supervision. A better integrated organization results, facilitating communication between units. In the end, the image and reputation of the organization may be improved. This, in turn, can influence or support a change in the organization's mandate in the form of greater trust, increased confidence, and perhaps an expanded mission.

Implementation of GIS and related technologies is often incremental, and for sound technical and fiscal reasons (Peuquet and Brocastow 1991). Systems integration usually comes about only after it is recognized that significant duplication in data gathering or information processing can be remedied by a better integration of analytical and mapping systems throughout different parts of the organization. Thus, systems integration can change an existing administrative, management, or analytical environment in much the same way as the original adoption of GIS technology and can have the effect of refocusing the nature of geographic information processing and product generation within an organization.

STRATEGIES FOR SYSTEMS INTEGRATION

The term "systems integration" has been applied to a variety of activities involved with geographic information processing. For example, data-processing professionals concerned with designing multi-user databases have a different view of what integration means than hardware vendors concerned with the compatibility of computing workstations and peripheral equipment. Consequently, the taxonomy decomposes the term "systems" into more specific forms of integration.

Taxonomy of Systems Integration

Due to the wide scope of integration concerns and in order to examine the expected benefits of various forms of integration, it is necessary to develop a taxonomy from which to proceed. The following are forms of integration, in increasing order of breadth. Each level assumes some development at the previous level.

- *Data Integration.* The ability to share access to data sources or access common databases. Various techniques for data modeling and view integration have been developed for this (Navathe 1985; Nyerges 1989a). Indeed, the database paradigm is built around the concept of integrated access to data in a non-redundant form to improve efficiency and consistency.
- *Applications Integration.* Bundling applications into larger applications. Two factors drive increasing integration of applications. First, increases in the performance of hardware and database systems over time enable more functionality to be built into existing applications. Second, market forces favor software competition on the basis of the functions a package can provide. In turn, vendor-supplied systems are employed in user applications by means of developing templates or writing macros.
- *Functionality Integration.* The integration of separate software and hardware systems for applications or products that combine many inputs. This is often what is meant with offhand references to "systems integration." "Functionality integration," as a term, is offered in a narrow sense (from Nyerges 1989b); the explicit objective is to make different information or analytical systems work together to provide information products.
- *Organizational Integration.* The integration of various units within organizations to achieve a common objective function. The objective function can be expressed in terms of the organization's mandate. Organizational integration can result from the needs of technological advances to achieve data, applications, and functionality integration.

- *Mission Integration.* The integration of organizations to achieve effective cooperation on programs and mandates. This last level lies principally in the political and societal realm, but certain enterprise benefits associated with other forms of integration can contribute to an environment that facilitates inter-agency coordination and contributes to an awareness of a common purpose.

At each level of integration the engine driving further development is the increased capacity to share data among increasingly interconnected analytical systems and organizations. Nyerges (1989b) terms this concatenated process of information sharing and systems integration "information integration" and presents a framework for its implementation. Our emphasis is on examining how various elements of this idealized typology coexist in several integration strategies undertaken by real organizations.

Integration, Expected Benefits, and Organizational Change

A successful strategy to achieve efficiency and effectiveness benefits combines data, and applications and/or functionality integration. As the effects of the technology become known throughout an organization, and as new opportunities owing to greater effectiveness are recognized, pressure for organizational change often arises to facilitate even greater efficiency and effectiveness gains.

Clapp et al. (1985) formulated a hierarchy of benefits that accrue with the adoption of geographic information technology. These begin within individual units experiencing "operational efficiency." "Operational effectiveness" and "program effectiveness" describe effects with respect to wider spheres of the use of information derived from the technology, although this is not discussed specifically with regard to organizational structures. A fourth category of benefit, "contributions to well-being," is meant to describe the social effects of better operations and programs enabled by the advances. From our perspective on enterprise benefits accruing to the wider organization, this notion is embraced if the wider organization is taken to include not only the adopting agency but its outside constituency as well. Thus, better integration of information resources throughout public agencies, for example, is meant to benefit public policy making and, ideally, constitute an important incentive for technological innovation.

It follows then, that reorganization is both a precursor and/or response to successful integration. Its primary objective is to capture enterprise benefits. Greater communication and information sharing within and among organizations is an important component of these benefits. Not all paths to integration are the same, however, and the mix of data, applications, functionality, organizational, and mission integration varies by organizational cultures and by the degree of resistance to the organizational change required to implement integration.

We have found at least three variations on this mix of integration objectives and their corresponding organizational costs in examining several Pacific Northwest case studies. Each example consists of several organizational units working to achieve a better data-sharing environment. However, the cases represent quite different institutional cultures and organizational responses. These examples constitute definite integration strategies, which can be described as:

- core database integration
- database constellation integration
- data marketplace integration

TRI-MET OF PORTLAND, OREGON

Integration Around a Core Database

Tri-Met is a provider of public transportation—bus, light rail, and paratransit—in the three-county Portland, Oregon, area. Various units within the organization employ geographic data for analysis and mapping. Tri-Met has developed an implementation plan for GIS to manage geographic data in an integrated environment. Specific applications for GIS include special-needs transportation (paratransit) scheduling and dispatching, route and system mapping and traffic revision notices for drivers, automated trip planning, engineering, and maintenance of facilities, bus schedule adherence monitoring, and service planning functions.

GIS is an appropriate tool to integrate data throughout an organization responsible for transportation because data concerning transportation supply and demand is characterized by location. Virtually all applications refer to the location of facilities or customers and depend on the ability to be used together. Specialized software is available commercially to perform certain scheduling and dispatch functions, but software from a single vendor is unlikely to address the full range of required applications. Moreover, turnkey vendor systems may be based on different representations of the underlying geography. Transit agencies find it difficult to integrate separate applications, each having some embedded GIS functionality. In addition, they must maintain the geographic databases independently, which usually results in inconsistent representations.

A needs analysis of Tri-Met applications revealed that a unified representation of the urban street network based on the U.S. Census

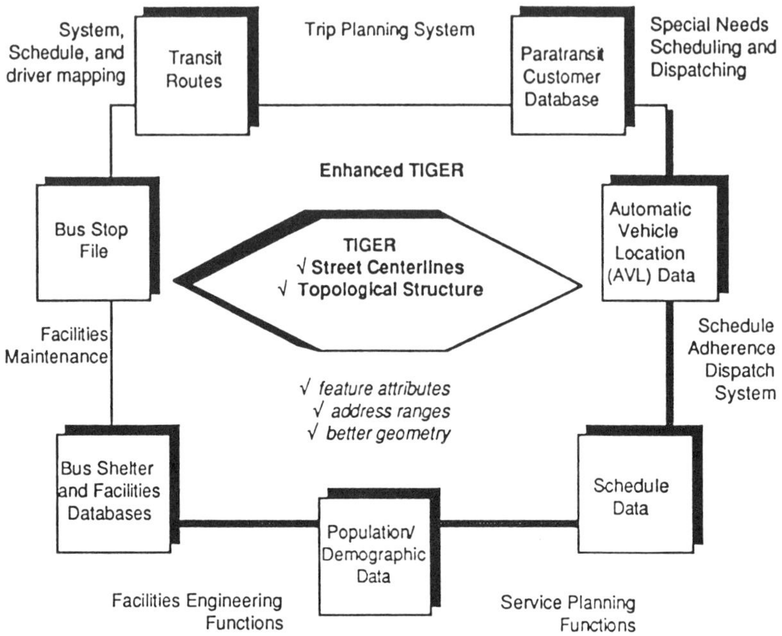

FIGURE 9.1
Tri-Met of Oregon's Core Database Integration Strategy employs a TIGER-centric approach

Topologically Integrated Geographic Encoding and Referencing (TIGER) files could be adopted to many transportation needs, although some enhancement and modification would be necessary (Dueker and Vrana 1991a). While an enhanced version of TIGER may serve as an integrated database, specialized databases derived from it would be required for applications that require additional attributes, or real-time transaction updating (Figure 9.1).

Adopting a core database integration strategy around TIGER facilitates the cooperative enhancement and maintenance of data and applications. This effect extends outside the organization as well. Data can be shared among regional organizations utilizing a common base map. In the Portland area, the Metropolitan Planning Organization is building and will maintain an enhanced TIGER database as a component of a regional land-information system. Tri-Met is a participant in this effort, supplying transit routes and facilities, cooperating in the initial development and enhancement, and receiving periodic updates.

Organizational Impacts at Tri-Met

Generally, a centralized data function evolves long after an initial project provides an entry point for geographic information technology.

Success on projects generates enthusiasm for more ambitious applications. Individual resistance to the technology has not emerged as an issue at Tri-Met, where the general level of computer use is already high and an enthusiastic data-processing department is eager to integrate GIS into its overall network strategy for distributed computing. However, progress was slow in establishing this initial entry point. The problem is that no single administrator of an application area will assume budgetary responsibility for shared database development and systems acquisition. Institutionalizing a core database integration strategy at Tri-Met requires top-level management support. An integrated core database GIS requires a prominent place in the organization. Without a champion emerging at the administration level to seed the growth of GIS technology, Tri-Met is contractually committed to a cooperative data acquisition venture until it develops GIS capability in-house.

It was posited that efficiency benefits would result from sharing development and maintenance costs of the core database, while effectiveness benefits would accrue from a horizontal integration where various applications work with consistent and accurate data. Vertical integration of data would provide managers and policy makers more accurate and timely data upon which to act, resulting in enterprise benefits (see Huxold 1991, 12-22, for a discussion of horizontal and vertical integration in a policy, management, and operations framework.)

Efficiency may be well served when measured by individual projects and applications with the core database approach, and further effectiveness benefits are expected with consistent data across applications. However, it remains to be seen if data integration results in enterprise benefits stemming from an expanded mission.

Organizational change may be needed to take advantage of GIS technology to achieve the advantage of horizontal and vertical data integration. It is not occurring rapidly in the existing organizational context at Tri-Met. Perhaps this is due, in part, to the inability of the current state-of-the-art to handle the temporal requirements of real-time control needed for vehicle dispatch and scheduling, a particular necessity for transit and paratransit operations.

BONNEVILLE POWER ADMINISTRATION

A Database Constellation Strategy

Facilities planning, environmental impact assessment, and substation and transmission line design are civil engineering applications that require geographic information at a variety of scales throughout the five-state area served by the Bonneville Power Administration (BPA). Geographic information technologies employed at BPA include:

- *GIS.* For small-scale regional mapping, environmental assessment of proposed transmission corridors, raster–vector integration for public presentation display, and visualization of transmission line and substation proposals.
- *CAD Mapping.* The workhorse technology for producing and updating hundreds of engineering drawings including large-scale mile-section plan and profile mapping of transmission line rights-of-way, and large-scale mapping to support engineering and construction of substations.
- *Photogrammetry (PHO).* To obtain planimetric data for large-scale mapping, data for digital elevation modeling, and timber and brush clearance, and danger tree monitoring to ensure the integrity of constructed powerlines.
- *Image Processing (IPS).* Scanning aerial photography for land-use classification, warping digital images to selected control points for integration with GIS coverages, and digital production of photomaps for right-of-way maintenance.
- *GPS and Digital Workstation Surveying Instruments.* For primary geodetic control, detailed mapping of transmission alignments and substation construction sites, and cadastral surveying for property and easement acquisition.

A transmission line project usually involves GIS analysis of proposed corridors using IPS inputs. Engineering design proceeds employing photogrammetric mapping and field survey methods, depending on terrain and cover for detailed design and tower spotting. This is then mapped with CAD software. These maps are useful to construction and maintenance crews, along with IPS photomaps, and photogrammetric analysis of tree clearing and conductor tolerances.

Each of these systems had been evolving along independent paths and involved individuals trained in highly specialized disciplines working in related but separate sections at BPA's Facilities Planning Branch. Recognizing the need for better integration among these units, management undertook a systems integration study to develop more efficient and effective geographic data in support of a broader organizational mission (Dueker and Vrana 1991b).

Information products and services provided by the Branch were analysed, and at least three kinds of customers external to the Facilities Planning Branch were found. These were other BPA units engaged in engineering design, environmental planning, and operations, maintenance, and construction (OMC). A series of three "database constellations" was proposed to organize production and provide a basis for data and functionality integration efforts based on resolution and scale requirements of the three application types.

The database constellation strategy recognizes the limits of a unified data model. It accommodates data sharing between constellations by providing for non-routine data exchanges between widely different systems. The constellations are identified by similar scales and analytical function and/or data resolution. In contrast to the single data model of the core database strategy, this approach emphasizes the data/work flow within and between units. Although the geographic extent of each constellation may be similar, each is tailored for the information processing needs that occur either on relatively incompatible systems or at different times in the project cycle. Figure 9.2 illustrates this strategy's clustering of databases and systems oriented by customer/product and by scale/format.

Of special interest is the T-Line Index Constellation. At BPA, no single data dictionary or indexing system existed to provide a clear picture of digital and analog data pertaining to the operating region. The T-Line index should produce efficiency benefits in terms of reducing the time and effort to assemble data for transmission line rights-of-way and identifying features that possess specified characteristics. Similarly, effectiveness benefits are anticipated in that organizational memory concerning line characteristics and data resources is diminishing with retirements and job reassignments. Enterprise benefits to the entire organization accrue when the T-Line Index is recognized as an agency data cross-reference tool, locating large-scale plan and profile drawings, environmental assessment documents, electrical transmission diagrams, photos, and digital imagery. All these data products are referenced to a common geographic base and accessed by a GIS for spatial searches of the organization's inventory and attribute searches with map display.

Organizational Impacts at BPA

At the outset of the Bonneville project, the integration approach concentrated on rationalizing data and work flow patterns. Part of the impetus for the integration study came from the need to make significant investments to upgrade expensive photogrammetry, image processing, and

CAD technologies. Some degree of organizational change was considered likely, even desirable.

Several factors in the institutional culture complicated the picture and created some turmoil and speculation. Coincident with the systems integration study, a productivity review was undertaken, mandated by an Office of Management and the Budget (OMB) circular A-76. In addition to a comprehensive review of individual productivity, this directive initially instructed management to identify activities for possible outsourcing to private ventures. Subsequent interpretation rescinded the outsourcing directive, but the internal productivity review continued, eventually prescribing a new, "Most Efficient Organization" for the Branch. All this occurred in the climate of a federal agency that has suffered hiring freezes and that already supplements its personnel needs with in-house contractors working alongside BPA employees.

Understandably, there was some resistance to reorganization at BPA. Although part of the initiative for the reorganization was external to the systems integration effort, to some degree individuals perceived the productivity review and integration study as threats to an existing order that worked well enough. In actual fact, the integration study revealed the high degree of interdependence of spatial-analytical systems and provided a rational means to integrate them. Thus, systems integration limited the desirability of outsourcing by increasing efficiency benefits in-house. This came at the price of some internal reorganization which, in turn, contributed effectiveness benefits as data are more easily accessed and reused in a life-cycle management context. Similarly, enterprise, and additional effectiveness benefits, are anticipated as other units of BPA can access more accurate and timely data.

BPA may be headed for a broader examination of GIS applications and opportunities for data sharing. The Facilities Planning Branch is a small cog in a larger organization, though it is playing the role of a GIS service center. But a service center functions best on an ad hoc project basis. This may be efficient for projects, but enterprise benefits require a more comprehensive and continuing program of applications and data sharing.

NORTHWEST LAND INFORMATION SYSTEM NETWORK

Data Marketplace Integration

The Northwest Land Information System Network (NWLISN) is an ad hoc consortium of federal agencies with regional offices in Seattle and Portland, along with Oregon and Washington state agencies with GIS or LIS interests. A central purpose of the network is to cooperate in the acquiring

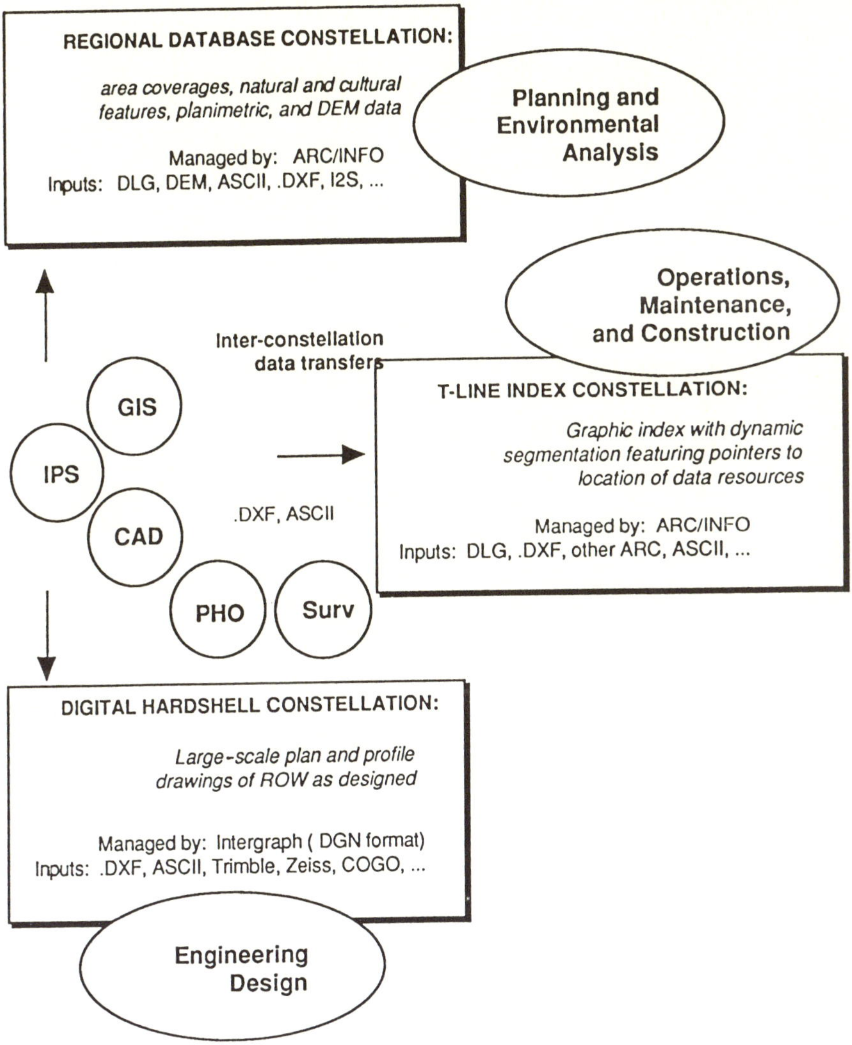

FIGURE 9.2
BPA's database constellation integration strategy optimizes geographic information processing for client units by similar scale and format. De facto graphic standards and ASCII file transfers handle less routine inter-constellation data exchange.

and sharing of cartographic and geographic data, reducing the duplication of digital data sets and establishing digital data priorities and joint mapping and analytical programs.

To facilitate data sharing among member agencies, NWLISN developed a Spatial Data indeX (SDX) to provide a central clearinghouse for information pertaining to the various digital spatial data holdings throughout the region. The SDX is not a central depository of the data, just a referral service to locate existing coverages and mapping projects. Users of the system call SDX, and a query is generated for the appropriate geographical extent and thematic subject of interest. The results of a query can be a listing of datasets by agency, a map of their spatial extent, and additional information on the system and format employed by the holding organization. The requesting organization then contacts the holding organization and can refer to the data of interest by name and native format. The benefits of such a system are twofold. Requesting agencies have "one-stop shopping" to locate potential sources of geographic data. Holding agencies catalog and report only once and get requests for specific items in familiar terms (Vrana 1992).

Though SDX does not physically integrate data holdings, a degree of systems integration is evident in its operation. The application is a metadata management system, one which must manipulate a variety of data from different organizations. To do this it is essential to formulate what constitutes a description of a data set. Individual data sets are reported by name, owning organization, completion status, thematic subject code, and other attributes, including a description of their geographic extent. But various agencies tile data by completely different schemes. SDX is a GIS application in part because it integrates quadrangle, township/range, and hydrologic unit basin (HUC) tilings for cross-referencing and query purposes. Figure 9.3 illustrates the integration of geographic tiling systems in use by SDX.

From an "information integration" point of view, this strategy is not comprehensive, but it has a wide potential impact. Data are a commodity, and sharing results when organizations trade with the aid of a market institution such as SDX. This strategy is appropriate for widely different organizations operating in coincident areas but not regularly cooperating on individual projects. This approach to data sharing reduces the number of one-to-one inter-agency agreements, therefore simplifying organizational adjustments required for multi-agency cooperation while simultaneously incorporating what may be considered enterprise benefits from an inter-agency "societal" point of view.

In contrast to the core database and database constellation strategies,

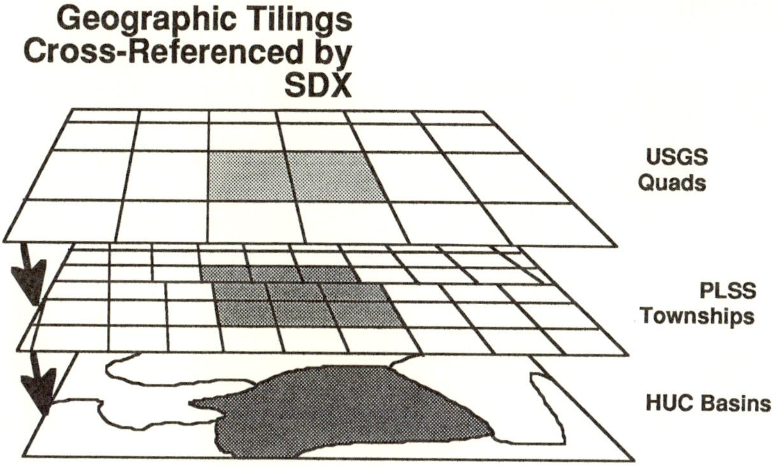

FIGURE 9.3
Metadata descriptions of geographic data can be cross-referenced for integrated search and retrieval.

this integration depends less upon unified schemas and transfer formats and more upon a means of describing data as a medium of exchange. Facilitating this process will require further standardization of spatial metadata (ASTM 1991). The emerging national Spatial Data Transfer Standard (SDTS; see USGS 1990) provides a framework for both content and data quality description. It is especially directed toward providing the metadata documentation to accompany a file transfer between organizations.

Organizational Impacts at NWLISN

Data sharing among organizations in a multi-participant land information system network is an example of efficiency benefits accruing to participants who share data sets. These organizations also realize effectiveness gains from their access to consistent and accurate data from other participants. Enterprise benefits may result in the form of an expanded mission pursuant to effectiveness gains. Since marketplace integration means better coordination between agencies, these take the form of more joint projects and co-development of digital data. In this case, the enterprise benefit accrues to the citizens and taxpayers, who no longer support the duplication of common base data themes and are served by governmental agencies that can cooperate in the geographic information technology-intensive issues facing the Northwest. Endangered species habitat protection, logging, anadromous fish migrations, and wetlands reclamation

TABLE 9.1

Benefits and Organizational Issues for Sharing Data

Type of Resource-Sharing Benefit	*Requirements*	*Organizational Change*	*Organizational Resistance*
EFFICIENCY BENEFITS			
Existing operations performed at lower per-unit cost	Data and systems standards	New positions specific to technology	Individual resistance to work rule or position description changes
	Data, applications, and functionality integration	Elimination of obsolete positions	
EFFECTIVENESS BENEFITS			
Enhanced capabilities for unit	Resources freed for additional analysis/services	Work flow changes	Distrust of data external to organization
		Interdependencies, complexities	
New products (timely and accurate data)	Applications, functionality, and organizational integration	Change in management and quality control functions	Resistance to coordination that challenges traditional hierarchies
ENTERPRISE BENEFITS			
New responsibilities	More complex analysis and more robust data models	Shift in relative importance of units within organization	Middle management reluctance leads to whole units resisting reorganization
Broader mission	Functionality, organizational, and mission integration	Role and importance of organization changes	Resistance in other organizations

are all examples of issues requiring regional cooperation of a number of federal and state agencies.

SDX does not directly impact the organizational structures of any of the participating member organizations. In order to become a more effective vehicle to direct and maintain the marketplace integration effort represented by SDX, the NWLISN must continue to function as an inter-agency consortium. This depends on management-level representatives with budgetary authority meeting on a regular basis to direct the operation of a technical working group level of representatives who in turn, meet and make recommendations for the management to act on as a steering committee. Voluntary participation of member agencies, each with a constrained budget and most overburdened with work in addressing the issues mentioned above, threatens the continued effectiveness of the consortium.

Meanwhile, a plan has been proposed to migrate the index service to the two states, half to be maintained by an Oregon Agency and the other half to be maintained by one in Washington. This organizational change could cause the two portions of the application to drift apart in terms of query, reporting, and updating procedures over time, further threatening the regional enterprise benefit of a coherent marketplace integration strategy.

FINDINGS: INTEGRATION AND DATA SHARING—A TWO-WAY STREET

Some Comments on Integration Benefits and Costs

In the case studies we have examined, data sharing is conditioned by type of integration strategy, which in turn is influenced by the type of benefits to be captured. Organizational costs, in terms of institutional resistance to change, also appears to be a function of expected benefits, although some external institutional factors may be influencing this as well. It should be emphasized that these case studies are merely representative of a range of organizational responses to data sharing and/or systems integration, and no attempt has been made to evaluate them systematically with respect to a control group. Still, the experience of working with all three organizations has afforded us the perspective of comparing several approaches to integration. This has led us to a number of conclusions and recommendations for further research concerning the relationship between data sharing and systems integration, and concerning the benefits and costs of achieving these aims.

Table 9.1 is a summary of benefit and organizational issues associated with various integration objectives. To the well-known delineation of efficiency and effectiveness benefits, we have added a third class of benefits that apply in a wider organizational context. We have related these benefit categories loosely to integration objectives according to several perspectives. The nature of these linkages is by no means definite, and further research may show enterprise benefits to be mainly second-round efficiency and effectiveness benefits, for example.

Along with benefit categories, we have looked at costs in terms of both the organizational changes necessary to achieve these benefits and the nature of the resistance to those changes. Generally, change becomes more complex with increasing benefits. To achieve efficiency benefits, little or relatively minor organizational change may be expected. Data integration, involving improvements to common databases, produces correspondingly minor institutional resistance. At the level of individuals this can be accounted for by factors such as personal loss of freedom, anxiety about acquiring new skills, and fear of the unknown. In fact, individual resistance may be due in part to latent dissatisfactions that only appear to be directly related to the implementation of new technology. In this case, GIS integration may be only a trigger that sets off a problem that may have been festering for some time.

Effectiveness benefits come at the cost of wider organizational changes to work flows and the accompanying management and quality control adaptations to these. The cases studies we have examined indicate a certain distrust of external data sources and a reluctance to coordinate new programs that upset traditional hierarchies. From the MIS literature we gain insight as to why. GIS service centers associated with the reorganization can become alternative power centers within an organization just as MIS managers were perceived as undercutting other departmental functions. This often takes the form of concerns within the organization about undercutting normal channels of communication and responsibility (Federico 1985). As GIS installations mature, however, the expertise migrates out of the project-centered service centers within an organization and becomes another analytical resource within departments.

Enterprise benefits can be impeded when resistance originates with managers who recognize that centralization and decentralization of information will occur with the new technology. This is particularly true in public-sector organizations (Somers 1989). Although the integration of separate systems might yield improved service to consumers (supplying bus schedule and actual arrival information to bus riders, for example) turf battles may inhibit the realization of these objectives. Since there is more organizational resistance to systems integration plans that emphasize enterprise benefits, they are riskier. However, we recognize that studying the relationship between organizational resistance and integration ambitions must control for variations of organizational characteristics such as size, level of central control, leadership characteristics, and other aspects of organizational culture. Even in the few case studies we have identified, significant outside influences have emerged.

Integration Strategies and Organizational Benefits and Costs

Integration and benefit issues are rarely confronted directly. Instead, they generally emerge in a pragmatic way, when prospective applications of GIS technology (or competing champions in different parts of an organization) have common needs for systems and geographic data. In exploring the potential for data sharing two issues become apparent:

- Efficiency savings are often the critical factor at the individual unit level. However, it may not be feasible to install separate and independent GIS in different parts of the organization, each of which would realize only efficiency benefits. Achieving effectiveness and enterprise benefits requires a coordinated approach to GIS implementation. Extensive organizational impacts are likely, as integrating systems and data require organizational change.
- Intra-organizational cooperation and data sharing to achieve effectiveness and enterprise benefits will require development of, and adherence to, systems and data standards. A major enterprise benefit of sharing common geographic data is the assurance of accuracy and consistency. However, achieving the common geographic framework for an organization means agreeing on update and maintenance procedures, and assigning those responsibilities. This means creating a new geographic database administration function, which has sometimes formidable organizational implications.

Table 9.2 illustrates some of the requirements of the three integration strategies we have identified. Surely there are many more approaches to the general problem, which can be succinctly stated as the need to mix integration objectives according to institutional requirements accounting for types and the extent of organizational change and resistance to that change.

The motivations for data sharing among organizations are twofold: to share the cost of common data and system resources, and to utilize the data of others for beneficial organizational purposes. Achieving cost reductions and gains in effectiveness by sharing data and system resources

TABLE 9.2

Integration Strategies and Expected Benefits

Strategy	*Technical Requirements*	*Organizational Requirements*
CORE DATABASE INTEGRATION		
Units access common database(s)	Shared conception of geographic domain	Centralized database administration function
Relatively uniform view of data	Unified data model supporting auxiliary databases	Intra-organizational data access with maintenance responsibilities defined
	In-house standards for data management	
DATABASE CONSTELLATIONS		
Like resolution/format databases clustered in optimal work-flow units	Proprietary/binary formats for within-constellation data transfers	Partly centralized database administration/data conversion function
	Standardized/ASCII formats for between-constellation data transfers	Records management and data library service function
MARKETPLACE INTEGRATION		
Market institutions to facilitate inter-organizational data exchange	Spatial metadata standards and transfer mechanisms	Oversight body responsible for data catalog, standards, and transfer monitoring
	Technical awareness and competence across several organizations	Participation in inter-agency data-sharing committees

will necessarily involve data and systems integration which, in turn, will require that organizational change be instituted. A concomitant effect of organizational change is resistance to change within an organization. This calls for expecting and mitigating the resistance to change rather than trying to avoid or minimize organizational change.

The appropriate focus for sharing data (and systems) resources is data and systems integration. Systems integration is a means by which data sharing can be achieved, and data integration is a compelling reason for sharing data. Integrating or combining data in a GIS increases its effectiveness and creates opportunities for wider enterprise benefits that accrue to entire organizations and constituencies. Further investigation of the incentives for, and impediments to, data sharing need to be made in the context of a conceptual framework that recognizes costs and benefits of various forms of systems integration.

This focus concerns the effect of integration strategies on work environments relative to the costs and benefits of the implemented technology. Important issues regarding information access, data quality, and the quality of the relations of production in the workplace are left unaddressed in this perspective.

Implications for Further Research

This investigation raises several issues that warrant further research involving the organizational impacts of the benefits of data sharing and systems integration.

What are the motivations within organizations for data sharing? Specifically, when can they be described as primarily efficiency benefit driven with effectiveness and/or enterprise impacts, and when are they motivated more with an effectiveness objective, with additional or unforeseen efficiency and/or enterprise impact?

Similarly, what are the motivations for systems integration, and when are these primarily efficiency, versus effectiveness, benefit driven? More systematic research into systems integration and how various strategies affect effectiveness and enterprise benefits is needed.

What can we observe from demand responses to changes in the supply functions for the availability of geographic data? The cost function determines efficiency benefits at the level of individual units. It also influences shifts in the demand function for geographic information, which spurs organizational change and realignment to capture effectiveness benefits. It would appear that the push to maximize certain benefits drives organizational change. But organizational change can involve up-front costs. How do these costs feed back to the costs of adopting new technology for sharing data and to subsequent rounds of organizational restructuring?

Finally, there is a need to study the motivations and characteristics of individuals in the technology adoption and data-sharing decision-making area. From our perspective, we believe that the relationship between organizational resistance and benefits, particularly enterprise benefits, is attenuated by individual differences. Persons championing enterprise benefits are often visionary and perhaps viewed by peers as impractical, whereas those who champion efficiency benefits may be viewed as practical but with limited long-range vision. Further research is needed on the risk-taking strategies of these individuals, and how these are influenced by integration objectives and organizational cultures.

REFERENCES

ASTM. 1991. Draft proposed specification for meta-data support in Geographic Information Systems. Version 10/91. ASTM D18.01.05 Subcommittee on Spatial Meta-data Standardization. Available from USGS-WRD, Reston Virginia.

Clapp, J., J. McLaughlin, J. Sullivan, and A. Vonderohe. 1985. Toward a method for the evaluation of multipurpose land information systems. URISA *Proceedings*, Ottawa, pp. 1- 10.

Dickinson, H. J., and H. Calkins. 1988. The economic evaluation of implementing a GIS. *International Journal of Geographical Information Systems* 2,4: 307-327.

Dickinson, H. J., and H. Calkins. 1990. Comment on "Concerning the economic evaluation of implementing a GIS." *International Journal of Geographical Information Systems* 4, 2: 211-212.

Dueker, K. J., and R. Vrana. 1991a. GIS applications in urban public transportation: a case study of Tri-Met, Portland, Oregon. GIS for Transportation Symposium, 1991 *Proceedings*, Orlando, Florida. Available from American Association of State Highway and Transportation Officials (AASHTO), Washington, D.C.

Dueker, K. J., and R. Vrana. 1991b. *Systems integration analysis and alternatives identification for facilities planning branch, BPA*. Task 3, Alternatives for systems integration, and Task 4, Recommendations and strategic plan. PSU Center for Urban Studies Publication #PR054, 66 pages, and #PR061, 45 pages.

Emery, J. C. 1971. Cost/benefit analysis of information systems. In E. Yourdon (ed.), *Writings of the revolution: selected readings on software engineering*. New York: Yourdon Press. 19-50.

Federico, P. 1985. *Management information systems and organizational behavior*. New York: Praeger Press.

Gillespie, S. R. 1991. Measuring the benefits of GIS use. ACSM Technical Papers, Fall meeting, Atlanta, Georgia.

Gramlich, E. M. 1990. *A guide to benefit-cost analysis*. Englewood Cliffs, NJ: Prentice-Hall.

Guptill, S. (ed.). 1988. *A process for evaluating Geographic Information Systems.* Federal Interagency Coordinating Committee on Digital Cartography, Technology Exchange Working Group Technical Report 1. USGS Open-File Report 88-105.

Huxhold, W. 1991. *An introduction to urban Geographic Information Systems.* New York: Oxford University Press.

Navathe, S. 1985. Important issues in database design methodologies and tools. In Albano et al. (eds.), *Computer Aided Database Design: the DATAID project.* Amsterdam: North Holland.

National Center for Geographic Information and Analysis (NCGIA). 1991. Progress report on Initiative 4 and plans for Initiatives 9 and 16. National Center for Geographic Information and Analysis Circular 9/9/91.

Nyerges, T. 1989a. Schema integration analysis for GIS database development. *International Journal of Geographical Information Systems* 3, 2: 153-183.

Nyerges, T. 1989b. Information integration for multipurpose land information systems. *URISA Journal* 1, 1: 27-38.

Onsrud, H., J. K. Pinto, and B. Azad. 1992. Case study research methods for Geographic Information Systems. *URISA Journal* 4, 1: 32-44.

Peuquet, D, and T. Brocastow. 1991. Organizational issues in the development of Geographical Information Systems: a case study of U.S. army topographic information automation. *International Journal of Geographical Information Systems* 5, 3: 303-319.

Prisely, S., and R. Mead. 1987. Cost-benefit analysis for Geographic Information Systems. GIS 1987 *Proceedings,* San Francisco. 29-37.

Seaborn, D. 1990. Enterprise GIS: management and organizational impact. Unpublished presentation at GIS/LIS 1990 conference, Anaheim. November.

Somers, R. 1989. Organizational change for successful GIS implementation. URISA 1989 *Proceedings,* vol. 2. Boston. 39-51.

United States Geologic Survey (USGS). 1990. *The spatial data transfer standard.* Draft standard, version 12/90. SDTS Technical Review Board, USGS, Reston, Virginia (FIPS 173, August 1992),

Vrana, R. 1992. Design and operation of a meta-GIS. GIS 1992 *Proceedings.* Vancouver, B.C.

Wilcox, D. 1990. Concerning "The economic evaluation of implementing a GIS." *International Journal of Geographical Information Systems* 4, 2: 203-210.

STEPHEN J. VENTURA

10 *Overarching Bodies for Coordinating Geographic Data Sharing at Three Levels of Government*

This chapter compares the development and potential effectiveness of overarching bodies charged with facilitating geographic data sharing at three levels of government – local (Dane County, Wisconsin: DaneLINC, the Dane County Land Information Consortium); state (State of Wisconsin: WLIB, the Wisconsin Land Information Board); and national (Trinidad and Tobago: Interministerial Committee on Land Information Systems). Observations are primarily focused on three groups of factors that appear to influence the process of creating the overarching bodies: technology introduction, practices and structures within organizations, and interagency relations. The process of formation and the form and authority of the overarching bodies will probably affect the success of data sharing among agencies, although it is too early in their implementation to document such effects. Observation of their formation process raises a series of questions to study through long-term observation of these and similar situations, which will provide evidence of factors influencing the success of coordinating bodies for geographic data sharing.

INTRODUCTION

In many jurisdictions at all levels of government, Geographic Information Systems and related technologies are being independently implemented without coordination or control. This is often noted with alarm, for fear that systems developed by various agencies and departments will be incompatible, and that some benefits of multipurpose systems will not be realized, especially benefits from sharing data and other resources. In some cases, the response to this perceived problem is to create an overarching body that can coordinate GIS implementation activities, promulgate standards, develop data- and cost-sharing agreements, and so forth. These bodies can take many forms, such as coordinating councils, interagency agreements, and even new organizations. The effectiveness of such bodies will likely depend on underlying institutional, organizational, and behavioral factors operating both during and after formation, as well as their explicit mandate and authority.

Like GIS itself, the creation and operation of overarching bodies to share geographic data is a relatively new phenomenon; their form, authority, and role are not yet completely clear. The commodity they deal with (geographic data) and their purpose (sharing) make them unique from an organizational perspective. Interagency cooperation and information exchange are not unique to GIS; this phenomenon is seen in areas such as emergency services, law enforcement, and social services. But the GIS aspect means there is technical complexity—hardware and software, standards, telecommunications, and so forth—that must be understood in addition to intra- and inter-agency interactions. The common thread between agencies is data, not a common service or activity or mandate. To be successful, overarching bodies need to resolve both technical and institutional challenges.

Some notable examples of overarching bodies are the IMAGIS project in metropolitan Indianapolis; New Brunswick, Canada's LANDNET (Coleman and McLaughlin 1988); RUIS in San Diego; and a private data service agency—POLARIS—in Ontario (Ministry of Consumer and Commercial Relations 1991). But these projects have all been started in the last few years, and detailed evaluations are not readily available. Their long-term viability has not been established, and their form, scope, and function continue to evolve.

Understanding Overarching Bodies

Multi-participant Geographic Information Systems and their overarching bodies have not been evaluated from an organizational perspective. It might be useful, therefore, to ask to what extent organizational and behavioral theory can help explain the formation and viability of overarching bodies charged with facilitating data sharing among organizations. Such studies could ultimately help practitioners guide implementation of GIS and formation of overarching bodies so as to increase the chances of successful data sharing. However, such studies are not without difficulties. Researchers have tended to look at a single organization at a time; the study of overarching bodies will require not only understanding the structure, behavior, and motivation of all participating organizations but their interactions as well. Researchers will also need to become conversant in technical issues surrounding data sharing so that they will know when technical issues are real barriers, when they simply reflect a lack of technical expertise, and when they are a mask for underlying institutional issues.

Formation of overarching bodies is a "real-world" process, so there is no such thing as a controlled experiment. Perhaps the most difficult obstacle to successful research on overarching bodies is the role of researcher. It is necessary to probe deeply into organizations to understand more than superficial interactions. In other words, a researcher must become familiar with personal traits and attend working meetings, not just read final reports and attend press conferences. However, this makes it extremely difficult for a researcher to remain a "detached observer." To make extemporaneous observations, rather than ex post facto studies, it will be necessary to develop a new research paradigm that recognizes the role of a "participant/observer."

Case Histories

The purpose of this chapter is not to develop that new paradigm. No attempt was made to remain detached in the cases presented herein. Our experience would suggest that organizations are hungry for information about GIS, and anyone that has enough knowledge to do this type of research will be perceived as an expert—someone asked to help with the process. Instead, three case histories are presented to show that direct observation provides useful lessons about organizational perspectives. Because of our "participant/observer" role, these histories cannot be considered case studies in the sense suggested by Onsrud et al. (1992). Whether general truths exist about overarching bodies for data sharing will remain an open question until a rigorous methodology is developed and more evidence is available.

The cases presented in this chapter do not have sufficient history to comment on the success of their ultimate goals (e.g., better data sharing among institutions). They represent part of a continuum of creation and operation of overarching bodies. The nation of Trinidad and Tobago has recognized the need for such bodies and has set some bureaucratic wheels in motion to figure out what to do; Dane County has gone through the wheel-spinning and recommended a structure for the overarching body, but it has not yet been adopted; the state of Wisconsin has adopted the structure for an overarching body, but the body has not yet focused its attention on state-level problems. In all three cases, there have already been improvements in data sharing. Some of that improvement can be attributed to the discussions of data sharing and institutional agreements, though it is difficult to sort the effect of data-sharing discussions from changes generally attributable to adoption of automated land information systems.

The chapter is divided into three additional sections: 1. an overview of the current situation—a description of the existing organizational structure and institutional relations in each jurisdiction, the GIS implementation activities prior to and during the initiative to create a coordinating mechanism, the process whereby a coordinating body was created, and the effectiveness of the mechanism to date; 2. observations on how and why institutional and behavioral factors influenced the creation of coordinating mechanism(s) and continue to influence data sharing and other cooperation among agencies; and 3. discussion of these observations in relation to other aspects of organizational and behavioral theory.

THE CASES: DEVELOPMENT OF OVERARCHING BODIES FOR DATA SHARING

The cases presented here can be considered typical of local, state, and national governments (at least typical of developing nations) from the standpoint of governmental structure and nascent interest in GIS and related technologies. It is worth noting that all three have relatively healthy economies in comparison to counterparts. This certainly contributes to their ability to invest in computer-based methods and might be an indirect reflection of governments with a vision for the future. All three are also considered to be progressive in comparison to equivalent or nearby jurisdictions. A brief overview of the jurisdictions and their governing structure follows. The agencies and actors involved in GIS implementation are introduced, along with the motivations or forcing functions compelling the use of GIS.

Overview of Jurisdiction and Data Sharing Histories

Dane County-wide Land Information Consortium (DaneLINC). Dane County is located in south-central Wisconsin. It has a land area of about 1,200 square miles and a population of nearly 400,000. It has a very stable economy comprised of agriculture, light industry, and government.

Dane County is set up as a "weak executive" county government, with an elected county executive and a large county Board of Supervisors. In reality, the executive has a great deal of power over day-to-day operations, with the board being largely reactive. County department heads are appointed by the executive with approval from the board. As such, department heads and other upper-level managers tend to be reasonably well-qualified in appropriate areas, rather than merely being "political" appointments. The board has standing committees that oversee the operations of departments but again, direction from the board is primarily reactive.

Land records modernization and environmental protection have been significant forcing functions in the adoption of automated land information technologies by county government. The County Land Regulation and Records Department and the Land Conservation Department have led the adoption of GIS. The largest municipality – Madison – has automated systems for infrastructure management and planning. Several utilities and special districts have or will soon have GIS or AM/FM (automated mapping/facilities management) systems.

Prior to the DaneLINC activities, county departments successfully shared data, equipment, and technical expertise among themselves. This is attributable to long-standing arrangements with researchers at the University of Wisconsin, who facilitated technology transfer and system implementation. A few "one-shot" exchanges of digital data have taken place between other agencies, including a large automation project done jointly by the city of Madison and a local utility.

Wisconsin Land Information Program. Wisconsin is on the western edge of the "rust belt." A diversified economy of industry, agriculture, and tourism has allowed it to escape severe impact from the most recent recession. It has a land area of about 40,000 square miles and about 4.5 million people.

The state has a typical bicameral legislature. The governor appoints department secretaries either directly or indirectly through appointed boards. Other upper-level agency positions are also appointed. These tend to be political, although generally not blatant mismatches between skills and responsibilities. Agencies prepare budget requests every two years; these are the major vehicle for change initiatives from within. The governor submits a combined budget recommendation to the legislature and then approves, with partial (line-item) veto power, the results of the legislature's budget. The final bill often is substantially different from agencies' initial requests, with change initiatives receiving a great deal of scrutiny.

At the state level, the Department of Natural Resources and the Department of Transportation are the only agencies with operational Geographic Information Systems. These systems were created primarily to meet internal needs. Other departments, such as Agriculture and Revenue, are exploring GIS through pilot projects and cooperative agreements.

When ARC/INFO was installed at the University of Wisconsin at Madison in 1986, a few "techies" from state agencies agreed to hold regular meetings. The "ad hoc state data-sharing group" has met monthly since then, successfully exchanging data, expertise, ideas, and plans. This has all been done without any official sanction or funding (or recognition from the Wisconsin Land Information Board after its creation). It was, and still is, essentially a group of technical-level staff who have realized the value of data sharing and figured out how to do it on their own.

Trinidad and Tobago. The nation of Trinidad and Tobago is comprised of the southernmost islands in the Caribbean; an eastern point is less than ten miles from Venezuela. Trinidad has one of the most robust economies in the region, primarily due to oil revenues. Tobago's economy depends on tourism and agriculture. The combined land area is about 2,000 square miles, with a population of about 1.2 million. Trinidad and Tobago has a parliamentary government modeled after the British system. Agencies are led by Ministers of Parliament; day-to-day management is by Permanent Secretaries, who generally come up through the civil service system. Ministers of Parliament are primarily oriented toward their political party, and so do not necessarily advocate for agencies in their portfolios. The Ministers' primary contact with agencies is through the Permanent Secretaries and other upper-level staff. Because the bureaucratic structure is rigidly hierarchical, any change initiatives must be routed through the Permanent Secretaries to Parliament, and so those individuals are key actors.

Several agencies have CAD or GIS pilot projects, including the Lands and Surveys Division (LSD), Town and Country Planning, Valuation Division, the Water and Sewerage Authority (WASA), and a quasi-governmental agricultural corporation. WASA is closest to having an operational system.

Several of the agencies learning about GIS have recognized the value of coordinating data collection and management, and WASA and Valuation are evaluating a data-sharing effort. However, most individuals will say that meaningful data sharing cannot take place until LSD automates its operation. LSD is responsible for both base and cadastral mapping, and so is considered to be the critical link in any coordination activities. There has been little effort to upgrade the decades-old system that LSD currently maintains.

Towards Formation of Overarching Bodies

Several common threads are apparent in the processes of forming the overarching body in each jurisdiction. In all three cases, professionals dealing with land records recognized problems related to data sharing (and lack thereof) for a long time before any attempt was made to resolve them. The push for change was from the technical level; it was not from upper-level decision makers, and it was not from citizens or groups affected by land information systems. In every case, a group of professionals began meeting, either formally or informally, to look at technical solutions—to begin data sharing with or without formal inter-agency agreements. These groups all found themselves unable to carry their efforts beyond some simple exchanges. They encountered institutional barriers that compelled them to seek a broader vision for data sharing and cooperation.

In all three cases, concepts of the overarching body came from University of Wisconsin experts, as did at least some of the organizing. The following section introduces the overarching bodies, the processes of their formation, and the role of University of Wisconsin technology transfer agents.

Dane County-wide Land Information Consortium. The Dane County-wide Land Information Consortium (DaneLINC), as this chapter is written, is a proposal in a published report (CLISC 1991). The report was released for public discussion and debate in late 1991. As of this writing, the proposers seem to be resting, perhaps waiting to see how upper-level officials react to their proposal before trying to move it through the legislative process.

For almost a year, a group of professionals—the County-wide Land Information Systems Committee (CLISC)—met weekly to develop a response to a perceived need for better data sharing. CLISC was co-appointed by the mayor of Madison, the Dane County executive, and the chancellor of the University of Wisconsin at Madison. These decision makers took this action at the urging of University of Wisconsin LIS/GIS experts and a few county staff with whom they had worked through the years. CLISC was ultimately joined by more than forty individuals representing federal, state, county, city, village, and town agencies; gas, electric, sewer and telecommunication utilities; real estate and title insurance interests; and consulting engineering firms. The major common interest among this group was better access to county land records, particularly land ownership and related cadastral data, though other types of data sharing came up during discussions.

The DaneLINC proposal calls for short- and long-term solutions. The short-term solution is essentially formalized institutional relations in the form of a new quasi-governmental body. The major tasks would be data cataloging, coordination between agencies, and provision of standards.

DaneLINC is proposed as a two-tier membership organization, with agencies that are major data producers having greater rights than agencies that are primarily interested in data acquisition. CLISC did not resolve how to include private organizations in this quasi-governmental structure. The resolution of this problem was left for the long term. The long-term structure also included additional services, such as the sharing of equipment and technical expertise, and specialized products and analyses.

Wisconsin Land Information Program. In the summer of 1989, the Wisconsin legislature enacted a law called the Wisconsin Land Information Program (WLIP). A year later, a funding mechanism was added, which made the law viable. The intention of the law is to foster local land records modernization and to coordinate the land-information activities of state agencies. Implementation of the law has been focused almost completely on counties, to the exclusion of effective enforcement of provisions for state agencies. There are a few individuals within state agencies who are trying to revitalize the state agency aspects of the law.

The WLIP was the result of many years of effort, initially by a small group of University of Wisconsin experts, and later by a broad-based professional organization that they founded: the Wisconsin Land Information Association (WLIA). WLIA is comprised primarily of professional staff of local and state government, of utilities, and of the land information enterprises with which they work. It evolved from the state Urban and Regional Information Systems Association (URISA) chapter, essentially as a grassroots lobbying force. WLIA finally succeeded when it found legislative allies—a property-tax reformer on one side of the partisan aisle and a land surveyor on the other side.

The WLIP is mostly directed toward local land information. A major provision requires the development of county-wide land information modernization plans. Another provision provides incentives for county programs through direct funding and grants-in-aid. The law is certainly not silent in expectations of state agencies. They are explicitly required to report on their land-information activities and are implicitly expected to coordinate them. However, the limited state staff charged with administering the program has not provided any guidelines or help to state agencies, and so there is effectively no improvement in coordination. Similarly, the statutorily created Wisconsin Land Information Board has been dominated by local interests and so has done little to excoriate state agencies. Limited exchange of data and technical expertise continues to take place through the ad hoc state data-sharing group.

Trinidad and Tobago. In 1989, the Minister of Food Production and Marine Exploitation and the Minister of Planning and Mobilisation co-appointed three committees to look at problems relating to land and development, including "Committee C" – a technically oriented committee that dealt with land information. Most major agencies dealing with land were represented on Committee C, including agriculture, forestry, housing, land taxation, and land-use planning, although government corporations such as utilities and the agricultural corporation were not represented. This committee reiterated many well-known problems in land information, including outdated base and cadastral maps, duplicative data collection and management, and cumbersome land transaction procedures. In essence, this group recognized the problems but lacked the political courage or the vision to propose any major needed reforms. The committee continues to meet and has been able to make some minor improvements because the meetings themselves serve as a coordinating mechanism.

In 1991, an international lending agency provided funding for a study of land rationalization in Trinidad and Tobago – to provide recommendations on how to diversify the agricultural economy. Several land information experts from the University of Wisconsin and elsewhere came to study the situation. They found that the lack of a functional land-information system was a serious impediment to agricultural reform. Their recommendations included a substantial investment in base-mapping–related activities and major reforms in land-information management, including GIS pilot projects and a permanent inter-ministerial body for coordinating land information.

The local professional staff who assisted in the study (essentially the same group as Committee C) embraced the consultants' recommendations. They started some reforms, such as developing a unique parcel referencing system and initiating discussions of data exchange standards. But substantive reform depends on funding from international lending agencies; a request for additional funding was included in the consultants' recommendations.

The overarching bodies conceived by these three jurisdictions cannot yet be considered fully functional. In Dane County, a temporary study committee has completed its work and provided recommendations about the nature and function of an overarching body, but it has not yet been implemented. In the state of Wisconsin the body exists, but the Wisconsin Land Information Board is presently preoccupied with county-level land records modernization. In Trinidad and Tobago, key actors have agreed on the need for a coordinating mechanism but are awaiting the impetus that external funding will provide before attempting to create it. In all cases, some data sharing continues to take place through informal mechanisms. And, in all cases, the technical staff involved in the data-sharing dialog are pushing decision makers to resolve the institutional barriers that constrain more effective exchange.

FACTORS INFLUENCING THE CREATION OF OVERARCHING BODIES FOR DATA SHARING

As suggested by the long list of possible explanations circulated with the paper that accompanied the National Center for Geographic Information and Analysis study on the use and value of geographic information (NCGIA 1991), many factors might explain why GIS is or is not used effectively within an organization. Many of these factors might also operate between organizations, in effect creating incentives or barriers to data sharing. Certain barriers are not hard to imagine operating at the institutional as well as individual levels—for example, "turf battles" and the protection of authority. These clearly will have some value in predicting the success of data sharing, especially if they are not addressed in the formation of the overarching body. There are, perhaps, more subtle influences on overarching bodies arising from the processes of their formation: actors involved (experts, staff, politicians); authority granted; funding provided; requirements and expectations of participants; and so forth. Though many others undoubtedly exist, this section discusses the following three general influences on the form and function of overarching bodies:

- *Technology Introduction.* The way in which agencies become aware of new technologies such as GIS can affect both immediate and long-term reactions to it. The agent(s) and their ability to convey data-sharing concepts and configurations might affect the nature and form of the overarching body.

- *Practices and Structures within and among Organizations.* Many factors can influence an agency's willingness and ability to accept the authority of, and work with, an overarching body. Initial evidence indicates that some of these factors influenced the cases presented.

- *Form of the Overarching Body.* In most cases where an overarching body is charged with facilitating data sharing, this authority is newly created. The group creating the body must consider many options, including the body's cross-jurisdictional authority, its mandates, and its funding sources. These choices are likely to affect the success of the body.

Technology Introduction

We argue that the way in which agencies become aware of new technologies such as GIS can affect both immediate and long-term reactions to it. The technology introduction process is influenced by the role of "agents of change," the type and specificity of information about the technology in relation to the audience to whom the information is presented, and the form and venue in which information is presented. Anecdotal and reported evidence suggests that these factors do affect the success of GIS implementation within an organization. The cases reported herein provide initial evidence that similar factors operate in conveying the principles and concepts (both technical and institutional) underlying a confederation of systems or an integrated multi-agency system.

In all three cases, the agent of change can be considered GIS experts from the University of Wisconsin at Madison. These experts started with the concept of a Multipurpose Land Information System (MPLIS) as presented by Chrisman and Niemann (1985) and Niemann et al. (1987). For the purpose of data sharing, the salient point of the concept is the notion of data custodians. Many agencies have mandates to maintain spatial data; if these "custodians" do so in a form and format compatible with other agencies, data sharing can be greatly facilitated. These concepts are now seemingly so generally accepted that they have been incorporated into the efforts of all three jurisdictions without hesitation.

In recent years, the MPLIS concept has been modified to accommodate the proliferation of multiple software and hardware platforms used by different agencies. The operational term becomes "systems." The concept is expanded to include a confederation of multiple independent Geographic Information Systems. And, the cooperation and sharing are data driven, not hardware or software oriented. This orientation is reflected in each jurisdiction, as follows.

Dane County-wide Land Information Consortium. Language about custodianship and independent, coordinated systems is found in the DaneLINC proposal (CLISC, pp. 1 and 7):

> DaneLINC is envisioned to be a centrally coordinated entity with distributed responsibilities for land information modernization.
>
> DaneLINC should be established as a Dane County intergovernmental land information consortium to implement, facilitate, and support the

> modernization and sharing of land-related information by all *public* and *private* organizations that produce, maintain, and use such information. (emphasis in original)
>
> Individual organizations will retain authority and responsibilities for their own records.

Wisconsin Land Information Program. Wisconsin Land Information Program (WLIP, 1991, p. A3) guidelines state that:

> information housed in one department or jurisdiction can be shared with other departments and jurisdictions at every level of government, utilities, and the private sector. This confederation of systems will be independently developed and operated and will be linked by formal and informal data-sharing arrangements.
>
> A primary objective of the program is the development of ways in which data can be organized and transmitted for sharing. It must be recognized that this is a *"data-driven system"* and is not hardware or software dependent. . . . [T]echnical specifications are focused on the data. (emphasis in original)

Trinidad and Tobago. The consultants' report for Trinidad and Tobago (IADB 1991) has engendered such thoughts in Trinidad and Tobago agency staff:

> There are no longer any substantive technical reasons for insisting on a massive centralized land information processing facility.
>
> With appropriate standards and coordination, it is possible to envision a confederation of independently maintained systems that still has the benefits of a centralized system.
>
> [A]gencies can maintain control over sensitive or proprietary data and still make other data available.
>
> [A]gencies can obtain data from the "custodian," the agency with statutory responsibility for particular data.

Recent University of Wisconsin technology transfer efforts have included coordinating mechanisms such as overarching bodies. This is a fait accompli at the county level under the WLIP: the program requires the establishment of county land information offices. The guidelines for county plan development include suggestions about what offices or functions should be included in the land information office. The program also created the Wisconsin Land Information Board (WLIB) to serve the same function at a state level. The board has done well coordinating with counties but only nominally with state agencies. The County-wide Land Information Systems Committee (CLISC) effort in Dane County has gone one step further: they not only have a county land information office to coordinate county department activities but also have the CLISC recommendations to work with a broader group of land information users and generators. In Trinidad and Tobago, the only coordinating mechanism is the inter-ministerial committee, though they recognize the need for a more formal mechanism that has super-ministerial authority.

There has also been a suggestion, in University of Wisconsin technology introduction efforts, that there should be two functional components to the overarching bodies: technical and policy. There needs to be continuing dialog between technical staffs over issues such as hardware and software compatibility, data exchange and quality standards, media, documentation, and so forth. However, technical staff will not or cannot resolve other issues, particularly concerning funding, cost recovery, data and equipment access, and system security. These issues must be dealt with by decision makers.

In the case of the WLIB, there was no explicit technical committee. The WLIB has relied so heavily on the "ad hoc state data-sharing work group" for technical support that work group participants recently drafted a policy paper essentially asking that WLIB create an official WLIB technical support sub-committee and name them as members. Among other reasons, this was requested so that work group members would get credit within their agencies for what was becoming extensive extra-agency efforts. The DaneLINC proposal specifically calls for technical committee(s) and a policy board. The Trinidad and Tobago "Subcommittee C" currently functions in both technical and policy capacities, though members have recognized that there are many issues that they cannot resolve, issues that must be brought to an authority that spans or supersedes ministries.

The imprint of the agent of change—University of Wisconsin experts in land records and GIS—is apparent in the concepts and forms of overarching bodies that are evolving in the three jurisdictions. In essence, it appears that the agent(s) and their ability to convey data-sharing concepts and configurations had a substantial influence. However, it is difficult to distinguish clearly these from ideas that would have emerged spontaneously from these groups or from infusion from other sources of information. These jurisdictions are at the forefront in developing such bodies, so the ideas must be at least partially attributed to the agents of change; few peer organizations exist from which they could be learning.

Practices and Structures within and among Organizations

Many factors can influence an agency's willingness and ability to accept the authority of, and to work with, an overarching body. Many of these factors are similar to those that influence adoption of GIS, including experience with other technological innovations and their implementation, the type and degree of centralization and control of data management systems within an organization, lines of authority and communication between key actors, and perceived risks and advantages. Some factors might be unique to the process of interacting with an overarching body, such as previous experience with similar bodies and the orientation of an agency regarding data use. For example, an agency such as a surveying department that provides data services for other agencies as a major portion of its mandate might have a lot to risk in a confederation; an agency that is primarily gathering data for its own use, such as a planning department, might have more to gain, and so have more incentive to promote a coordinated approach. Again, initial evidence indicates that some of these factors influenced the cases presented.

Experience with GIS within an organization clearly influenced participation in the formation of overarching bodies. Not surprisingly, the leaders in efforts to create overarching bodies for data sharing have come from agencies that have already implemented Geographic Information Systems. Dane County and city of Madison staff involved with GIS went up the chain of command to convince their leaders to create CLISC. Individuals from Wisconsin state agencies had leadership roles in the Wisconsin Land Information Association and Land Information Board. In Trinidad and Tobago, the planning agency and the land taxation agency became leaders in the consultant's evaluation and organizing efforts, though they officially had no role in the consultant's scope of work. Staff in all these agencies already had to promote and explain GIS to decision makers and used the advantages of data sharing as a selling point. The creation of an overarching body became an extension of the implementation process—one necessary to realize the full potential of the technology.

Not surprisingly, many of the impediments to GIS implementation that occur within an organization also present barriers to the creation of overarching bodies. Individuals from data-processing shops worried about data access and security. Surveyors worried about data accuracy. Turf battles within and between organizations were expressed in many ways during discussions of overarching bodies. The cost of data sharing and the price of data became surrogates for deeper issues of control and authority. In all cases, the formation of an overarching body independent of existing agencies was at least partially attributable to the unwillingness of participants to entrust that function to an existing agency. The County-wide Land Information Systems Committee (CLISC) specifically sought some mechanism under state statutes to create a new kind of entity. Wisconsin leaders had a great deal of explicit discussion about the appropriate institutional home in the development of the Wisconsin Land Information Board. Trinidad and Tobago has recognized the need for super-ministerial authority to mitigate turf struggles.

Evidence is mixed as to whether an agency's perceived risk affects its willingness to participate in an overarching body. Most of the agencies or departments that had both a great deal to gain through data sharing and a substantially increased burden as data providers were active proponents of the concept. However, agencies such as the Dane County Regional Planning Commission, whose data-gathering needs could be greatly facilitated by an overarching body, were nominal participants in CLISC. This can probably be attributed to institutional inertia; they have not adopted GIS, and so have not made organizational changes that would allow them to take advantage of a data sharing environment. Likewise, the Trinidad and Tobago Lands and Surveys Division, which seemingly had a lot at stake as a major provider of data, embraced the concept of an overarching body as a necessary component (and cost) of its move toward automated mapping technology.

Form of the Overarching Body

In most cases where an overarching body is charged with facilitating data sharing, this authority is newly created. A new body must be created because there is usually no single agency with a mandate to coordinate geographic data among agencies. This requires initiative from a level of authority above individual agencies, such as a legislative body. The "authority above" must define the mission of the overarching body, create a funding mechanism, and grant the body some authority. The authority of the overarching body over participating organizations can range from a situation of entirely voluntary participation to one with substantial enforcement mechanisms.

A number of questions about mandates and authority arose from observation of the formation of the overarching bodies presented herein. These questions include:

- Do the nature of the edict and the authority granted affect success?
- Do the source(s) and amount of funding affect success?
- How do technical staff convey issues and potential solutions to legislators, and what effect does that have on the edict?
- How do inter-agency relations (or lack of relations) interfere with the new authority?
- How do bureaucratic and political allegiances affect the outcome?

Resolution of such questions about how the form of the overarching body will affect its success must await the passage of time. These distinctions among the three cases and their respective coordinating bodies might provide interesting comparisons in the future.

Dane County-wide Land Information Consortium. The DaneLINC proposal calls for a quasi-governmental body created under a statute originally promulgated to allow the formation of multi-jurisdictional emergency medical service coordinating bodies. There is no direct funding; it relies on contractual arrangements between participants. Part of the CLISC activities included a fairly detailed data inventory; the DaneLINC proposal calls for ongoing data cataloging.

CLISC is designed as a voluntary mechanism, though once individual agencies join, there is enforcement through economic means.

Wisconsin Land Information Program. The WLIB is attached to an existing agency—the Department of Administration—and has its own direct funding mechanism. Local interests control the board, so no funding has been directed to help state-level coordination. Data-reporting requirements are not effectively enforced for state agencies, so potential data resources still are not known. WLIB envisions an on-line data inventory but has done little to promote it.

Participation in the Land Information Program is mandatory. Related laws concerning information technology could be used to enforce provisions.

Trinidad and Tobago. The Trinidad and Tobago concept calls for a committee with super-ministerial authority, although previous attempts to mandate this sort of thing (e.g., the National Computing Authority) have had mixed success. At this point, Trinidad and Tobago is counting on international lending agencies for at least initial funding. The consultants' work included a data inventory; their proposal calls for a more detailed inventory and needs assessment.

The issues of authority and enforcement mechanisms have been debated often but not yet settled. There is apparent mistrust of "superagencies."

Time will tell if any of these options are more viable, though many competing variables will also have to be considered.

DISCUSSION IN RELATION TO ORGANIZATIONAL AND BEHAVIORAL RESEARCH

Although the work reported herein does not adhere to a rigorous case study methodology (the author has not been a neutral, uninvolved observer), it is possible to discuss the cases in relation to topical areas developed for the I-9 Specialist Meeting Report (Calkins et al. 1991) and in broader organizational and behavioral terms. The observations of institutions and actors provide anecdotal insight into incentives and impediments to the sharing of geographic information from a perspective deep inside the process. This perspective has provided some evidence that supports, and some evidence that contradicts, prevailing organizational and behavioral hypotheses as applied to geographic data sharing in the I-9 report. The following is a brief discussion of some of this evidence; in essence, it is organizational and behavioral observations in addition to the three discussed in detail in the previous section.

Technical Capacity

Discussions on the feasibility of geographic data sharing typically begin on the assumption that participants must have more than rudimentary technical skills. The I-9 report suggests that "certainly it is necessary that two organizations possess similar technical capacities and capabilities in order for information exchange to proceed efficiently." This apparently is not true for the formation of overarching bodies designed to facilitate data sharing. In all three cases, technical proficiency of participants in discussions about the formation of overarching bodies ranged from agencies just beginning to learn what GIS was during the discussions to agencies with operational systems and years of experience. And, almost as wide a range existed in agencies' anticipated functionality and their expectations about how data would move between agencies. The implication of this observation from a research standpoint is that it may not be appropriate to "assume" equivalent technical proficiency among organizations.

Rather, proficiency should be treated as another variable for measurement or observation in order to characterize organizations, and ultimately to control or eliminate technical impediments as competing explanations for data sharing success.

Organizational Issues

Calkins et al. (1991) presented a sample of five organizational factors that could impact institutional data sharing. The cases indicated that at least the first four of these factors are useful in explaining some of the observed behaviors.

Bureaucratic practices and standard operating procedures. The most striking evidence that institutional inertia limited participation in data sharing came from a planning agency in Dane County and utilities (government corporations) in Trinidad and Tobago—agencies that potentially could benefit from improved data sharing. One explanation for their general lack of interest in GIS and specific limited participation in coordination activities could be their substantial investment in manual cartographic techniques: they are set in their ways because the existing system seems to be good enough and/or it would be upsetting to adopt new methods.

Degree of cross-functional cooperation. A good example of situations where agencies' predisposition to operate in a vacuum has affected their ability to work cooperatively at data sharing comes from the lack of effective participation of local offices of federal agencies in either Dane County or the state of Wisconsin. The internal organization of agencies such as USDA—Agricultural Conservation and Stabilization Service is not set up to facilitate cooperation at the local level. Although attempts have been made to include them and other federal agencies in discussions, they apparently do not have the flexibility to adapt to local conditions and become officially involved in overarching bodies. To their credit, some federal agencies have participated in an advisory or ex officio capacity in local and state efforts.

Organizational structure. The diversity of agencies involved in these three cases suggests that if the individuals involved are willing, structural impediments can be resolved. There is evidence throughout this discussion that structural impediments do exist, including organizational inertia and inflexibility; but organizational structure does not seem to be a strongly explanatory factor in participation in discussions about overarching bodies.

Corporate culture. Barriers to data sharing under the rubric of "corporate culture" can also be labeled "disciplinary bias" or "attitude problems."

Under any name, it is apparent that the technology transfer process must include an educational component. Different "corporate cultures" use different terminology, have different standards, and have different expectations of cooperative efforts. In all three cases, a great deal of time was spent developing a common understanding of terminology and needs. Issues such as appropriate standards for spatial data accuracy still remain unresolved, though at least now the participants understand basic time/cost/accuracy trade-offs. Research on attitudes and expectations with regard to overarching bodies should build on earlier work by Portner and Niemann (1983) about land-information systems in general.

Political environment. There is no overt evidence that agencies involved in the formation of overarching bodies were using the process to gain influence or resources. It would take more investigation to determine if these were underlying motivations.

Behavioral Issues

Calkins et al. (1991) presented three examples of behavioral issues that might affect the sharing of data. Casual observation of participants in the discussions about overarching bodies lend credence to these explanations. For example, turf battles and the role of technology champions were discussed previously. But there is little evidence that their other issue, concerning "individual differences," operated at least to the level of constraining data sharing. Participants in the discussions recognized their differences but generally worked hard to overcome them, rather than developing an "us versus them" mentality. Even at the political level, decision makers who could have allowed personal or organizational enmity to stand in the way overcame that urge for the sake of better government. This was an unexpected observation, given the predilection for agencies to look narrowly at their own mandate in most circumstances.

CONCLUSIONS

In these three case observations of overarching bodies, technical and institutional issues constrained the development of effective agreements. In each case, however, the jurisdictions have continued to move forward. Even where there is not yet official sanction for the activities, data sharing has improved merely by the process of meeting on a regular basis. Discussions about data sharing and building overarching bodies have built awareness of GIS-related activities in other organizations and have provided a mechanism for technology transfer agents to enlighten the group. This has been a very positive outcome, and indeed many other jurisdictions follow their progress with interest.

Coordinating councils, inter-agency agreements, and similar bodies have been initiated for a variety of reasons. But the cost and volume of geographic data have made data sharing, and consequently overarching bodies, an imperative. It almost goes without saying that the form and function of overarching bodies will be influenced by the structure and behavior of organizations and individuals that participate. The tenets of organizational theory and related studies of institutional behavior should be useful to help understand these influences and ultimately to foster successful developments. However, these realms of thought have been developed primarily from observations within organizations. Our challenge now is to extend these notions to inter-organizational processes. This chapter has presented some of the many influences on form and function of overarching bodies—the technology introduction process, the practices and structures within organizations, the history of interaction between organizations, and the mandates that move organizations toward cooperation.

REFERENCES

Calkins, J., E. Epstein, J. Estes, J. Onsrud, J. Pinto, G. Rushton, and L. Wiggins. 1991. Sharing of geographic information: research issues and a call for participation. Anonymous ftp at ncgia.ucsb.edu:/pub/initiatives/I-9/call-for-participation, and Chapter 2, I-9 Specialist Meeting Report.

Chrisman, N. R, and B. J. Niemann. 1985. Alternative routes to a multipurpose cadastre: merging institutional and technical teasoning. AutoCarto 7 *Proceedings*, pp. 84-94.

Coleman, D. J., and J. D. McLaughlin. 1988. The LANDNET project in New Brunswick, Canada: moving from research to reality. URISA *Proceedings*, vol. 1, pp. 80-90.

County-wide Land Information Systems Committee (CLISC). 1991. *Report of the County-wide Land Information Systems Committee.* Dane County, Wisconsin. 62 pages.

Inter-Americas Development Bank (IADB). 1991. Technological options for a modern multipurpose land information system in Trinidad and Tobago. Report #10, Trinidad and Tobago Land Rationalization Project. 43 pages.

Ministry of Consumer and Commercial Relations. 1991. Partnership between Ontario government and private sector to create jobs. *News Release Communique.* February 15.

National Center for Geographic Information and Analysis (NCGIA). 1991. Testing thirty technology transfer theories: a GIS case study research project. 26-page handout, anonymous ftp at ncgia.ucsb.edu:/pub/initiatives/I-4/case studies.

Niemann, B. J. Jr., J. G. Sullivan, S. J. Ventura, N. R. Chrisman, A. P. Vonderohe, D. F. Mezera, and D. D. Moyer. 1987. Results of the Dane County land records project. *PE&RS* 53, 10: 1371–1378.

Onsrud, H. J., J. K. Pinto, and B. Azad. 1992. Case study research methods for Geographic Information Systems. *URISA Journal* 4, 1: 32–44.

Portner, J., and B. J. Niemann. 1983. Belief differences among land records officials: implications for land records modernization. URISA *Proceedings*, pp. 121-135.

Wisconsin Land Information Board (WLIB). 1990. Developing the Wisconsin Land Information Board's recommendations and requirements for county-wide plans. Wisconsin Department of Administration. 47 pages.

Nancy Tosta

11 The Evolution of Geographic Information Systems and Spatial Data-Sharing Activities In California State Government

Continued expansion in the use of technology for mapping and other geographic data activities has generated a growing demand for digital spatial data. California state agency use of GIS technologies in the late 1970s was limited to approximately six individuals in as many agencies but by 1992 had spread to more than 100 users in at least twenty-five agencies. Interest in digital spatial data saw an even more dramatic rise as agencies came to understand the potential usefulness of the technology in addressing numerous management concerns. Most agencies realized the value of sharing digital geographic data because of the costs of data capture and maintenance, but actual instances of sharing, particularly in data creation, were relatively rare. The evolution of GIS technologies in California state government between 1977 and 1992, and the accompanying interest in, practice of, and constraints to data sharing are examined in this chapter. Comments are offered on factors that influenced the ability to share digital geographic data including politics, individual behavior, organizational structures, and changes in technology. Various approaches to facilitate data-sharing activities are identified.

BACKGROUND

From approximately 1977 to 1992, the use of Geographic Information System (GIS) technologies rose significantly within California state government agencies. The number of actual users of the technology (individuals with access to hardware and software from their desks) rose from approximately half a dozen in as many agencies in the late 1970s to more than 100 in approximately twenty-five agencies by 1992. At least 200 additional people were also taking advantage of the technology in 1992, through the use of consultants, universities, and service bureaus.

As of 1992, roughly 10 percent of all California state agencies, including all of the major land and natural resource management agencies, were using GIS for geographic analysis and policy purposes. As examples, GIS was used to assess endangered species habitat, design transportation corridors, plan emergency responses, monitor air and water resources, track toxics, study disease patterns, and analyze population growth.

Successful use of GIS is dependent on access to accurate, complete, and current geographic data in digital format. Interest in sharing digital geographic data grew as the use of GIS technologies spread throughout state agencies. Aside from the availability and use of Landsat data in the late 1970s, digital spatial data in use were primarily created within agencies through digitization by contractors or individual users of the technology. These data sets were primarily raster files, representing specific geographic features in limited areas, primarily of interest to the creating agencies. By 1992, representations of numerous "layers" of spatial data for the entire state were available in common digital formats. These were often derived from national digital data sets created by the U. S. Geological Survey.

For the most part, GIS use evolved independently within a variety of state agencies. Overall, the California state government infrastructure consists of more than 200 departments, agencies, boards, and commissions, most of which are led by Governor appointees. In the late 1970s, the agencies using computer mapping tools (the terminology in use at the time) were primarily those with natural resource management responsibilities, including the Air Resources Board, the Energy Commission, and the departments of Forestry, Conservation, Agriculture, and Fish and Game.

The presence of NASA Ames Research Center and the Western Mapping Center of the U. S. Geological Survey (located in close proximity to each other in California's Silicon Valley) spawned some of the interest in mapping and spatial technologies in California state agencies. During the late 1970s, NASA was engaged in an aggressive technology transfer program to establish valid uses for Landsat imagery. The departments of Forestry and Water Resources, and the Governor's Office, all became involved in technology transfer projects related to the use of remotely sensed data. In the final outcome of most of these projects, however, the knowledge about how to actually manipulate the technology, as well as an understanding of the hardware and software involved, continued to reside with the federal agencies or their contractors, and not with the state cooperators.

The departments of Transportation and Water Resources began using computers in the 1970s for engineering applications, including Computer-Aided Design (CAD). Digital databases were created for limited areas from large-scale maps for water and transportation construction projects. Also in the late 1970s, The Nature Conservancy (TNC) began its national efforts to build rare and endangered species databases. In California, unlike other states in the nation, from the start this work was initiated with the intent of using computers for inventory and analysis. The Department of Fish and Game was responsible for implementing the project, along with TNC staff. This was called the California Natural Diversity Data Base (CNDDB).

Through the 1980s and into the early 1990s, knowledge about the use of GIS permeated nearly every state agency. By 1992, at least twenty-five agencies had actually installed hardware and software, while close to that number were in some phase of GIS technology justification or implementation. GIS use evolved from primarily natural resource applications to include emergency response activities, transportation, and health and social services. The GIS tools in use included all major commercial software packages and several home-grown systems running on almost every hardware platform available, including personal computers, workstations, minicomputers, and mainframes. There was no overall plan or architecture for implementing GIS in state government in the late 1970s nor was there in 1992. Similarly, there were no official policies or plans to address the production, maintenance, and dissemination of digital spatial data.

GEOGRAPHIC DATA ENVIRONMENT

Political Factors

California experienced three governors from 1974 to 1992. Jerry Brown (D) was Governor from 1974–1981, George Deukmejian (R) from 1982–1990, and Pete Wilson (R) took office in January 1991. Each change in administration resulted in new programs and policies and new leaders in most departments and agencies. Each change obviously brought new agendas and priorities and "adjustments" in programs that affected data-sharing efforts.

During the Brown administration, the use of technology was not prevalent, but idealism about what it might be able to accomplish was high. The administration had visions of the potential of technology, and the Governor appointed a former astronaut (Rusty Schweickart) as his Science and

Technology Advisor. During this time, concurrent with great growth in both the population and economy were increasing concerns about a variety of environmental issues. Many state agencies began to experiment with information technologies for environmental analyses although, for the most part, these efforts were technology demonstrations rather than applied use of technology for environmental problem solving. Funding was available for a variety of information technology projects, even if specific applications and expected results were ill-defined. For the most part, agencies undertook technology projects internally and independently from other agencies.

In the Governor's Office, Brown established an Environmental Data Center (EDC), which began to set the stage between 1978 and 1980 for a number of data-sharing efforts. First, an inventory of sources of data was compiled. This included tabular as well as spatial data such as maps, photos, and images, and identified local, state, and federal data creators and collectors. Secondly, several committees were formed to address data analysis needs for specific resources. For example, committees existed on wildlife, rangelands, recreation, and vegetation. By 1980, there was also a committee on computer mapping that involved the agencies using mapping technologies for environmental assessments, those working on technology transfer projects with NASA, and proponents of the CNDDB. Although the committees had no formal control or authority, discussions about technology and data development were informative for all participants. Data sharing was discussed in the computer-mapping committee, but many agencies were unable to share data given proprietary software and lack of common transfer formats between systems.

In 1982, the populace elected George Deukmejian as the new governor, and several changes were very quickly initiated. The position of Science and Technology Advisor and the EDC were eliminated. Shutting down the EDC had several effects on technology development and data-sharing activities. The communication infrastructure and focal point of knowledge related to spatial data were lost just at a point in time when GIS technologies were beginning to take hold in state agencies. That loss, and concurrent changes in departmental executive staff to individuals with little knowledge about technology uses and data requirements, slowed the momentum of many agencies in use of GIS. Champions in a few agencies continued to argue the value of technologies for processing digital spatial data, but for several years after the change in governors, communication among departments about GIS and support for the technology were extremely limited.

In general, the Deukmejian administration was characterized by pragmatic experimentation of technology to solve functional problems. Departments tended to isolate issues and activities, tackling data development projects in a singular fashion. Agencies concentrated on inventories and assessments in database development efforts, with much less emphasis on "system" approaches to environmental problems. One exception to this was the San Joaquin Valley Drainage Program. The discovery of toxic levels of selenium in the San Joaquin Valley raised the level of awareness of agricultural drainage issues. A cooperative project team was formed among eight state and federal agencies to create a common, comprehensive geographic database, using technologies such as GIS. This effort was brought to a close in 1991, and the comprehensive database that was created was "mothballed" because there was no mechanism in place to cooperatively manage and maintain the component data sets.

Toward the end of the Deukmejian administration, as other environmental issues such as growth management, endangered species, and toxic waste site contamination became obvious, task forces of agency directors were formed that cut across traditional agency authorities. These interagency efforts were often mandated by the legislature and not the administration. Data availability and sharing were usually the first topics of discussion in task force meetings. Knowledge about GIS and the role that digital spatial data could play in environmental problem solving began to spread from technical to executive staff.

By the time the Wilson administration took office, the state was facing a severe budget deficit. Hiring and contracting "freezes" were immediately imposed, significantly affecting agencies' ability to obtain needed staff and technology resources. At top organizational levels, there was belief in the potential of GIS technologies although still limited knowledge about what was required to realize that potential.

Executive staff became concerned about getting "control" over technology use within and among their agencies. In some cases, "software battles" erupted between agencies, with directors and executive staff arguing the merits of particular GIS packages in which their agency had invested, in efforts to establish control over other agencies' GIS activities.

Technology

The evolution and implementation of information technology in general, and specifically as used for geographic information analysis, has been significant within California state agencies. Between 1977 and 1992, mainframes and minicomputers running proprietary operating systems and home-grown GIS software, and supporting users on "dumb" terminals, gave way to personal computers and workstations running software from a score of vendors and supporting "open" architectures. Databases that were supported in only one location, by data-processing gurus, were fractured, split, modified, and duplicated a hundred times over in the processors on desks of "end users."

Changes in electronic networking kept pace with hardware platforms. State-agency use of networks evolved out of the offices of data-processing staff connected to mainframes to electronic mail connections within most departments, and worldwide Internet connections in many agencies with research programs.

Information Technology Policies

Implementation of technology in California state government requires the development of a Feasibility Study Report (FSR) that is reviewed and approved by the Office of Information Technology in the Department of Finance. Justification of technology in an FSR requires an analysis of five years of costs and benefits. Agencies are encouraged to look for personnel savings (i.e., decrease staff) through the use of technology. All GIS technology FSRs written up until the late 1980s underestimated the costs of data development or data acquisition.

The primary focus in all FSRs is on hardware and software justification. Most agency GIS feasibility studies were written by individuals who had never used digital spatial data and who had never performed geographic analyses. Data needs were often ignored or treated with a "we'll cross that bridge when we come to it" attitude. The costs of digital spatial data were often underestimated and the assumption often made that data would simply be available and accessible when needed.

Budget procedures and policies force agencies to justify most technology implementations project by project and department by department. There is little flexibility to examine cooperative and integrated technology alternatives. Agencies actually "compete" for funding from the state's general fund. Aside from multi-agency contracting with one of the state's centralized data centers, where costs for technology services are spread across users, cooperative technology projects were difficult to accomplish and were consequently rare.

Organizational Structures

How and where GIS technologies were implemented within state organizations changed over the course of fifteen years. From the late 1970s to the late 1980s, interest in GIS was almost always initiated by individuals in programs with specific applications that could be served by GIS. These initiators were usually research analysts or scientists, several levels down in the organization, who had specific problems to solve. Data-processing staff were not often involved. If GIS was funded and implemented, a small pocket of GIS expertise would evolve to meet a specific program's needs.

In 1987, after many years of departments' ignoring the activities of others, the Resources Agency noted that four departments that fell under its authority had programs that were independently implementing GIS technologies and, in some cases, digitizing the same information. A recommendation was made to create a "committee" to discuss common requirements. From the beginning, the "technologists" involved in the committee understood that every agency used different software and hardware, selected initially for meeting specific application needs. Once an agency had selected a system, debates about "the best" system were fruitless, as no agency was willing to admit that the software other than that in which they had invested was better. However, the group also understood that there were common data needs, and coordination might focus on how best to share data. The initial group of four individuals grew steadily over the years to more than forty regular attendees in 1992, including federal and local government representatives. The group consisted primarily of GIS users, a few managers, and several "wannahaves."

The Computer Mapping Coordinating Committee (CMCC), as the group came to be called, began as and continued to be an ad hoc entity with no official responsibilities. The primary reason for continuation was its value as a communication forum for data sharing. Participants would meet and exchange information on current projects and available data. CMCC became the starting point of digital geographic data-sharing activities among many state and federal agencies. A subcommittee on remote sensing and vegetation mapping evolved and provided a forum for discussion and activity associated with creation of a statewide cover-type map.

The "program-by-program" approach to GIS implementation began to change around 1989 as more people became interested in the technology and demands for access to GIS spread across agencies. Also, around that time, more data-processing staff began to see GIS as a technology that they should be part of, if not in control of. Departments began to create internal "task forces," usually mandated by executive staff, of all potential users of GIS in the organization, as well as data-processing staff. Feasibility studies were initiated that began to examine multiple program (within a department) needs for GIS, interdependencies on data, and roles of data-processing staff in supporting GIS technologies. However, budget constraints continued to limit agencies' ability to actually implement any department-wide GIS projects.

Also in the late 1980s, two of the state's data centers that provide cost-recoverable data-processing services, predominantly on mainframes, began to examine the possibility of offering GIS services. Their efforts were opposed by many of the agencies already engaged in GIS activities because of the potential threat of being forced to do business at the data centers and losing control over their own computing. Teale Data Center evolved a GIS service offering that was based on user-supported distributed processing, with a shared library of digital spatial data. The Health and Welfare Data Center experimented with offering GIS services on a mainframe to integrate and display geographically the primarily social, economic, and health data that they maintained.

The data centers were able, in many cases, to provide access to technology tools that would not otherwise be available to some agencies, given procurement and staffing constraints. By 1992, more than fifty local, state, and federal agencies were sharing and using data through Teale. Eight departments had individually established data library access over a wide-area electronic network and were also able to communicate with each other via high-speed data networks.

A GIS user group evolved from the Teale Data Center activities, and subgroups were formed based on interests in specific thematic data layers. The user group coordinated its activities with the CMCC but tended to focus on needed standards for documentation, attribution, and spatial data maintenance.

As stated earlier, in about 1991, inter-agency task forces became a means to establish communication, particularly at an executive level, on specific issues. A Task Force on Geographic Information was legislatively mandated on January 1, 1992 to examine needs for and opportunities to share development and use of geographic data among federal, state, and local governments, and the private and academic sectors. The Governor's staff took an interest in the task force and, through an Executive Order, established the Director of the Governor's Office of Planning and Research as task force chair. Directors and representatives of many local, state, and federal agencies, as well as representatives of the surveyors' associations, were asked by Governor Wilson to serve on the task force and make recommendations to his office and the legislature on improving access to, and use of, geographic data in California. The task force report is under development as of this writing.

DATA SHARING

Constraints

Several constraints or impediments to sharing geographic data became obvious over the course of GIS evolution in California. Initially, there were many technological constraints, although these decreased significantly as software and hardware vendors began to consider requirements for interactions among different systems and means for common data transfer.

Organizational and bureaucratic constraints exist in the areas of procurements, contracts, and sharing of resources among agencies. For example, the process of developing FSRs, focusing agency by agency without considering interactions among agencies, and justification of systems based on decreases in personnel, contributed to unrealistic estimates of costs for data.

The bureaucratic policies and procedures that must be addressed to implement new technologies force agencies to underestimate data development efforts. The cost to one agency to capture all of the digital geographic data that might be required to support GIS use would be impossible to justify. As mentioned previously, agencies tend to "assume" data exist and may identify a cost to access or purchase, but seldom do they include the true costs of data capture in an FSR. There are no easy ways for agencies to identify partial funding that might be matched by other agencies to create new spatial databases. For this reason, data sharing after data are developed is an easier process than sharing in data creation. This is why many state agencies look to the federal government for data.

Agencies that have mandates and/or funding for data creation are often less willing to compromise on data characteristics in data sets to be created than they are in accessing existing data. Agencies are often willing to "take what they can get" if the price is right but will not commit funds if new data are to be created that do not specifically meet their mandates or requirements. For example, many efforts to cooperatively generate a covertype map for the state fell apart because no one agency had adequate funding to accomplish the entire job, and agencies that each had small amounts of funding could not come to agreement on an appropriate classification system or minimum mapping-unit size. Several agencies continued to use the U. S. Geological Survey Land Use/Land Cover data base despite the fact that it was nearly two decades old and the classification was very general, because agreement could not be reached to create a more current digital coverage.

The unwillingness to compromise in data creation activities is indicative of underlying behavioral constraints that play a major role in data-sharing activities. Although technological and institutional constraints are often more obvious, human nature and behavior are likely more central to issues of data coordination.

Data sharing implies an interaction between individuals or among organizations comprised of individuals. How individuals behave towards each other can significantly affect the success of data-sharing activities. How people feel about each other, how secure they feel about their own knowledge and roles, how they perceive their organizations, and whether they see themselves personally benefiting from a data-sharing activity may influence their response to data interactions. These emotions and modes of behavior are relevant in all interactions, not just those related to geographic data sharing obviously, and seem to be derived from basic human characteristics: desire for power, desire to do good, fear, self-preservation, and morale.

In the course of many data-sharing discussions and meetings, conflict and tension related to desires for personal recognition, power, or control are often present. The ways in which such emotions are expressed are usually subtle rather than blatant displays of anger, for example. The ideas of certain individuals will be ignored or overruled, compromise related to specific mapping issues such as what classification system should be used cannot be reached, and agreements made in meetings will later be broken.

Budget constraints create situations where individuals are concerned about the viability and continuation of their programs (and jobs) and, in their own self-interest, are resistant to sharing knowledge (or data) that might make someone else's program appear more successful. Several excuses for not sharing are offered, such as concerns on the part of the data creator that the data are not accurate enough for all uses that someone might want to put them to (someone might find errors in "my" data)—or, alternatively, that other people do not have the knowledge (or the right?) to correctly use the data and might misinterpret or draw incorrect conclusions based on it. Sometimes the fear also seems to exist that someone might think of ways to use the data that the creator had never envisioned and hence make "better" use of the data.

Individuals act in their own best interests unless given reason to do otherwise. Public servants are expected to serve the public's interests, although oftentimes, and particularly during transitions in leaders, those interests are muddled in bureaucracy and are difficult to discern. Complexity in issues, and conflicting agendas among agencies, can undercut morale and contribute to an attitude of "I'll just do my own thing." Budget cuts and state employee layoffs during the most recent administration have created fear among many about their own livelihoods. Although tight budgets can, in some instances, encourage data sharing because resources are limiting to all agencies, other factors—such as self-preservation—may actually contradict sharing incentives.

As described earlier in the Political Factors section, various executives may have their own agendas in interacting with other organizations that transcend interests in creating sharable databases. The databases of their own organization represent sources of authority or "power." Often this behavior seems to occur in individuals within organizations with less-defined areas of responsibility, or with responsibilities that overlap other agencies. Missions may be less clear-cut than exist in specific resource protection or regulatory agencies, for example.

The behavior of vendors also plays a role in the ability of public agencies to share data. Vendors are often anxious to establish a handhold in the state bureaucracy, particularly if another vendor has tended to dominate. Understandably, winning a bid and establishing a software and/or hardware site is usually more important to a vendor than the institutional data-sharing architecture of which they are trying to become a part. There were several instances of vendors lavishing attention, including free hardware/software or database development activities, on a prospective client with little conscience about how that client might function in the larger data-sharing picture with other state agencies when the vendor is no longer offering free support. Of course, most clients appreciate the attention, and for an agency that needs technology tools and has given little thought to where data will come from, the vendor involvement is welcome. The client is ultimately responsible for establishing data parameters to guide vendors, but the lack of client expertise in many cases limits their ability to do this.

FACILITATORS

Various activities can facilitate the sharing of geographic data. These include providing credit and recognition for ongoing GIS programs, establishing means for communication, developing standards and quality control, defining specific responsibilities for data production and management, and developing processes for integrated, anonymous access to geographic data sets. Many of these activities serve to counteract basic behavior that impedes data sharing.

The creation of the CMCC was one positive step, providing a forum for discussion about various data activities. At a minimum, CMCC meetings offered an opportunity for individuals to discover ongoing GIS and data development projects in other agencies while also offering a place to begin discussions about potential data development projects. The committee, as an ad hoc entity, also survived changes in administrations, providing a certain amount of stability and continuity, although no real power in terms of influencing decision makers. About three years into the life of the CMCC, a decision was made to rotate the chairmanship to allow various individuals the opportunity and responsibility of running the meetings. This action tended to give more people ownership in the process as well as allowing leadership through a variety of viewpoints. More people were given recognition for their views and acknowledgment that they were officially part of the GIS community within state government.

Given that requirements for geographic data often cross program and agency authorities, the challenge exists to develop a vision of shared database development, use, and maintenance. State institutional structures do not commonly fit this model but rather function hierarchically and independently. This is one area where the involvement of the state data centers in GIS activities became something of an advantage and facilitated data sharing.

The Teale and Health and Welfare Data Centers provide computer-processing and technology services to numerous state agencies. Traditionally, an agency's data were not of concern to the data centers other than in understanding how much data were being collected and what sort of manipulations were required such that adequate hardware and appropriate software for data storage and management could be provided. However, as GIS services were examined, Teale, in particular, began to realize the advantage of providing a neutral site for cooperative geographic data integration and access across state as well as other governmental agencies. This presented a challenge, however, in that many agencies were somewhat antagonistic toward the data centers, particularly as the use of personal computers supplanted and replaced mainframes.

When Teale first began development of GIS services, reaction of most of the GIS community was negative, and many justifications were developed for not using Teale for processing. However, as staff expertise grew and the ability to provide distributed processing services expanded, many agencies came to depend on Teale for GIS services and for access to data. Teale represented a neutral site of expertise with no specific agenda of its own, accessible to anyone. Once network connections were established, access to data was at the end user's discretion; no one needed to ask permission to use data. To some extent, data sharing could become "institutionalized" as more agencies considered meeting standards required to communicate electronically to access digital data and to provide access to their data as well. The concept of "owning" data began to yield to a belief that "you don't have to own it to use it."

The Teale GIS user groups and CMCC began to lay the foundation of multi-agency shared databases. Specific discussions on data quality expectations, data needs, documentation standards, and collection responsibilities provided the beginnings of a framework for data sharing. Users of the data library at Teale were assured that the data in the library met some minimum standards, or at least included labeling that attested to the quality of the data. The quality of the data actually improved as layers of information were integrated and registered to each other. Individuals had less basis to argue for "their own thing" when their activities had an obvious place in the overall data architecture of multiple agencies.

Attempts to refine a data-sharing vision and establish a "code of ethics" related to geographic data development were just beginning in 1992. How successful the efforts will be will depend to a major extent on the ability of agencies to procure technology tools to meet specific analysis needs and on the willingness of executives to share in the vision of integrated geographic data management.

Also in 1992, the expansion of electronic networks began to change the way individuals within and among organizations communicated with each other. Network technologies began to change everyone's perception of the "bigger picture." Two major effects of electronic networks appeared to be at least a partial breakdown of traditional hierarchies and a democratization of technology use. In some state agencies, all members of the organization, from entry-level office assistants and technicians to directors, were connected by electronic mail. Whereas staff might be reluctant to call a director personally, electronic mail seemed to provide a safer and more accessible means for reaching across hierarchies.

Electronic mail also became an initial "hook" for previously non-technical individuals to begin to realize the possibilities of sharing data as well as mail. Executive staff who had relatively little experience with technology began to understand the potential through simple mail systems. Many departments and agencies initiated discussions about corporate databases and business process redesign projects to create departmental data sets, accessible electronically by all agency personnel. As of 1992, the discussions had not been extended to geographic data sets, but directors and other decision makers were beginning to realize the possibilities of "viewing" the activities of interest to their departments. For example, resource management agencies could track fire protection activities; regulatory agencies could view sites of various pesticide applications or air quality violations; and service agencies could analyze patterns of collection and distribution activities.

CONCLUSIONS

The evolution of GIS use in California state government has been influenced by political, bureaucratic, technological, and behavioral constraints. Initially, technological constraints, such as hardware and software expense and compatibility, controlled growth of the technology and interactions among agencies. However, as use of the technology spread, growth became at least partly a function of availability of data. Geographic data sharing became a concern of all agencies using the technology tools. Over a fifteen-year period, various agendas and activities have constrained and contributed to the ability of agencies to share geographic data. The critical components of successful data sharing appear to include recognizing various individual contributions to geographic data activities, developing a vision for common benefits and a framework for geographic data sharing that includes standards, establishing neutrality in data control and access, and institutionalizing processes of data sharing through the use of technology such as electronic networks.

The passage of legislation to create a Geographic Information Task Force raised the awareness at the Governor's level and among executive staff of issues of geographic data sharing and use of GIS technologies. A hoped-for outcome of the task force work is creation of a more permanent, officially endorsed and funded structure for developing geographic data sets, coordinating among all levels of government and the private sector within California.

Bruce Rocheleau

12 *Computers and Horizontal Information Sharing in the Public Sector*

Implementation of Geographic Information Systems requires a great deal of sharing between municipal departments, but there is little information available on the nature and frequency of shared information. This chapter reports on the rate of sharing of information between police and other city departments in Illinois. It reveals that the overall rate of sharing is infrequent – less than monthly. The highest rate of sharing is with the administrative and budget departments. Qualitative data suggest that much of sharing is due to factors such as hierarchical authority and accountability requirements rather than task-oriented reasons. Statistical analysis shows that the use of electronic sharing does encourage a higher overall rate of sharing between departments. Other variables including hardware configurations, departmental size and autonomy, and crime-related variables, were not found to be significant. The implications for the implementation of Geographic Information Systems are discussed.

INTRODUCTION

Do computers facilitate the exchange of information in public-sector organizations? Will the implementation of Geographic Information Systems (GIS) spur inter-departmental exchanges in cities? There have been few systematic studies of interdepartmental information sharing in the public sector. Researchers such as Laudon (1986) have looked at the construction of national police databases that have enabled police to share data more easily. Indeed, one of the basic themes of these studies is the fear that computers make the sharing of police files too easy. From the point of view of information exchange, these studies are useful but limited because they have focused almost completely on exchanges of information among police agencies; exchanges with non-police departments have not been a subject of inquiry.

There is a substantial amount of qualitative evidence that public-sector organizations have difficulty in exchanging information horizontally. For example, a recent General Accounting Office (GAO) report found that one of the major problems in combating illegal drug use in the United States was the inability to share computerized drug-related information among the agencies involved (GAO 1991). U.S. Army information systems specialists have found that the Army has primarily a "vertical" information system in which it is impossible, for example, to exchange data between personnel and logistics offices (Bass 1989). A recent study of the feasibility of using administrative data for medical research found that veterans' hospitals have very limited sharing capabilities (Agency for Health Care Policy 1991, 3-4):

> Indeed, even within a single hospital, one department's automated information system quite frequently cannot exchange patient information with the system of another department.

The Department of Agriculture has also found that its information flows are almost entirely vertical (Rogers 1991). As a result, their employees have difficulty in solving issues that cut across departmental boundaries.

Kraemer et al. (1974) long ago pointed out the possibility of fruitful exchange of information among municipal departments. For example, fire, police, clerk, and building departments can often usefully exchange information and this exchange, they argue, could be facilitated by computers. However, the attempt to build integrated municipal databases that would cut across department functions failed (Kraemer and King 1979). One of the major hopes for GIS systems is to encourage inter-departmental information to meet the problem of "higher level" functions such as economic development and planning (Eichelberger 1986). Nolan and Nyerges (1991) recently reported that 88 percent of organizations responding to a questionnaire in the Puget Sound Region had engaged in some sort of sharing of land-related information. But recent comprehensive reviews of GIS implementations found resistance to sharing (Crosswell 1989; Craig 1990).

In summary, there is substantial evidence to show that the relationship between computers and the sharing of information, both inter-departmental and inter-organizational, is one of the most significant but neglected issues in research on public-sector information management. The reports and research that do exist on information sharing, such as that cited above. suffer from three major limitations and weaknesses:

1. They are largely anecdotal in nature with little quantitative data provided on how extensive or limited sharing is;
2. They are largely based on the study of single organizations;
3. Much of the literature was gathered during the mainframe era; thus, we have no assessment of what impact, if any, the development of microcomputers and local area networks have on inter-organizational sharing.

The goal of this chapter is to develop and test some preliminary hypotheses concerning factors that influence inter-departmental sharing based on a study of sharing of information between police and other municipal departments in Illinois. We study the nature, frequency, and some potential causes of information sharing. Although our study does not directly examine the impact of GIS systems, the results clearly will have important implications for researchers and practitioners planning such systems.[1]

CONCEPTS AND HYPOTHESES

We use the term "sharing" to refer to the exchange of an organization's information residing in its information management system such as files containing budgetary, personnel, operational, and other sorts of information. In this study, we have defined the term "information" broadly for several reasons. Since this is an exploratory study, we wanted to get a comprehensive view of all kinds of information exchanged, whether that information is considered mundane or crucial. We framed our questions to cover both manual and computerized information sharing. In later studies, we may be able to categorize different types of information. Currently, other than obvious functional categorization (e.g., budgetary versus non-budgetary), it would be difficult to make such distinctions until we have a good inventory of the types of information being exchanged. Such an inventory has not existed until now. Also, preliminary interviews with municipal officials suggested that the amount and frequency of information exchanged was limited, and attempts to define the term narrowly would risk missing the little that occurred.

Based on qualitative studies of information sharing cited above as well as theoretical literature in the information management field, we developed the following hypotheses for preliminary testing.

H1: *Information sharing among departments in cities will be infrequent.*

There are a number of task-oriented reasons why organizations would like to share information. For example, the police and fire departments share overlapping functions with respect to responding to emergencies and investigating arson. Police–fire sharing could help improve their performance on these important functions. However, based on our review of the literature cited above, as well as organizational and sociological studies (e.g., Gouldner 1959), we believe that there are stronger forces that encourage organizations (and individual departments within cities) to attempt to maintain autonomy over their information. Laudon's (1974, p.143) case studies of police and welfare agencies support this argument:

> Control over information functions serves organizations as an important boundary-maintaining device against potentially disruptive inquiries and it's a requisite for self-evaluation and planning. . . . Organizations resist any threat to impair . . . control over information.

Numerous other studies (e.g., Dery 1981; Malvey 1981) suggest organizations prefer to keep tight control over their information systems and databases. For example, release of files to outsiders could result in reanalyses of the data that could embarrass and/or reflect negatively on them. A good example of such an occurrence concerns the nation's hospitals, which are required to give the Department of Health and Human Services (DHHS) mortality and other data. DHHS then analyzes this data and releases reports that identify certain hospitals as having excess mortality rates according to DHHS analyses. This data can obviously cause harm to the reputation of the hospitals (Martinsons 1987).

The GIS literature also contains numerous examples of the desire to retain autonomy over information. Craig (1991) found reluctance to share information due to a number of reasons, including fear that sharing might lead to analysis of controversial racial data. Craig also points out that organizations place highest priority over their core tasks.[2] They often discard files and data that are not of central value to them and resist spending time on meeting the needs of others. Guthrie (1991) found resistance to a GIS system due to a total lack of interest and incentives to pool databases. Guthrie also cites the fear of misuse, sometimes justified, if data were made available to others. Van Buren (1991) found several reasons why towns in Vermont resisted GIS systems developed by the state. One fear was that the state would use the system to get at town tax information and compared it to allowing "your neighbor access to your bank account." Kishor and Nieman (1990) show that the desire to retain control over information is common to other countries, such as India, and is tied into territorial and survival imperatives. In summary, despite the presence of computer systems, we expect to find a limited amount of sharing.

> H2: *Sharing will take place most often among police and the manager/mayor and budgeting offices. Sharing with other departments, such as the fire department, will be less frequent.*

Due to the forces constraining inter-departmental sharing, we would expect that strong forces would be necessary to overcome this natural resistance to share data. Consequently, we would predict that the frequency of sharing will be higher with departments such as the mayor/manager, which exert hierarchical authority over police and can influence crucial factors such as budgets and staffing, than with other departments (e.g., fire) with which they share closely related and sometimes overlapping functions, because these latter departments exert little or no control over important contingencies for the police.[3] For example, Kling (1978) found that United Way was able to achieve exchange of client data among human service agencies involved in serving the same clients by offering financial incentives to the agencies involved; but Kling discovered that the sharing stopped when the funding incentive was removed. Also, we would expect the frequency of information exchange to be high with the Budget/Finance Department since accountability and reporting requirements mandate that the police share, at a minimum, budgetary data for reports to city councils. In short, our expectation is that most sharing will be due to hierarchical authority and accountability requirements as opposed to task-oriented sharing, such as the desire to perform better, solve common problems, and develop new programs.

> H3: *Computers lead to the increased exchange of information between police and other departments.*

> H4: *Departments with more sophisticated police computer systems will share more information than departments with computer systems with less capability.*

Both of these hypotheses reflect the viewpoint of the "technological imperative" (Markus and Robey 1988,) which assumes that greater technological sophistication of the departments will improve all aspects of information management including information sharing between organizations. The successes achieved in the private sector, such as sharing of information among airlines and banks, would support this point of view (Veith 1981; Vincent 1990; Synnot 1987).

H5: *Organizations using mainframes will be more likely to share information than those based on microcomputers.*

There are many (e.g., Lindberg 1979; Allen 1983; Chandler 1990) who argue that microcomputer technology tends to fragment and compartmentalize information systems and, thus, makes sharing more difficult. The balkanization of computer systems can result when end-users of stand-alone microcomputers control databases. If this is correct, we would expect greater sharing among organizations that have mainframe-based systems (or local area networks) than those using only stand-alone personal computers.

H6: *Police organizations that have greater autonomy over their information systems will be less likely to share information with other departments.*

As Kraemer et al. (1989) have shown, there are wide differences in the approaches of cities to information management, and these differences have important implications. In some cases, departments may be given autonomy over decisions about hardware and software while in other municipalities, these decisions are centralized. It would seem logical to expect more sharing in those cities where autonomy of individual departments like police is less.

METHODS

Measurement Issues

The above hypotheses are based on a combination of technological, organizational, and environmental factors that are likely to influence sharing.

Dependent variable(s). We asked questions concerning how frequently the police shared information (manually or electronically) with other city departments and whether any sharing was done electronically. We formed a summary measure of average frequency of sharing across departments. Finally, we asked for specific examples of the types of information shared. In addition to information sharing, we also gathered information on other variables, such as overall satisfaction with the police information system.

Computer hardware and software. We asked for the specifics on what types of hardware and software they were using. We grouped the hardware into analytical categories for the purposes of analysis (e.g., whether they had a mainframe or not, mainframes only, personal computers only, PC networks, and combinations of hardware). We also asked the police to rate the capability of their system with regard to a variety of factors, including flexibility, reporting capacity, mapping ability, and others. We formed a summary average index of this capability variable. We also used another "proxy variable" to measure capability: the size of their computer budget, the assumption being that the larger the budget, the greater the capability. Also, we treated one of our dependent variables, whether the departments engaged in electronic sharing or not, as an independent variable in order to determine if electronic sharing spurred greater overall frequency of information sharing.

Organizational factors. We asked the police who participated in the decision making concerning the acquisition of a police information system. Thus, we are able to compare departments in which police played an exclusive role versus those in which other departments (e.g., mayor/manager's office, fire, data processing, and other departments) participated. The assumption is that those departments in which outside departments played a role have less autonomy than those in which only police participated. Organizational size was also studied, based on the number of sworn officers in the department. As noted above, we tested the importance of hierarchy and accountability incentives to share indirectly by predicting that sharing with the administrative and budget offices would be higher than with departments (especially fire) with which they shared overlapping tasks and functions.

Environmental variables. Researchers have pointed out the importance of studying potential impacts of environmental variables on internal information systems (e.g., Bozeman and Bretschneider 1986; Huber and Daft 1987). Consequently, we studied the effect of crime rates, changes in crime rates, number of arsons, city population, geographical location, and total crime index on the dependent variables. City population and total crime indexes are highly inter-correlated with departmental size. However, the effect of the environmental variables on sharing is not clear. High or rising crime rates may spur investments in computer technology to control crime but not necessarily spur exchanges between police and other departments. However, we would expect a high number of arsons to spur exchanges between fire and police departments.

It should be noted that there are many other variables that are likely to influence sharing that were not studied in this research. Future research should address other variables that may influence information sharing, such as organizational culture, privacy laws, and individual factors, which may all play a role in facilitating or inhibiting information exchange. By focusing on a single department in a single state, we helped to hold fairly constant some of these potentially confounding variables (e.g., privacy laws and organizational culture).

Questionnaire and Qualitative Data

We selected local police organizations as the focal department for studying computer impact on sharing for a number of reasons. As Laudon (1986) and others have shown, police organizations have the demonstrated capability to share information electronically because they have been successful in sharing computerized information with other police departments. Also, police departments have traditionally been one of the earliest departments in cities to computerize (Gurwitt 1988). There are a number of reasons why police organizations would possibly need to share information with other departments, such as those outlined in Figure 12.1 below. On the other hand, while sharing of information between police and other departments may be desirable in many cases, it is usually not mandatory. (By way of contrast, for example, it may be required that departments would have to provide budgetary data to the Administration or Finance departments.) Thus, there is likely to be a substantial amount of potential variability among cities in regard to the amount of sharing done by police. Consequently, the amount of sharing by police is likely to be more sensitive to the impact of computers than sharing with the administrative or finance departments.

Our theoretical population of interest included all Illinois police departments in cities with a population of 10,000 or higher (based on 1980 census figures). We limited the study to these cities because there are more than 600 local police agencies in Illinois, and the smaller ones are much less likely to have computerized systems and a full set of other city departments with which to share. This yielded 178 departments, of which 149 responded to our survey (83.7 percent). Sixteen (16) of these responding departments had no computer at all other than a dumb terminal to enter data into the State Law Enforcement Agency Data System (LEADS). We compared the 149 responding and twenty-nine non-responding agencies and found no statistically significant differences on the following dimensions: (1) size of the department in terms of sworn police officers and number of civilians; (2) crime index, rates, and number of arsons; and (3) population size and geographical location of the police agencies.

The survey was endorsed by the Illinois Association of the Chiefs of Police, which helps to explain our excellent response rate.[4] This association, as well as selected local police officers, were asked to review the preliminary questionnaire, which was revised based on their comments. Several questions were asked to form scales of the department's computer system capability and the frequency of sharing with other departments. Measures of the internal consistency of the scales (Cronbach's alpha) were .85 and .93, respectively. On a few response items (e.g., number of sworn officers in the department), we were able to compare their survey responses with data recently provided by the state; they were very closely correlated. A number of phone calls were made to individual departments (about one-quarter of the respondents), primarily to clarify responses about hardware and software. These calls allowed us to check on the general accuracy of the surveys, which proved to be excellent. The surveys were mailed to the police chiefs, who filled it out themselves in 15 percent of the cases but gave it to someone else to fill out in other cases. We analyzed the data on all of the key dependent and independent variables but found no significant impact of the position of the person filling out the survey (e.g., chief versus other police, police versus civilian, and so on).

Prior to doing the above survey, we conducted qualitative research including site visits or semi-structured phone interviews with officials from more than twenty-five cities and villages. We have reported on the site visits in a previous work (Rocheleau 1988). We interviewed city managers, police, and other departmental staff concerning general issues related to information management. These site visits and phone interviews helped to develop the hypotheses we explore here and also assist in interpreting the results.

Note that we report tests of statistical significance below despite the fact that our survey represents a census of all departments (excluding, of course, non-respondents) in our sampling frame. We use these tests as an indicator of the strength and importance of the statistical associations. Our study is based on cross-sectional data. Consequently, our hypotheses need to be tested in the future using a longitudinal design.

RESULTS

The Nature of Information Sharing

It is instructive to study the substantive nature of information sharing done by the police departments in our study. Above we summarize the different kinds of information that police reported sharing with other

Department/Types of Sharing

Administration
reports on crime rates and other data, arrests, tickets, incidents, personnel, complaints, background checks, budgets, vehicle tags, abandoned vehicles, special projects, calls for service

Accounting/Finance
budgets, purchase orders, operational costs, invoices and other bookkeeping, payroll data, fines, liquor license background checks, credit union information, parking ticket data, attendance, false alarms, bills, damage to village property, overtime data, scheduling, utility information, resident information, animal registration

Engineering/Transportation
activity data, traffic accidents (historical record and by location), citations, information concerning transportation growth needs, fire calls, ambulance calls, spot speed survey data, planning data, occasional inquiries, vehicle checks, repair data, new equipment installation information, vehicle overweight permits

Fire
dispatching/incident information, resident data, inspection data, fire ambulance runs, calls for service, 911 case reports, specific investigations, resource listings, arson reports, operational information, fire data including distributions

Public Works
roadway activities, street maintenance data, lighting, violation locations, service requests, vehicle maintenance data, citizen problems, hazards, water break data, fleet maintenance data, building maintenance information, snow removal information, signs down, lights out, work order information

Building Investigation
investigations, permit violations, data about specific addresses, annexation and developments—information, crime statistics by zone, building ownership information, city property information, light maintenance

Social and Human Services
activity and arrest data, domestic incident data, community service information, referrals, youth information including adjudications, missing person data, employee program information, child abuse reports and inquiries, name checks, complaints and service requests, crime statistics, nursing home violation reports

FIGURE 12.1
Police information sharing with other departments

Department/Types of Sharing

Planning/Economic Development

crime and other activity information by location and by zone, roadway information, resident data, traffic accident data, community service information, annexation-development information

Recreation/Park

dispatch information, calls, incidents, and crimes by location, officer assignment information, enforcement activity data, special event information, injury data, problems in park building information, damage to park information, gangs and graffiti information, park inspection reports

Clerk's Office

registration for park information, business license inquiries, vehicle licenses and checks on licenses, liquor license reports, amusement machine requests, city stickers, personnel problem information, water bill data, arrest and ticket data, budget data, court docket information, parade routes

FIGURE 12.1 (CONTINUED)
Police information sharing with other departments

departments (see Figure 12.1). Note that the list covers all sharing done, whether manual or computerized.

This list represents a summary of the different types of sharing reported by police with other departments. No individual police department did all of the types of sharing shown in Figure 12.1. Indeed, as we will see below, many departments did little or no sharing. The list shows that police departments have substantial possibilities to exchange information horizontally with other municipal departments, and much of it could be shared with the aid of computers. It demonstrates that police department sharing is not necessarily limited to fire, budget, and administrative departments but can occur with all the departments in the municipality including recreation, human services, building and others. This list is useful as a starting point in classifying different types of information exchanged. If it were possible, it would be good to attempt classification of the types of information such as between routine, operational, and transactional data elements versus decision-support type of information. However, such classification will not be easy because these distinctions depend on the perception of the users of the information. For example, data on the number of accidents at an intersection could be viewed as routine information by police but could be the key component of a Transportation Department decision support model on the need for road improvements.

The Frequency of Sharing

In our survey, we asked how frequently sharing, manual or computerized, occurred with other departments in the city. The data for the frequency of sharing is summarized in Table 12.1 below.

Table 12.1 shows that, overall, about 95 percent of the police departments report some form of information sharing with other city departments. Since we expected that most departments would, at a minimum, need to share budgetary data with the finance department, we break out the data into two additional categories: (1) police sharing with the city manager and/or budgeting/finance office; and (2) police sharing with other city departments. About 92 percent reported sharing with the manager/budgeting office versus 79 percent with other departments. The overall average rate of sharing for all departments was 2.8 (2 = annual, 3 = monthly) which shows that, overall, interdepartmental sharing is infrequent in a majority of cities. However, there is substantial variability among cities. More than 25 percent of the cities report an overall sharing rate of a weekly or daily basis. Seven departments (5.6 percent of the respondents) report sharing daily with all of the other city departments. But more than one-third report a sharing rate of less than once per year.

TABLE 12.1

Occurrence of Sharing Between Police and Other Departments

Nature of Information Sharing	*Percent*
Sharing with any other department	95.0
Sharing with administration/budget departments	91.7
Sharing with departments other than administration/budget	79.2
n = 120 (valid cases)	

Overall Frequency of Sharing Between Police and Other Departments in Municipality

Mean = 2.8
Standard Deviation = 1.2
(1 = not at all; 2 = annual; 3 = monthly; 4 = weekly; 5 = daily)
Percent < 2 (2 = annual sharing): 34.2%
Percent > 4 (4 = weekly sharing): 25.8%
n = 120 (valid cases)

TABLE 12.2

Electronic Sharing Between Police and Other Departments

	Percent
Did any electronic sharing take place?	
No electronic sharing	81.3
Electronic sharing	18.7
n = 125 (valid cases)	
What types of electronic sharing occurred?	
Tape	5.0
Disk swapping	15.0
Network	10.0
Via mainframe	55.0
Disk and network	5.0
Other	10.0
n = 20 (valid cases)	

The electronic sharing of information is the exception rather than the rule. In Table 12.2, we show that only 18 percent of the departments with computer systems report doing any kind of electronic sharing with other departments. Table 12.2 shows that, of the electronic sharing reported, about half occurred via a mainframe computer while the remaining occurred via computer tapes, disks, and networks.

Table 12.3 shows the frequency of sharing done by the police with each department. The two highest average ratings are for the administration (mayor/city manager) and accounting/finance departments. This fact, combined with the overall low rate of sharing revealed in Table 12.1, supports the argument that hierarchy and accountability requirements are the paramount reasons for sharing. It is harder to ignore information requests from the city manager or mayor who has hierarchical authority over police. Discussions with police and managers suggested that much of this sharing is for the purposes of presentations to city councils, not for the purpose of the manager or mayor intervening in police operations. Likewise, it is necessary for most departments to share budgetary data with the budgeting/finance departments. The high ranking of sharing with the clerk's department, according to our interviews, results from the necessity of many departments to provide them with information concerning parking tickets and other legal matters. Overall, the average frequency of sharing is less than monthly, except with the administrative department.

TABLE 12.3

Frequency of Sharing by Police with Other Departments

	Department	*Mean Value*	*% Daily*	*% None*	*Number of Cases*	*Standard Deviation*
V1	Administrative/ Managerial	3.52	30.8	11.7	120	1.32
V2	Budget	2.96	19.8	33.7	101	1.58
V3	Fire	2.90	30.8	39.3	107	1.73
V4	Clerk	2.81	27.1	43.8	96	1.74
V5	Public Works	2.69	25.5	45.9	98	1.72
V6	Human Services	2.32	15.4	51.3	78	1.57
V7	Building	2.17	11.8	56.5	85	1.49
V8	Transportation	2.15	5.4	50.5	93	1.33
V9	Parks	1.86	8.6	66.7	81	1.37
V10	Economic Development/ Planning	1.71	2.4	62.7	83	1.05

(1 = not at all; 2 = annual; 3 = monthly; 4 = weekly; 5 = daily)

WILCOXON MATCHED-PAIRS SIGNED-RANKS TEST

Variable Pair			*2-Tailed Probability*	*Cases*
V1	vs.	V2	.001	100
V2	vs.	V3	.743	90
V3	vs.	V4	.669	86
V4	vs.	V5	.386	84
V5	vs.	V6	.808	72
V6	vs.	V7	.103	64
V7	vs.	V8	.738	74
V8	vs.	V9	.479	68
V9	vs.	V10	.314	69

Table 12.3 also shows what percent shared information daily versus those who did not share at all. While many police departments do not share information at all with fire or other departments, those that do share with them may share frequently. Nearly 31 percent of the police report daily sharing of information with both the manager and fire departments. Daily sharing of information is also fairly common with the public works (25.5 percent) and city clerk (27.1 percent) offices, whereas less than 20 percent (19.8 percent) report daily sharing with the budget offices. Accountability types of sharing, such as with budget offices, may require only periodic (e.g., monthly or quarterly) rather than daily exchanges.

In some cases, the police and fire department actually shared their main computer systems. In several other cases, the police and fire departments at least shared a computer-aided dispatch system. Still, it is notable that 39 percent of the police departments report no information exchange at all with the fire department, and the average frequency is less than once a month. This is surprising in view of their overlapping functions (e.g., arson and responses to emergencies). We checked on this further by studying the relationship between number of arsons and average amount of sharing. Perhaps sharing takes place only when there are a significant number of arsons and a threshold is passed, making exchange more likely. Further analysis demonstrated that there was no relationship between the two variables.[5] Informal discussions with police and fire also reveal that, despite their overlapping functions, police and fire departments often remain quite remote from each other, if not in competition. This remoteness seems to be reflected in the data.

Impacts of Computer Technology and Organizational and Environmental Variables

Table 12.4 shows the results of a stepwise multiple regression analysis with the overall average frequency of sharing as the dependent variable.[6] The standardized (beta) coefficients are used to facilitate comparison of the relative importance of the independent variables on sharing. Only one variable, the use of electronic sharing, reached the level of statistical significance. Organizations that use electronic sharing tend to share somewhat more frequently than other police departments. This finding supports the argument for the technological imperative that computers facilitate information exchange. However, the police rating of their system's capability had no relationship to sharing. Nor did the configuration of the hardware (i.e. stand-alone PCs versus mainframes only versus combinations of hardware).

The positive findings concerning the impacts of electronic sharing are moderated by the modesty of their impact: the adjusted R-Square for the independent variables was only .04. The limited impact of electronic sharing is not surprising in view of the fact that only 18 percent of the departments used electronic sharing as noted earlier in Table 12.2. Thus, although computers may facilitate sharing where departments choose to use them, clearly they have not been a very powerful force for exchange up until now. The technological imperative may work quite well for sharing between police departments. For example, police told us that the establishment of mobile terminals in police cars had enabled them to conduct far more computerized investigations than previously by facilitating

TABLE 12.4

Regression Analysis with Average Sharing Frequency as Dependent Variable

Variables	*Beta*	*T Value*	*Level of Significance*	*Valid Cases*
Variable Entered into the Equation				
Use electronic sharing	.217	2.36	.02	113
Variables Not Entered into the Equation				
Presence of mainframe	.132	1.40	.16	113
Mainframe only	- .13	-1.373	.17	113
Capability of police system	- .02	- .204	.84	113
Police computer budget	.148	1.410	.162	92
Stand-alone PCs only	.067	.707	.481	113
Autonomy	.177	1.90	.063	113
Size	.083	.88	.38	113
Crime rate	- .01	- .107	.915	113
% Change in crime rate	- .012	-.131	.90	113
Total crime index	-.009	-.107	.915	113

use of the state LEADS system. But if technology does assist inter-police sharing, it appears that the technological imperative is less effective in breaking down department boundaries within cities. This interpretation follows Bozeman and Bretschneider's (1986) argument that public information linkages tend to be more external than internal.

Another possibility is that, although departments do not directly use computers for electronic exchanges, computers may assist information exchange indirectly by allowing them, for example, to easily retrieve and reorganize the data in the form that the requesting agency needs.[7] However, our findings did not find any relationship between the capability of the system (including its flexibility, reporting, and mapping capacity) and the frequency of sharing.[8] It is likely that organizational barriers are the reason for the lack of relationship. For example, police agencies providing information may fear that too much responsiveness to demands by requesting agencies would lead to a spiraling of requests, taking them away from police-oriented activity.

There is no confirmation that mainframes encourage sharing or that stand-alone PCs discourage sharing. Autonomy, whether or not other city departments were involved in selecting the police computer system, also fails to show a statistically significant (p = .06) impact, but the results warrant an attempt at more direct measure (e.g., survey questions on the degree of autonomy individual departments exert over their information systems). Neither organizational size nor environmental variables has any impact on sharing. In the future, it will be important to study the reasons for sharing directly. Overall, the low power of the model suggests that it may exclude other crucial explanatory variables (e.g., the nature of the informal organization and organizational climate.)

The partial correlation coefficient (controlling for organizational size) between the average frequency of sharing and police satisfaction with their computer system was -.02 (p = .42). The lack of a correlation is not surprising in view of the fact that satisfaction with information systems is likely based on how happy they are with its performance of crucial departmental functions, not sharing with other departments. This shows that sharing is independent of departmentally oriented outcome measures (e.g., measures of user satisfaction and other departmentally defined outcomes). Since interdepartmental sharing can contribute to the integration and productivity of the city operations as a whole, it points out the importance of the role of generalist managers (e.g., city managers) in information management. They need to take a proactive stance in encouraging information sharing because it is not likely to be a priority of departments. It is an activity that is not likely to flourish without strong incentives.

CONCLUSIONS

Our study of information sharing has both practical and theoretical implications. Overall, we found sharing to be relatively infrequent. We found that computers have had a small but significant impact on the extent of information sharing between police and other departments in cities. Although interdepartmental sharing of information is facilitated by computers, most departments still do not exchange information electronically despite the presence of computers. Hierarchical authority and accountability requirements appear to be the strongest incentives to share information in cities, with mayor/manager and finance/budget offices being on the receiving end of the information. Findings did not show any major impact of organizational size or autonomy on information sharing. However, the latter finding should be further studied with more direct measures of organizational autonomy.

Individual factors and other organizational factors may also be important and need to be studied. For example, Schermerhorn (1976) found that openness to inter-organizational cooperation was most likely when administrators considered their organizations to be high performers. Informal relationships, for example whether the police and other department personnel meet frequently on a friendly basis, may help to encourage interchange. Studying sharing by all departments, not just the police, is clearly desirable. Police and their information systems have unique aspects, and we need data on other departments to see if our results hold for others.

It would appear that there are substantial possibilities for productivity improvement if police and other departments were to engage in more electronic sharing. Many new functions of government (e.g., environmental and economic development planning) require information drawn from many municipal departments. But our research confirms the fact that the information flow of most cities is vertical, thus inhibiting the performance of these new functions.

Many cities and other local governments are now in the process of investing huge amounts of money in GIS technology. However, to make full use of this new technology, there would have to be extensive cross-departmental sharing of information as Toregas (1991) has noted. Our research suggests that achieving the necessary integration to use GIS systems across departments would require a major change in the nature and structure of most city information systems. Somers (1989) has suggested a number of ways of spurring possible lateral links including liaison individuals, inter-departmental committees, task teams, and coordinators. Although these mechanisms should help, our research suggests that more powerful incentives such as hierarchy, funding, and accountability requirements may be necessary. Unlike private agencies, city departments are not faced with competition as a motivating force to share information. Craig (1991) argues that bureaus change only "when hit hard" and that the best hitters are elected officials. A state law such as the one passed by Wisconsin that requires departments to prepare plans for data sharing would give generalist managers a strong tool to develop sharing (Brown 1991). A growth management act that requires governments to develop public facilities concurrently with new development may put pressure on cities to develop shared information (Hoover and Goff 1990). The need to develop data to support the imposition of impact fees on developers can also serve as an incentive for governments to force interdepartmental sharing (Sternagel and Enos 1990).

Although information exchange may be forced through laws, it is important that departments accept the legitimacy of the exchange or the quality of data provided may be harmed. Organizational communication research shows that exchanges are less likely to flourish when they are asymmetrical in nature: departments feel like they are giving and never receiving (Thorn and Connolly 1987). It is important that the department providing the information receives some reciprocal benefit. The benefit could be receiving useful information in return from other departments. Freund (1990, 223) has made this same point concerning sharing in highway infrastructure systems:[9]

> One of the most common reasons that information management systems fail to meet users' expectations is that the systems are designed to accommodate *only one direction of information flow* [emphasis added].... If the working level cannot realize the use of their data to impact their own operations, they have no incentive to ensure that it is either accurate or timely.

The same is likely to be true for sharing between departments. If it is not possible to reciprocate, it would then be important for the receiving agencies to provide feedback to the providing department to show that the information they provided has some useful result. In our research, we did not distinguish between providing and receiving information. Our interviews and the types of sharing cited in Figure 12.1 suggest that police are primarily providers, not receivers, of information to non-police departments. In future research, it will be important to distinguish between the two to see if exchanges are two-way or primarily asymmetrical in nature.[10]

Another interpretation of the lack of sharing of information in cities is that police (and most other departments) have a very limited, reactive view of the role of information systems. A study of a computerized crime mapping system in Chicago (Maltz et al. 1991) found most officers, young and old, skeptical about the system, and they rarely used it for the exploratory analysis of crimes. The officers placed high value on case narratives that were not included in the system. They also preferred their own memory as a way of identifying crime patterns (Maltz et al. 1991). In another paper, this author (Rocheleau 1991) found that police view their computer systems primarily as vehicles for responding to requests for reports by outsiders. Much of the GIS literature assumes that public managers view information systems as decision support systems and that shared data will be employed to influence operational, management, and policy issues. Although some police officers and departments do use systems this way, it does not appear to be the modal pattern.

As a consequence of the development of PCs and local area networks, we found that the structure of the information systems of even small city departments could be very complex. The amount of inter-departmental information sharing and the ratio of horizontal to vertical information sharing may be a good indicator of the structure of the information system and degree of integration of the city's overall information system. One of the potentially most important impacts of GIS may be to spur more interdepartmental exchanges. In short, we believe that information sharing represents an important new focus for research on public information systems.

Notes

1. We asked questions concerning software and hardware, and none of the responding departments mentioned having operative GIS systems. Several of the cities and villages are just in the process of beginning to build GIS systems. A few police departments do have a mapping capability for crime, and this was one of the variables included in our index of the system capacity. Studies such as ours will help to provide a baseline to compare the rate of sharing after GIS systems have been implemented.

2. Organizational theorists such as James D. Thompson (1967) have held that organizations act to protect their core technology.

3. Note that we would also predict that, when departments do share information, they are more likely to give other departments specific, discrete data and reports than to let the requesting departments have access to their databases. However, we did not test this hypothesis in this study.

4. A copy of the complete survey is available upon request.

5. Stepwise regression (using frequency of sharing with fire department as the dependent variable) revealed no relationship between arsons and police–fire sharing.

6. Note that we tested for autocorrelation (Durbin-Watson test) and multicollinearity and found no problems with either.

7. Indeed, some would classify sharing in which computers are used to compile the responses (e.g., through word processing, database, or spreadsheet programs) as constituting "computerized exchange" even if the actual exchange is done manually. We think that there is a major difference between exchanges that are implemented via computer media and those that are not in terms of speed and productivity.

8. Note that we tested in stepwise regression analyses each of the capability variables including flexibility, reporting capacity, and sharing with neighboring and state police departments individually, as well as using the aggregate measure of system capability.

9. Although Freund (1990) makes the point concerning vertical integration, it would seem to hold, if anything, more strongly with respect to horizontal integration where the worker is not subject to the direct authority of the people being provided the data.

10. The only GIS article (Bennett 1990) to cite any quantitative data sharing found that the planning and assessment departments were the chief providers of data in regard to land systems.

REFERENCES

Agency for Health Care Policy and Research. 1991. *The feasibility of linking research-related data bases to federal and non-federal medical administrative data bases.* Washington, D.C.: U.S. Department of Health and Human Services, AHCPR Pub. No. 91-0003.

Allen, B. 1983. An unmanaged computer can stop you dead. In L. Salerno, ed., *Catching up with the computer revolution.* New York: John Wiley and Sons. 104–121.

Bass, B. 1989. A way to share data "horizontally, not vertically." *Government Computer News,* November 27: 64.

Bennett, P. G. L. 1990. Identifying and integrating business functions to develop a GIS needs assessment. Urban and Regional Information Systems Association 1990 Conference, Edmonton, Alberta, August 12-16, *Proceedings* 3: 1–13.

Bozeman, B., and S. Bretschneider. 1986. Public management information systems: theory and prescription. *Public Administration Review* 45 (November, special issue): 475–487.

Brown, K. 1991. State legislatures promote GIS use. *URISA Journal* 3, 1: 61–66.

Chandler, D. 1990. Interview with Forest Baskett. *Unix Review* 8, 1: 42–49.

Craig, W. J. 1990. URISA '89 advances information systems research agenda. *URISA Journal* 2, 1: 79–82.

______. 1991. Empty promises for small area data: monitoring our neighborhoods using operational records. Urban and Regional Information Systems Association 1991 Conference, San Francisco, California, August 11-15, *Proceedings* 4: 220–229.

Crosswell, P. L. 1989. Facing reality in GIS implementation: lessons learned and obstacles to be overcome. Urban and Regional Information Systems Association 1989 Conference, Boston, Massachusetts, August 6-10, *Proceedings* 4: 85–92.

Dery, D. 1981. *Computers in welfare: the mismatch.* Beverly Hills: Sage Publications.

Eichelberger, P. 1986. Land/structure/occupancy database design: handling an increasingly complex urban reality. Urban and Regional Information Systems Association 1986 Conference, Denver, Colorado, August 10-14, *Proceedings* 2: 1–12.

Freund, D. M. 1990. Vertical integration of highway infrastructure management systems: possibility or pipe dream. Urban and Regional Information Systems Association 1990 Conference, Edmonton, Alberta, Canada, August 12-16, *Proceedings* 1: 220-230.

General Accounting Office. 1991. *War on drugs: information management poses formidable challenges.* Washington, D.C.: General Accounting Office, IMTEC-91-40.

Gouldner, A. 1959. Reciprocity and autonomy in functional theory. In L. Gross, ed., *Symposium on sociological theory*. New York: Harper and Row, 241-270.

Gurwitt, R. 1988. The computer revolution: microchipping away at the limits of government. *Governing* 1 (May): 35–42.

Guthrie, J. 1991. Linking disparate data sources in a large organization's GIS. Urban and Regional Information Systems Association 1991 Conference, San Francisco, California, August 11-15, *Proceedings* 3: 39–51.

Hoover, M., and T. Goff. 1990. Concurrency management and the Orlando/Orange County GIS project. Urban and Regional Information Systems Association 1990 Conference, Edmonton, Alberta, Canada, August 12-16, *Proceedings* 4: 44–56.

Huber, G. P., and R. L. Daft. 1987. The information environments of organizations. In F. W. Jablin, L. L. Putnam, K. H. Roberts, and L. W. Porter, eds., *Handbook of organizational communication*. Newbury Park, CA: Sage Publications, 130-164.

Kishor, P., and B. J. Niemann, Jr. 1990. Geographic Information Systems in India: a "look before you leap caveat" to technology transfer. Urban and Regional Information Systems Association 1990 Conference, Edmonton, Alberta, Canada, August 12-16, *Proceedings* 1: 55–66.

Kling, R. 1978. Automated welfare client-tracking and service integration: the political economy of computing. *Communications of the ACM* 21, 6: 484–493.

Kraemer, K. L., and J. L. King. 1979. A requiem for USAC. *Policy Analysis* 5, 3: 319–349.

Kraemer, K. L.; J. L. King; D. E. Dunkle; and J. P. Lane. 1989. *Managing information systems*. San Francisco: Jossey-Bass, Inc.

Kraemer, K. L.; W. H. Mitchell; M. E. Weiner; and O. E. Dial. 1974. *Integrated municipal information systems*. New York: Praeger Publishers.

Laudon, K. C. 1974. *Computers and bureaucratic reform*. New York: John Wiley and Sons.

______. 1986. *Dossier society*. New York: Columbia University Press.

Lindberg, D.A. B. 1979. *The growth of medical information systems in the United States*. Lexington, MA: Lexington Books.

Maltz, M. D.; A. C. Gordon; and W. Friedman. 1991. *Mapping crime in its community setting*. New York: Springer-Verlag.

Malvey, M. 1981. *Simple systems, complex environments: hospital financial information systems*. Beverly Hills, CA: Sage Publications.

Markus, M. L., and D. Robey. 1988. Information technology and organizational change: causal structure in theory and research. *Management Science* 34, 5: 583–598.

Martinsons, J. N. 1987. Services put mortality data in perspective. *Hospitals* 61 (October 20): 78.

Nolan, T., and T. Nyerges. 1991. Land information systems network in the Puget Sound region. *URISA Journal* 3, 2: 25–34.

Rocheleau, B. 1988. The organizational context of the new information technology. *Public Productivity Review* 12, 2: 165–178.

______. 1991. Evaluating public sector information systems: satisfaction versus impact. Paper delivered at the 1991 Meeting of the American Association on Evaluation, Chicago, October 31.

Rogers, B. 1991. Agriculture rethinks how its information flows. *Government Computer News* 11, 12: 69.

Schermerhorn, J. R., Jr. 1976. Openness to interorganizational cooperation: A study of hospital administrators. *Academy of Management Journal* 19, 2: 225–236.

Somers, R. 1989. Organizational change for successful GIS implementation. Urban and Regional Information Systems Association 1989 Conference, Boston, Massachusetts, August 6-10, *Proceedings* 2: 39–51.

Sternagel, J. A., and R. D. Endos. 1990. Impact fees: designing and implementing a program to fund the impacts of growth. Urban and Regional Information Systems Association 1990 Conference, Edmonton, Alberta, Canada, August 12-16, *Proceedings* 4: 29–43.

Synott, W. R. 1987. *The information weapon*. New York: John Wiley and Sons.

Thompson, J. D. 1967. *Organizations in action*. New York: McGraw-Hill.

Thorn, B. K., and T. Connolly. 1987. Discretionary databases: a theory and some experimental findings. *Communications Research* 14, 5: 512–528.

Toregas, C. 1991. Perspective: giving something back to the world. *Geo Info Systems* 1, 8: 16–23.

Van Buren, T. S. 1991. Rural town regional information systems: issues in integration. Urban and Regional Information Systems Association 1991 Conference, San Francisco, California, August 11–15, *Proceedings* 3: 136-151.

Veith, R. H. 1981. *Multinational computer nets*. Lexington, MA: D.C. Heath.

Vincent, D. R. 1990. *The information weapon*. Homewood, IL: Dow Jones-Irwin.

IAN MASSER AND HEATHER CAMPBELL

13 *Information Sharing: The Effects of GIS on British Local Government*

The arguments for information sharing are discussed in the context of British local government. The case for more corporate approaches that maximize the benefits to be obtained from information sharing is evaluated and an alternative case that highlights the operational advantages of departmental approaches is considered. The advantages and disadvantages of these two approaches are examined both in overall terms and in relation to the preliminary findings of a comprehensive survey of Geographic Information System (GIS) implementation. The findings of this survey show that there is a fifty–fifty split between corporate and departmental approaches in current British practice.

INTRODUCTION

It is often argued that the greatest benefits from GIS implementation come as a result of the pressures that are placed on organizations to share information. The findings of the joint Nordic Project, for example, suggest that the returns on investment in GIS range from 1:1 where the automation of basic mapping functions is involved, to 2:1 where GIS is applied to planning activities, and are as high as 4:1 where they involve the collective use of databases common to all users (Tveitdal and Hesjedal 1989). It can also be argued that these benefits will be accompanied by considerable cost savings, particularly where a lot of time is spent on inter-agency map exchanges. For example, a study of six agencies in Santa Clara County estimates that overall savings of between 42 and 54 percent of current costs could be obtained by the development of decentralized and centralized shared AM/GIS (automated mapping/Geographic Information Systems) facilities respectively (Finkle and Lockfield 1990).

Nevertheless, the theoretical assumptions behind arguments such as these can be questioned from both economic and political standpoints. In economic terms, they assume that geographic information can be treated as a public good (i.e., something accessible to everybody, which can be shared without loss to any individual). Yet in many respects geographic information must be regarded as a private good, as a commodity that is traded at a price with access being determined by ability to pay (Bates 1988, Openshaw and Goddard 1987). It is important also to note that the commodification of geographic information is not restricted to the private sector given the increasing pressures on public-sector agencies to recover their costs (for example, Hepworth 1990; Maffini 1990).

Similarly, these arguments do not take account of the political economy of geographic information, that is, the extent to which geographic information is used to maintain the existing balance of power within late capitalist society (see, for example, Mosco and Wasko 1988). In this perspective geographic information must be regarded as a tool to be used in turf battles between different interests. Moreover, differences in access to information are likely to result in a growing divide between the information-rich and the information-poor within cities and regions while creating circumstances that lead to the emergence of surveillant societies (Pickles 1991).

Even where the public good dimension of geographic information is paramount and there are no conflicts between the agencies involved, the success of information sharing is likely to depend on a wide range of organizational and institutional factors. There have been a number of attempts in recent years to set down maxims or conditions that promote effective implementation (see, for example, Croswell 1991; Levinsohn 1990; Masser and Campbell 1991; Wellar 1988). However, these are largely directed toward single organizations and must be modified in the context of information sharing.

Some of the key organizational issues involved are highlighted in the reviews by Croswell (1991) and Masser and Campbell (1991, 1992). Croswell's content analysis of thirty-nine articles describing users' experiences demonstrates that "obstacles related to the technical side of system implementation operation are considered minor compared to organizational and institutional problems" (Croswell 1991, 46). To increase the chances of success in system implementation, Croswell suggests ten organization-specific maxims:

- carry out an initial evaluation of organizational risk;
- get commitment from management;
- assign a GIS manager early in the project;
- adopt a structured approach to system development;
- involve users in system design;
- formulate a goal-oriented plan and schedule;
- develop a project organization that encourages cooperation and consensus;
- allocate sufficient staff time;
- keep users, managers, and constituents informed; and
- provide education and training at all implementation stages.

Drawing on the work of the Irvine group (see, for example, Danziger et al. 1982; King and Kraemer 1985) on the use of computers in local government, Masser and Campbell (1992) contend that three necessary and generally sufficient conditions must be met for the effective utilization of GIS in organizations:

1. the existence of an overall information management strategy based on the needs of users in the agency and the resources at its disposal;
2. the personal commitment of individuals at all levels in the organization with respect to overall leadership, general awareness and technical capabilities; and
3. organizational stability with respect to personnel, administrative structures and environmental considerations.

Each of these maxims or conditions must be extended when information sharing between organizations is involved. In the case of Croswell's maxims, for example, it will be necessary not only to carry out an initial evaluation of organizational risk but also to demonstrate that the risks are equitably distributed between the organizations involved. Similarly, it will be necessary not only to get the commitment of the management of all organizations involved but also to maintain a similar degree of commitment from all the organizations throughout the implementation process.

The effects of moving from a single organization to a multi-organization environment is particularly marked with respect to Masser and Campbell's three conditions. Effective utilization will depend, first, on the existence of an overall information management strategy in each of the organizations concerned that takes account of the needs of each set of users and the resources at the disposal of the respective organizations. This assumes a high degree of compatibility between the priorities that are given to these operations where sharing is required and the resources that are set aside by each organization for these purposes. The second condition presupposes a high degree of personal commitment at all levels in all the organizations. For example, given the importance attached in the literature to the role of particular individuals or champions in ensuring effective implementation (see, for example, Beath 1991), it will be necessary to think in terms of supra-organizational champions in the context of information sharing. However, the biggest obstacle to effective implementation is likely to be with respect to the third condition, regarding the need for organizational and environmental stability where information sharing is concerned, given that the failure of even one organization in this respect may threaten the success of the whole operation.

It may be argued that the above discussion is unduly negative in tone and that information sharing in most cases involves largely technical operations that are affected to only a very limited extent by economic, political, or organizational considerations. It may also be argued that in many cases the benefits from information sharing are sufficiently great to justify special efforts to satisfy the conditions set out above. With these considerations in mind, this chapter examines the effect of GIS on British local government with particular reference to information sharing. It examines a case where the circumstances are especially favorable for information sharing (i.e., the local authority environment where information sharing takes place between different departments within the same organization rather than between different organizations). Under these circumstances, given the common corporate objectives of the organizations to which the departments belong and the number of common factors associated with organizational stability, it may be argued that the adoption of GIS should be associated with corporate information management strategies that maximize the benefits to be obtained from information sharing.

This chapter first considers the thesis that is implicit in such a standpoint, then examines the antithesis that draws attention to the strengths of a departmental standpoint. The next section of the chapter presents some of the findings from a comprehensive survey of the impact of GIS on local government in Great Britain, which are discussed in relation to the assumptions of both the thesis and the antithesis. Finally, a number of general conclusions are drawn from the evidence with respect to information sharing as a whole, and a number of suggestions are made regarding the further research that is required.

THESIS AND ANTITHESIS

Thesis

The basic thesis underlying the case for information sharing in local government revolves around the need to regard information as a capital asset. To exploit this asset an integrated approach is required that maximizes the potential for resource sharing among users. This, in turn, creates economies of scope in the form of multiple applications.

The case for a corporate approach to information resource management in British local government can be developed in a variety of ways. For example, Hepworth (1990) calls for greater efforts by local authorities to mobilize their information capital to improve the territorial management public services. Similarly, Gault and Peutherer argue that "to manage change in the future, local authorities need to develop and make better use of the information sources at their disposal" (Gault and Peutherer 1989, 2). They also argue that GIS represents a useful tool for strategic information management that adds value by integrating diverse data sets. In their view, "a corporate view is essential if the synergy is to be realized which will assist both the management of change and increase operational efficiency" (Gault and Peuther 1989, 13). The Audit Inspectorate also note the importance of data integration in their advice to local authorities on preparing an IT strategy (Audit 1990). They view with concern the disorganized and uncoordinated approach to acquisition that exists in many local authorities as a result of a preoccupation with short-term considerations. "Users have invariably been able to demonstrate the short-term benefits of providing their own solutions and have only later come to appreciate such disbenefits as the inability to share corporate data and the consequence of need for expensive duplication" (Audit 1990, 4).

The advantages of a corporate approach are highlighted in many empirical studies. For example, Coulson and Bromley's detailed assessment of user needs in Swansea district demonstrates "the impracticability of introducing an isolated departmental GIS" (Coulson and Bromley 1990, 215). In their view, four types of benefits would accrue to the local authority as a result of adopting a corporate approach.

1. *Maps:* "Access to up-to-date urban mapping is a bonus to any organization for reasons of completeness and common usage with other members of the organization."

2. *Data:* "The main benefits of GIS are due to better access to data . . . the corporate approach encourages a steady reduction of the artificial barriers that have grown up between departments over the years."

3. *Operations:* "The combination of better mapping and data improvements brought about by a GIS should result in better operations within the local authority, in particular better resource targeting and better forward planning."

4. *General benefits:* "The most significant is the idea of great coordination, but only if a corporate GIS is adopted. While the management of corporate GIS is clearly more difficult than a departmental GIS, the benefits far outweigh the disadvantages" (Coulson and Bromley 1990, 216).

The Antithesis

It is also necessary to consider the antithesis to these claims. The case embodied in the antithesis revolves largely around matters relating to control and continuity. It can be argued that those most directly concerned with the provision of specific local authority services, that is the individual line departments, need to retain control wherever possible over the technical facilities and the information that they need to carry out their tasks. This presents local government with a fundamental dilemma: "maintaining sufficient centralization of control in city government that enables oversight, efficiency and accountability, while providing sufficient decentralization to deal adequately with the problems of size and complexity and the demands of neighborhoods and the special interest groups" (Kraemer and King 1988, 28).

It can also be argued that the importance of continuity in terms of administrative routines and practices is often underestimated in practice. British local government is highly departmentalized, and many departments have their own professional cultures that are reflected in the distinctive character of their administrative practices. Under such circumstances, movements between departments within local authorities are relatively rare, and it can be argued that the most successful technological innovations are likely to be those that reinforce existing departmental practices.

Evaluation

The arguments underlying the thesis and antithesis can be expressed in terms of pressures towards more corporate approaches and more departmental approaches, respectively. The main advantages and disadvantages of these two approaches are summarized in Figure 13.1, which is based on Campbell and Masser (1991). This shows that the respective advantages of corporate approaches in terms of information sharing, more informed decision making and increased efficiency, which were incorporated in the

CORPORATE APPROACH

Advantages

Integration of data sets

Increased data sharing

Improved access to information

More informed decision making

Increased efficiency due to reduced duplication leading to time, staff, and cost savings

Disadvantages

Variations in priorities between departments

Differences between departments in their abilities to exploit GIS facilities

Differences in the level of awareness and spatial data-handling skills between departments

Inter-departmental disagreement over:
- access to information
- leadership
- data standards
- equipment
- training

DEPARTMENTAL APPROACH

Advantages

Independence

Control over:
- priorities
- the form and accessibility of information
- technical specialists
- equipment

Clear lines of responsibility

Continuity

Disadvantages

Departmental isolation

Lack of support in terms of:
- finance
- technical specialists
- training

Absence of authority-wide benefits, including lack of system and data compatibility

FIGURE 13.1
The main advantages and disadvantages of corporate and departmental approaches to implementing GIS

thesis, are often offset by corresponding disadvantages in practice. For example, against the benefits to be derived from data integration, there are the potential costs associated with variations in priorities between departments. Similarly, the advantages of improved access to information may be offset by differences between departments in terms of their abilities to exploit the potential offered by GIS.

Furthermore, the benefits to be derived from corporate approaches in terms of more informed decision making may be outweighed by differences between departments in terms of the level of awareness and the spatial and data-handling skills at their disposal. In addition, the potential savings to be derived from increased efficiency in terms of the reduction of duplicated efforts may be offset by inter-departmental disagreements with respect to access to information, professional leadership, data standards, equipment and training needs.

On the other hand, the advantages of a departmental approach in terms of independence, control over resources, clear lines of responsibility, and continuity that underlie the antithesis may be offset in practice by the corresponding disadvantages. For example, the benefits to be derived from independence may be outweighed by the costs incurred as a result of the department's isolation within the authority. Similarly, the advantages of retaining control over priorities, information content, and information processing facilities may be offset by the lack of support in terms of finance and technical expertise. In addition, the benefits to be derived from clear lines of responsibility may not outweigh the lack of system and data compatibility that would enable authority-wide benefits to be derived from departmental activities. Under such circumstances, the advantages to be gained from continuity may be offset by inertia, which seriously impedes the realization of the potential offered by the new technology.

The most important conclusion to be drawn from Figure 13.1 is that both approaches have their advantages and disadvantages. The extent to which the advantages overcome the disadvantages is likely to vary considerably in practice according to the organizational context within which GIS is being introduced. The extent of these variations is explored in the next section with respect to preliminary findings of the study of GIS on British local government that is being carried out by the authors (Campbell and Masser 1991).

THE EFFECTS OF GIS ON LOCAL GOVERNMENT IN GREAT BRITAIN

Key Features of British Local Government

Local government in Great Britain at the present time is largely based on the two-tier structure introduced in 1974 in England and Wales and in

1975 in Scotland. In England and Wales this resulted in an upper tier of forty-seven shire counties outside London and the six other main metropolitan areas with a lower tier of 333 shire districts. The responsibilities allocated to shire counties include education, highways and transport, strategic land-use planning, emergency services, and social services, whereas the shire districts look after local land-use planning, housing, public health, and refuse collection. From 1974 until 1986 a two-tier system also operated in Greater London and the other six metropolitan areas. However, since the abolition of the upper-tier authorities in 1986, the sixty-nine metropolitan districts have become unitary authorities.

A slightly different system operates in Scotland that involves nine regions, fifty-three districts, and three Island Councils. The regions have a broader range of functions than their counterparts south of the border.

The primary reason for the reorganization of British local government in 1974 and 1975 was that of the perceived managerial efficiencies associated with the economies of scale. Consequently, the average populations of the shire districts, metropolitan districts, and shire counties in England and Wales in 1990 were 96,600, 259,800 and 684,000, respectively, while those of the Scottish districts and Scottish regions at the same date were 95,100 and 559,000. It is worth noting that the smallish shire district in England (Teesdale with a population of 24,800 in 1990) would be regarded as relatively large in comparison to its North American or continental European counterparts.

Research Strategy

The findings discussed below form part of a comprehensive survey of British local government that is being undertaken by the authors. A combination of research methods has been adopted as the basis for this project, including a telephone survey of all 514 local authorities in Britain and detailed case study investigations. This discussion concentrates on the results of the telephone survey that was undertaken between February and June 1991. Given the large number of postal questionnaires that have been sent to local authorities and their often partial coverage, it was decided to undertake a telephone survey. The 100 percent response rate to the survey supports the adoption of this method and also provided valuable subsidiary information during the interviews. This method also removes the ambiguity that exists in some surveys regarding the definition of GIS, as the capabilities of the software and the precise nature of its use in the authority can be related to operational definitions. The researchers adopted a broad interpretation of GIS, including automated mapping/facilities management and land information systems but excluding thematic mapping and computer-

TABLE 13.1

Plans for GIS in Local Authorities in Great Britain

Number	*Plans for GIS*	%
Already have GIS facilities	85	16.5
Firm plans to acquire GIS within one year	44	8.6
Considering the acquisition of a GIS	227	44.2
No plans to introduce a GIS	158	30.7
TOTAL	514	100.0

aided design (CAD) packages. A more detailed discussion of the survey findings with respect to the adoption and implementation of GIS in British local authorities can be found in Campbell and Masser (1991).

Some Findings from the Survey

The extent of current interest in GIS in local government in Great Britain can be seen from Table 13.1. This table shows that 356 out of the 514 local authorities (69.3 percent) were considering introducing GIS in one form or another. However, the introduction of GIS is at an early stage in most authorities. Only eighty-five authorities, accounting for 16.5 percent of the total, had already acquired a GIS, while an additional forty-four authorities (8.6 percent) had firm plans to purchase one in the twelve months following the survey. It is likely, therefore, that a quarter of all authorities in Great Britain will have acquired a GIS by the middle of 1992.

Table 13.2 shows the distribution of GIS in relation to the type of local authority. From this it can be seen that more than 59 percent of all county-

TABLE 13.2

Plans for GIS by Type of Local Authority

Local Authority	*Already Have GIS*		*Total*
	NUMBER	%	
Shire districts	24	7.2	333
Metropolitan districts	22	31.9	69
Shire counties	32	68.1	47
Scottish districts	4	7.5	53
Scottish regions	3	25.0	12
TOTAL	85	16.5	514

TABLE 13.3

Level of GIS Adoption by Authorities and Systems

Local Authority	*Number of Authorities Possessing GIS*	*Number of GIS*
Shire districts	24	24
Metropolitan districts	22	22
Shire counties	32	44
Scottish districts	4	5
Scottish regions	3	3
TOTAL	85	98

and regional-level authorities in Great Britain already have GIS facilities. In contrast, only 7.3 percent of shire districts and Scottish districts have GIS facilities. The proportion of metropolitan districts with GIS falls midway between the two extremes (31.9 percent). It should be noted, however, that the number of authorities with GIS in each of these categories is broadly similar. Systems have already been adopted by twenty-eight shire and Scottish districts, twenty-two metropolitan districts, and thirty-five county and Scottish regions.

The units of study in this analysis are systems rather than authorities. A system is regarded as a distinct piece or combination of software which one or more departments within a local authority are implementing. Table 13.3 indicates that a total of ninety-eight systems have been purchased by eighty-five local authorities. Shire counties are the most likely to have more than one system within an authority, as there are forty-four systems in thirty-two counties. The only other authority to have adopted more than one system is a large Scottish district.

The presence of more than one system in many of the shire counties indicates the extent to which departmental-based approaches are still prevalent in some areas of British local government. This finding is borne out by Table 13.4, which distinguishes between departmental geographic information systems and those being introduced by several departments. The most striking feature of these findings is the virtually even split between departmental and corporate approaches. Given this broad trend, there are distinct variations between the different categories of authority. It is nearly twice as likely that a system of a county or a region will have been implemented departmentally rather than corporately. In contrast, more than three-quarters of the systems in Metropolitan Districts and nearly 60 percent of

TABLE 13.4

Approach to GIS Implementation

Local Authority	*Departmental*		*Corporate*		*Total*
	NUMBER	%	NUMBER	%	
Shire and Scottish districts	12	41.4	17	58.6	29
Metropolitan districts	5	22.7	17	77.3	22
Counties and regions	31	65.9	16	34.0	47
TOTAL	48	49.0	50	51.0	98

those in shire and Scottish districts were, to some extent, corporate in nature.

One important factor influencing the type of GIS adopted in many authorities is the division of responsibilities for funding. Table 13.5 shows that, although funding is more or less evenly divided between central and departmental sources, there are important differences in emphasis among the different types of authority. In the case of the counties and regions, nearly 47 percent of funding came from departmental sources. In contrast, nearly 45 percent of funding in shire and Scottish districts came from central sources. Nevertheless, it appears from this table that the introduction of GIS into local authorities frequently requires a significant commitment to resources by departments even in cases where a corporate approach is adopted.

TABLE 13.5

Sources of Funding for GIS
(in %)

Local Authority	*Shire and Scottish Districts*	*Metropolitan Districts*	*Counties and Regions*	*Total*
Central	44.8	36.4	27.6	34.7
Departments	24.1	31.8	46.8	36.7
Central and Departments	24.1	31.8	19.1	23.6
TOTAL	6.9	–	6.4	5.1

TABLE 13.6

Number of Departments Involved in GIS Implementation Where a Corporate Approach Has Been Adopted

Number of Departments †	*Shire and Scottish Districts (%)*	*Metropolitan Districts (%)*	*Counties and Regions (%)*	*Total (%)*
ALL	23.5	5.9	6.2	12.0
7	5.9	–	–	2.0
6	–	5.9	12.5	6.0
5	11.8	5.9	6.2	8.0
4	29.4	17.6	43.7	30.0
3	11.8	47.0	18.7	26.0
2	17.6	17.6	12.5	16.0
	(n = 17)	(n = 17)	(n = 16)	(n = 50)

† Departments are defined by the respondent and vary in size and type among authorities.

Table 13.6 shows that there are considerable differences among the number of departments involved in the various types of authority where some form of corporate approach has been adopted. Generally, system implementation involves no more than three or four departments, and more than 70 percent of corporate systems are being developed by less than five departments. Only 12 percent are being introduced in line with the classic corporate approach, which would entail all departments in an authority participating in the implementation of GIS. Smaller authorities such as shire districts are most likely to favor this framework.

Table 13.7 shows that planning is most likely to be the lead department, followed by information technology/computer services, where some form of corporate approach has been adopted. The findings suggest that, even in instances where a top-down structure is favored, the lead is taken by the central computing department rather than by the Chief Executive's department. However, only in the case of one in four systems is the key coordinating role within the authority taken by one of the central departments. If it is assumed that the lead department is the one with the strongest commitment to GIS, the findings presented in Table 13.7 suggest that this is particularly pronounced in the technical and property-related departments. Consequently, the introduction of GIS appears to reflect the bottom-up demand for such facilities rather than a perceived corporate

TABLE 13.7

Lead Departments Where a Corporate Approach Has Been Adopted for GIS Implementation

Lead Departments †	*Shire and Scottish Districts (%)*	*Metropolitan Districts (%)*	*Counties and Regions (%)*	*Total (%)*
Planning/Development	47.0	26.5	6.2	27.0
Info.Tech./Computer Services ††	17.6	17.6	25.0	20.0
Highways/Engineering	5.9	8.8	31.2	15.0
Combined Technical Services	–	26.5	6.2	11.0
Chief Executive	–	11.8	6.2	6.0
Property/Land Management	–	8.8	6.2	5.0
Legal/Related Services	11.8	–	–	4.0
Community Services	5.9	–	–	2.0
No Lead	11.8	–	18.7	10.0
	(n = 17)	(n = 17)	(n = 16)	(n = 50)

† The activities of departments have been interpreted as appropriate given the information provided by respondents.

†† Information Technology/Computer Services includes sections in Treasurers' departments.

need by those at the center. The contribution of planning is most pronounced in the case of the shire/Scottish and metropolitan districts, where 47 and 26.5 percent of systems, respectively, are led by planning departments. However, planning departments are far less significant in the case of systems being developed by county and regional authorities, with highway departments and information technology/computer services most likely to take the lead. There is also a high proportion of systems in these authorities for which responsibilities are shared by all participating departments.

The general trends noted above with the respect to the departments are confirmed by the findings presented in Table 13.8, which includes all departments involved in system implementation. The interest in GIS among planning and technical services departments such as highways is again clear. The results also suggest that while property-related departments such as estates and architecture show a reluctance to lead, they are often keen to participate in such projects. The central computer services section is also likely to be involved with the development of GIS. As a result, three-quarters of the total number of departments participating in implementing corporate

TABLE 13.8

Departments Involved with the Corporate Implementation of GIS

Departments	*Shire and Scottish Districts (%)*	*Metropolitan Districts (%)*	*Counties and Regions (%)*	*Total (%)*
Planning/Development	25.0	17.0	20.7	20.7
Property/Land Management	20.8	18.9	20.7	20.1
Highways/Engineering	12.5	18.9	20.7	17.6
Info.Tech./Computer Services	12.5	15.1	20.7	16.3
Community Services	8.3	7.5	3.4	6.3
Combined Technical Services	4.2	11.3	1.7	5.7
Chief Executive	2.0	5.7	6.9	5.0
Legal/Related Services	10.4	3.8	–	4.4
Environmental	4.2	1.9	1.7	2.5
Emergency Services	not available	not available	3.4	1.2
{All Departments (number of systems)}	{4}	{1}	{1}	{6}
	(n = 17)	(n = 17)	(n = 16)	(n = 50)

systems are accounted for by planning, property, and highways management, engineering, and computer services, with this figure rising to around 80 percent when the combined technical services-type departments are included.

These findings suggest that active interest in GIS has not permeated all local authority activities. Large community-service–type departments such as housing, education, and social services tend not to have become involved. This may be a reflection of the considerable administrative pressures faced by these departments and the limited resources that are available for new activities associated with data handling.

Departmental systems are shown in Table 13.9 to be associated with very similar types of activity to the geographic information systems that are being implemented corporately. Again, technical services such as property management and planning are prominent, with a particular tendency for departmental systems to be developed by the engineers and highways authorities. The absence of the central computer service sections from this list reflects their corporate role within local government.

TABLE 13.9

Departments Implementing Their Own Geographic Information Systems

Departments	*Shire and Scottish Districts (%)*	*Metropolitan Districts (%)*	*Counties and Regions (%)*	*Total (%)*
Highways/Engineering	33.3	40.0	29.0	31.2
Planning/Development	50.0	20.0	12.9	22.9
Property/Land Management	16.7	20.0	25.8	22.9
Combined Technical Services	–	20.0	12.9	10.4
Emergency Services	not available	not available	9.7	6.2
Community Services	–	–	3.2	2.1
Other	–	–	6.2	4.2
	(n = 12)	(n = 5)	(n = 31)	(n = 48)

Overall, the findings of the survey indicate that there are very few examples of centrally led fully corporate arrangements for the implementation of GIS in Great Britain at the present time. In practice, around half the systems are being developed by a single department, while collaborative projects most often involve no more than three or four departments, with a technical service-type department taking the lead. There is also evidence to suggest that within the counties, in particular, there may be several systems being developed within one authority.

CONCLUSIONS

The findings from the survey of GIS adoption in British local government must be interpreted from two different perspectives reflecting the thesis and antithesis that were set out in the previous section. In terms of the thesis, the fact that half the systems currently involve more than one department points to the extent to which the benefits of information sharing have been recognized in British local government. In this respect, it may be argued that the adoption of GIS is having a marked impact on existing administrative practices, particularly in the shire districts/Scottish districts and metropolitan districts.

At the same time, the findings of the survey also indicate the diversity of current arrangements within local authorities where more than one department is involved in GIS implementation. The classically corporate approach led by the central chief executive or computer services department and involving most of the departments in the local authority is being implemented in only a relatively small number of cases, primarily small shire districts. More typical are the pragmatic arrangements made among three or four departments in the larger shire districts and metropolitan districts. These have usually been instigated by planning or property-related departments (see Campbell 1993).

The argument behind the antithesis is evident in the number of departmental applications of GIS that are being implemented in British local government. This is particularly apparent in the shire counties and Scottish regions. It is especially pronounced with respect to departments such as highways, which already possess in-house technical expertise. Under such circumstances, the advantages to be derived from information sharing as a result of inter-departmental collaboration appear to be offset by the benefits of retaining control over GIS operations.

With these considerations in mind, it can be argued that these two contrasting standpoints reflect the technological and organizational context within which GIS is being implemented. On the one hand, the facilities and technical experience of the county highway departments make it relatively easy for them to adopt and implement GIS without having to make major changes in administrative practices. Under such circumstances, there are strong pressures toward a departmental rather than an inter-departmental or corporate approach. On the other hand, the position in many shire districts and metropolitan districts is entirely different. The more limited experience of the line departments, together with their more limited resources, tends to favor some form of collaborative venture. In cases of this kind, the impetus to innovation often comes from the top initially, even though it may eventually be implemented by computer services or the planning department rather than by the chief executive.

A critical factor in both cases is the nature of the geographical information that is being used. Apart from basic map information, it can be argued, for example, that a large proportion of the information required by highway engineers is operational information that is specific to the department itself, and the use of strategic information related, for example, to population and social trends is relatively limited in this case. Under such circumstances, the case for information sharing is relatively weak by comparison with the advantages of retaining control. On the other hand, it can be argued, especially in small authorities, that property-based maps are a key component in the activities of several departments and that there are considerable advantages to be gained by combining efforts. This is often reflected in existing practices prior to the adoption of GIS whereby one department—for example, planning—already provides a mapping service for other departments. In cases of this kind, it is not surprising that collaborative and/or corporate approaches to GIS will be implemented.

To explore these questions in greater depth, a number of detailed case studies are currently being carried out by the authors. In addition to providing valuable information about the technological and organizational context within which geographic information systems are being implemented in different authorities, these case studies will also throw light on the dynamics of the GIS implementation process. One limitation of the present survey is that the findings give only a cross-section of experience at one point in time. By means of the case studies it will be possible to study the impacts of GIS on organizational practices over time.

Until the findings from the case studies are available, only interim conclusions can be drawn with respect to the effect of GIS on British local government. However, there is a great deal of evidence from the survey to suggest that the adoption and implementation of GIS has led to the emergence of a variety of corporate approaches that seek to exploit the advantages of information sharing, especially in the shire districts/Scottish districts and metropolitan districts. On the other hand, there is also evidence of the continuation of department-based approaches, especially in the shire counties, despite these apparent advantages. As a result, it must be tentatively concluded that the findings from British local government point to the extent to which organizational and political factors apparently offset, in many instances, the theoretical benefits to be obtained from structures that seek to promote information sharing.

ACKNOWLEDGMENTS

The research described in this paper was funded under the Economic and Social Research Council/Natural Environment Research Council Geographic Information Handling Project.

Portions of this material appeared in I. Masser and H. J. Campbell, "Information Sharing and the Implementation of GIS: Some Key Issues," in M. F. Worboys, ed., *Innovations in GIS: Selected Papers from the Forist National Conference on GIS Research UK* (United Kingdom: Taylor & Francis, Ltd., 1994). Permission of Taylor & Francis to reprint this material is gratefully acknowledged.

The authors wish to thank Kluwer Academic Publishers for granting permission to include in this derivative article extensive verbatim excerpts from I. Masser, "The Diffusion of GIS in British Local Government" (pp. 99–115) and H. J. Campbell, "GIS Implementation in British Local Government," contained in I. Masser and H.J. Onsrud, eds., *Diffusion and Use of Geographic Information Technologies* (Dordrecht, the Netherlands: Kluwer Academic Publishers, 1993). Reprinted with permission of Kluwer Academic Publishers.

REFERENCES

Audit Commission for Local Authorities in England and Wales. 1990. *Preparing an information technology strategy: making it happen*. London: HMSO.

Bates, B. J. 1988. Information as an economic good: sources of individual and social value. In V. Mosko and J. Wasko, eds., *The political economy of information*. Madison: University of Wisconsin Press.

Beath, C. M. 1991. Supporting the information technology champion. *MIS Quarterly*, 355–372

Campbell, H. 1993. GIS implementation in British local government. In I. Masser and H. J. Onsrud, eds., *Diffusion and use of geographic information technologies*. Dordrecht, the Netherlands: Kluwer.

Campbell, H., and I. Masser. 1991. Impact of Geographic Information Systems on local government. *TRP 101*. Department of Town and Regional Planning, University of Sheffield.

Coulson, M., and R. Bromley. 1990. The assessment of user needs for corporate GIS: the example of Swansea City Council. EGIS 1990 Proceedings, Utrecht, the Netherlands.

Croswell, P. L. 1991. Obstacles to GIS implementation and guidelines to increase the opportunities for success. *URISA Journal 3*: 43–57.

Danziger, J. N., W. H. Dutton, R. Kling, and K. Kraemer. 1982. *Computers and politics: high technology in American local governments*. New York: Columbia University Press.

Finkle, R. W., and F. M. Lockfield. 1990. Major cost savings in multiparticipant AM/GIS: the Santa Clara County case study. JMM Technical Publishing Service, J. M. Montgomery, Inc.

Gault, I., and D. Peutherer. 1989. Developing GIS for local government in the UK. Paper presented at the European Regional Science Conference, Cambridge.

Hepworth, M. 1990. Mobilising information capital in local government. *Local Government Policy Making* 16: 42–48.

King, J. L., and K. L. Kraemer. 1985. *The dynamics of computing*. New York: Columbia University Press.

Kraemer, K. L., and J. L. King. 1988. The role of information technology in managing cities. *Local Government Studies* 14: 23–47.

Levinsohn, A. 1990. Institutional issues in GIS implementation. ISPRS Commission IV 1990 *Proceedings*, Tsukuba, Japan.

Maffini, G. 1990. The role of public domain databases in the growth and development of GIS. *Mapping Awareness* 4, 1: 49–54.

Masser, I. 1993. The diffusion of GIS in British local government. In Masser, I., and H. J. Onsrud, eds., *Diffusion and use of geographic information technologies*. Dordrecht, the Netherlands: Kluwer.

Masser, I., and H. Campbell. 1991. Conditions for the effective utilization of computers in urban planning in developing countries. *Computers Environment and Urban Systems* 15: 55–67.

______. 1992. Geographic Information Systems in organizations: some conditions for their effective utilization. In M. de Lepper, H. J. Scholten, and R. M. Stern, eds., *The added value of Geographic Information Systems in public and environmental health.* Dordrecht, Netherlands: Kluwer.

Mosco, V., and J. Wasko, eds. 1988. *The political economy of information.* Madison: University of Wisconsin Press.

Openshaw, S., and J. B. Goddard. 1987. Some implications of the commodification of information and the emerging information economy for applied geographic analysis in the United Kingdom. *Environment and Planning A* 19.

Pickles, J. 1991. Geography, GIS, and the surveillant society. Applied Geography 1991 Conference *Proceedings*, Toledo, Ohio.

Tveitdal, S., and O. Hesjedal. 1989. GIS in the Nordic countries—market and technology, strategy for implementation: a Nordic approach. GIS 1989 *Proceedings*, Vancouver, Canada.

Wellar, B. 1988. Institutional maxims and conditions for needs sensitive information systems and services in local governments. URISA 1988 Conference *Proceedings*, 4: 371–8.

Section III

Economic, Legal, and Access Issues in Sharing Geographic Data

Economic, Legal, and Access Issues in Sharing Geographic Data

The first two chapters in this section both address economic theories relative to the selling and pricing of geographic data by government, and both address the "free rider" problem. John King begins his chapter by describing a dilemma being faced by many municipalities across the United States and elsewhere. A city is being asked by private parties under authority of the jurisdiction's "open records" law to release its entire Geographic Information Systems (GIS) database at the cost of reproduction. Elected city officials view such releases as tantamount to giving away city assets. After discussing the far-from-resolved legal situation, King focuses on the complicated problem of arriving at appropriate pricing for GIS data and products. He concludes with a series of recommendations to local governments currently being confronted with this very real problem.

The problems addressed by Richard Taupier are very similar to those addressed by King but within a different governmental context. Taupier begins by describing initiatives within state government in the Commonwealth of Massachusetts for the sharing of geographic information among state and regional agencies. He then examines those initiatives through applied economic theory in relation to public goods concepts, the value of geographic information, and the goals of public organizations. Similar to King, Taupier argues that economic theory has a contribution to make relative to understanding the behavior of geographic data sharing.

The chapter by Harlan Onsrud provides an overview of the potential role and limitations of the law in facilitating the sharing of geographic information. He argues that the sharing of large data sets, such as those being amassed within GIS, gives rise to numerous public concerns. He also argues that GIS developers, as purveyors of a powerful social resource, need to identify the consequences of use of the technology and consider those consequences in the light of their general social effects. Such social effects include the widening versus narrowing of social and economic gaps among various segments of society, and equitable opportunity to be informed about and participate in governmental decision making. He provides several examples of comparative case studies of already implemented GIS sharing arrangements that might provide guidance in addressing some of the legal and social concerns.

The chapter that follows by Earl Epstein echoes several of the same themes present in the Onsrud chapter but goes on to discuss, in depth, the issue of control of public information. Epstein takes the advocacy position that distribution of public information by public agencies at dissemination costs remains good public and economic policy in the age of computers.

The chapter by Patrick McGlamery moves on to pragmatic considerations in increasing the ability of citizens to access geographic information. He argues that the library system as an institution is both a highly appropriate model and vehicle for the effective distribution of spatial data. The current status in handling of digital spatial data in libraries is discussed, and future potential roles for libraries in aiding the sharing of geographic data are described.

Finally, the chapter by Jack Dangermond expands upon the theme of broadening public access to geographic data and, by so doing, empowering people with knowledge. Dangermond describes some of the public issues associated with the distribution of data, advocates connection of the public to the databases that governments use in making policy and conducting government operations, and describes some of the benefits that increased citizen participation in government might be expected to produce.

As a whole, the chapters in this section address a range of complex issues affecting the sharing of geographic information. In many senses, the issues of economics, law, and access to information are inextricably intertwined. Similarly, the resolution of these issues, or progress on them, is substantially affected by the behavioral and organizational issues raised in the preceding sections of the book. The chapters in this section provide a useful introduction to at least a sampling of the economic, legal, and access issues likely to be encountered in geographic data-sharing environments now and in the future.

John Leslie King

14 *Problems in Public Access Policy for GIS Databases: An Economic Perspective*

The leadership of the city of Buena Vista faces a dilemma following the successful implementation of their new Geographic Information System. The GIS, which cost the city more than $5 million, is desired by various organizations, including for-profit firms, who argue that the city is compelled by state "open access" laws to release the entire, intact GIS database at cost for reproduction. Elected city officials believe this constitutes a "gift of public funds" and argue that compliance with these requests is equivalent to giving away city assets. Not surprisingly, the debate quickly turns away from arguments on principle to the practicalities of establishing a reasonable price for access. But establishing an appropriate price turns out to be an extraordinarily complicated challenge. The dilemma faced by the city of Buena Vista confronts many local agencies that are struggling with the economics of their Geographic Information Systems in light of fiscal constraints.

INTRODUCTION

Consider the following scenario. The city of Buena Vista has developed a highly sophisticated Geographic Information System, with an extensive database covering the entire jurisdiction with many layers of detail. The GIS took eight years of heavy investment by the city in talent, time, and trouble. Most important, the GIS cost the city a great deal of money—more than $5 million. The GIS went live for routine use by city agencies in late 1991. Recently, Buena Vista was approached by several companies seeking access to the GIS database. These companies, which include two powerful regional public utilities, have insisted that, under state law, Buena Vista is required to provide full copies of the GIS database "for the cost of reproduction." The companies further have argued that the cost of reproduction can include only the price of several magnetic tapes and about $200 of staff time. Three members of the city's seven-member city council, after hearing about the demand, told the Buena Vista city manager they would consider such "release" to be a giveaway of public property worth millions of dollars and issued a strong warning against such action. The city has no policy regarding release of electronic records of this kind, and the city manager never faced a situation like this before. What should she do?

Buena Vista is in this difficult position because it is a leader in application of information technology to city operations. In particular, it is far ahead of most cities in creation of comprehensive, computer-based Geographic Information Systems. These applications of information technology have, as a natural part of their development, resulted in the creation of electronic databases of information essential to city operations and management. These databases are remarkable artifacts. They cost a great deal of money to build and maintain. They also have current or potential value beyond their contribution to the city's operations (*Economist* 1992).

To the extent these databases provide value to citizens in the furtherance of public policy objectives established by the elected leadership of the city, over and above their utility in city operations and service delivery, they constitute in themselves a kind of public service provided by the city. In this scenario, they are no different from other services the city provides, and existing policies and protocols should cover the issues of access adequately. Similarly, to the extent that the databases developed by Buena Vista are of value to other governmental jurisdictions, existing policies and precedents governing intergovernmental data sharing should cover access issues adequately.

A special problem arises, however, when these databases have the potential to be used for commercial gain by individuals and companies. Policies that govern the cases noted above are not likely to cover completely the issues of access when an individual or company requests databases for the purpose of producing products and/or services to be sold for profit. The conflict, in brief, arises from two basic principles that must govern the city's leadership in dealing with city resources. On one hand, the leadership of the city is charged legally and held accountable politically to guard the public trust. Any "giveaway" of that trust, either outright or through sale at a price unreasonably below legitimate market value, would constitute dereliction of duty. On the other hand, the city exists to provide services to its citizens and cannot, without good cause, deny citizens (including its corporate citizens) access to these public services.

The crux of the problem is this:

> *Are the databases created by the city part of the public trust, like land, equipment, and other assets acquired in the process of developing and providing services to citizens? Or are these databases themselves services to be provided to the citizens on request?*

It would be a simple matter for the city if appropriate legislative or judicial agencies provided an unambiguous answer to this question. However, no such answers are available for the city at this time. This is uncharted territory for local governments in Buena Vista's state and, indeed, for many local governments nationally. There are no complete answers because the questions themselves are very new. Buena Vista, as a leader in the creation of such databases, is one of the first local governments to enter this confusing new territory. In the absence of firm guidance, the city has only two choices.In one option, the city can ignore the problem and attempt to forestall dealing with it until such guidance is in place. Alternatively, the city can develop its own policies with the expectation that they shall be successful in dealing with the issues and will be sustained by appropriate legislative and judicial agencies if challenged. For several reasons, the first option is probably foreclosed at this point. Requests for access to databases have already been made on the grounds that such databases are public services. The city must choose how to respond, but the choice is problematic. A decision to provide access under the public service model is likely to preclude or weaken future attempts to treat the databases as part of the public trust, and value them accordingly. Similarly, a decision to treat the databases as part of the public trust without thinking carefully about the public policy benefits of treating certain kinds of access as a public service could deprive the citizens of the city of a valuable service the city is well able to provide.

We return to our opening question: What should Buena Vista's city manager do? Curiously, the answer to that question is found in the ambiguity of its situation, which is both an asset and a liability for all parties to a disagreement. Unclear issues need not be resolved in any particular fashion. Thus, the city does not have to accept any specific definitions of the "truth" about this difficult question, simply because those definitions are strongly stated. Rather, public policy governing access to electronic databases will be negotiated in the absence of clear and lasting direction. It is recommended that city leadership gain the upper hand in the discussions over how database access issues are to be handled. This is done through understanding the complexities of this issue, the reasons why the issue will not simply solve itself, and the basic facts about the databases and the city's investment in them that will serve as a basis for negotiating resolutions to the problem.

This chapter reviews the economic rationale necessary to build that understanding in a way that would give Buena Vista's city leadership the upper hand in negotiating resolutions to the problem.

UNDERSTANDING THE ISSUES

A fair question at this point is "Why is this problem any different from those that other local governments have faced regarding the protection of the public trust and access to public services?" The answer lies in the peculiar legal and economic environment surrounding electronic information. We will first review the legal situation and then address the economic situation.

The Legal Situation

We begin by asking whether Buena Vista should be able to protect the value of, and in some cases recover the costs of, creating valuable databases of public information. In today's legal climate there is little agreement among legislative, administrative, or judicial authorities whose judgments in this matter are binding on Buena Vista. The issue is particularly sticky because many people of different opinions have reasonable arguments to back up those opinions. There is a big difference between reasonable opinion and public policy, although one would hope the latter is backed up by the former. Reasonable opinions might vary, but public policy must be applied consistently. Therefore, the problem of public access will be defined and confused by a set of competing views until and unless there is a decisive rendering of how the issue will be resolved by binding authorities above the city's jurisdiction.

There is, at present, no coherent and sustained set of public policy declarations at the federal or state level to guide local governments in making policy decisions for public access to their electronic databases (Dansby 1992). The trend in this area appears to be almost completely reactive, evolving in response to specific problems brought to light in the legislative process or by litigation. The only solid base of social fact on this matter is the established principle that governments, as a general rule, cannot refuse public access to information gathered by the government in the course of providing public services. That is, the long-standing assumptions governing traditional record keeping are assumed to apply to electronic records, including GIS. The exceptions are relatively rare and circumscribed, usually having to do with personal privacy or law enforcement concerns. Among the states, only Alaska has undertaken a comprehensive rewrite of its open-records laws to balance cost recovery, openness of government, privacy, and so on (Dansby 1992). Beyond that, there is not much specific guidance other than the principle that the public must have access.

Perhaps the most problematic area of vagueness concerns what "access" means under the general principle that the public must have access to government information. Court decisions and statutes recognize that information can be stored in many different formats and on many different kinds of media, and all formats and media must be recognized as included in the general principle of access (OTA 1986). But it is also clear that legislation, and particularly certain court decisions, recognize that "information" does not automatically mean information in a particular format or on a particular medium. Thus, it appears that a request for information could be met by a variety of responses that some would judge to constitute complete compliance with the general principle of access, while others would judge them to be an evasion of the duty to comply with the general principle (Onsrud 1992). The obvious case in point in the GIS case concerns a request for the "information in the GIS database." The information could be handed over on a computer disk or tape, stored in the exact formats required for loading into GIS application software; or it could be handed over in printed form in binary code on a long paper tape. It is the "same" information . . . or is it?

Along the same lines, it is not clear what "information" means under the general principle that the public must have access to government information. Public access to government records is fairly well established, but the relationship between records and information that was once moot because the records by themselves constituted at best raw data sources has become more ambiguous. Records usually are archival or documentary, containing information pertaining to vital factual data (e.g., vital statistics), recorded results of critical decisions (e.g., votes by legislatures, verdicts of juries), and narratives of events (e.g., court transcripts). They are kept, in most cases, to provide a trace or auditable trail of decisions and actions and the critical discourse that led up to those decisions and actions. Records in themselves do not constitute the "living" systems of information that are used increasingly for routine, real-time, ongoing operations and management.[1] The move from paper-based information management to electronic information management has been contemporaneous with a move from records-oriented systems to "information systems." This is a very significant move, because it reveals that the reality of information handling in government organizations has changed in degree *and* in kind. Records that once had relatively limited utility, and thus were made available for whatever utility they did provide, have become much more important as inputs to larger and more important systems. Guidelines established to deal with paper-based records systems are not necessarily applicable to the contemporary era.

When access protocols call for payment by the recipients of information held by governments, there is no clear guidance on how to price access to that information. Even the most strict reading of the duty to provide public access does not require that the government incur a financial loss in doing so. Rather, the city has a right to collect appropriate charges for providing such access. But it is not clear what "appropriate" means in any of these instances. It could mean anything from a token payment that does not even cover the marginal costs of reproduction to whatever the market will bear. And even in cases where charges must be limited to "costs incurred in providing" the information, one must ask what the term "provide" means. Does it mean the full cost of producing the information in the first place? If so, an "appropriate" charge could produce a very high price. And if one cannot recover all the costs of producing the information, what proportion of those costs can one recover?

The position of local governments could be improved dramatically if their respective states were willing to issue new laws defining the cities' rights to restrict access and recover costs associated with producing the information found in databases. But this is not likely for a number of reasons, the most important being the fact that there are powerful interests that do not want their current or future claims to the city's valuable information eroded or constrained in any way (OTA 1986). Why should they stand by while the legislature increases their costs of doing business? These interests have demonstrated their willingness and ability to lobby against legislation that would strengthen city control over such information. And the arguments they can bring to bear should not be underestimated in terms of their intellectual quality or their political appeal, whether they are of the "we the people already own these data," the "don't let government stifle free enterprise," or any other flavor.[2] It might be best not to engage the merits of the different arguments in such a politically charged public forum if they can be resolved effectively, to the benefit of the citizens and to the satisfaction of those who seek access to information, through more friendly and cooperative means. The same admonition holds for the option to have the issues resolved through litigation. It is not clear that the city's best interests will be served through such a mechanism.

In summary, the legal situation at present is confused and inadequate as a guideline for most local governments dealing with access questions regarding GIS data (Dansby 1992). Like many innovations, Geographic Information Systems are fraught with institutional and organizational complexities that confound their diffusion and adoption (Campbell 1991). It is highly likely, moreover, that the legal guidance necessary for local government to feel comfortable in handling such issues will evolve slowly, and only after many confrontations and litigious actions. The issues are complicated, and there are no absolutely clear principles to guide the formation of public policy. The most promising, or at least the most popular these days, is to apply some kind of neo-classical market rationality to the problem—that is, to "privatize" the issue by bringing the concept of price to bear in regulating supply and demand. Aside from the fact that it is not clear in many cases what authority local governments have to set prices for access, it is not even clear just how pricing for access to information should be handled. The next section addresses this problem.

Economic Considerations

An economic perspective on the problem is useful because fundamental assumptions of economics underlie many sound public policies regarding access to public resources (Hyman 1973; Mishan 1976). To the extent that a solid economic rationale for governing access to information is available and applicable to the problem at hand, the task of developing public policy becomes much easier. As we shall see, the economics of information serves mainly to illustrate the complexity and difficulty of the problem; it does not provide ready solutions. But understanding the complexity and confusion enables a much clearer perspective on why conventional remedies will not work and why new approaches are desirable.

In recent years, it has become popular to describe information as the newest of the "factors of production" in organizations. Traditionally, economists have described three basic factors necessary for production of all goods and services: land, labor, and capital. All organizational assets or resources were classified under these categories, and the management of any of the factors was influenced by economic theories specific to the categories. Managers and economists have long recognized the importance of information in production; indeed, no organization could operate without information. But information was always seen as a kind of invisible commodity, not readily classifiable under the three factors of production. In the 1960s scholars began to think of information as an asset or resource in its own right and to talk about information as a fourth factor of production (McDonough 1963; King and Kraemer 1988). As a euphemism to illustrate the importance of information in production, this concept was not a problem. However, a problem emerged when economists began to think seriously about the nature of information in light of economic principles.

The most important and troublesome feature of information from an economic perspective concerns the issue of *scarcity*. In classical economics, the scarcity of something is a basic determinant of its value (Hirschleifer 1980). In Adam Smith's classic interpretation, high demand and high scarcity produce high prices, while something with no demand, or something that is in great abundance, will command low prices. Gold is priced high because it is in demand and it is scarce. Air is in great demand, but it is priced low (in fact, so low that it is usually not priced at all) because it is so abundant. Many trace elements are scarce, but there is no demand for them, so they can be obtained at low prices. Scarcity is a feature that accompanies a commodity at more or less the same level over time, unless demand disappears or a large new source is discovered. Gold is valued, and has always been valued. The price is established by the market—the negotiations between sellers and buyers. It goes up and down but usually within some relatively narrow range. The same holds for most products and services that can be traded or sold.

Information is obtained through the expenditure of effort and resources. It takes an investment to get information, and in this way, information has initial scarcity. But once information is obtained, it can be given away to literally everyone at extremely low cost, and moreover, still be retained by its original producer. It loses its scarcity in the process, and in the process, many of the economic principles governing other things of value lose meaning for information. This interesting feature of information is captured nicely in a paradox noted by Nobel Prize-winning economist Kenneth Arrow (Arrow 1962).

Information is an input to overall economic production, but unlike most inputs, information is neither used up nor depreciated in the production process. It is expensive to produce, but its distribution is cheap, so users enjoy the benefits of the information without having to pay the costs of its production. This is a classic "free rider" problem associated with public goods,[3] and the standard argument about public goods is that social optimality requires that they be distributed to all potential users for the low, marginal costs of reproduction. However, this means the producer will not be compensated by the users for producing the information in the first place, and there will be no incentive to produce the information. It is optimal for everyone to wait until someone else produces the information. In this case, something of great general value will never be produced because no one will sacrifice their economic welfare by taking the first step.

The Arrow observation reveals the problem in establishing access policy for electronic databases in Buena Vista. The central difficulty can be understood via the concept of an essential mechanism of exchange markets: pricing.[4] The city has already made the large investment in producing the information contained in the databases. Producing in this sense includes not only obtaining the information but organizing and codifying it in a manner that facilitates its subsequent use. Having done that, the city has three options regarding access: it can deny access, it can provide access to the information at zero price, or it can provide access at some price. The city could have legitimate public policy reasons to deny access or to provide free access; for example, where issues of personal privacy are involved. But denial can be justified only in special cases, and is not a general remedy for the issue of public access. Indeed, "sunshine" laws and freedom of information laws in many states make it impossible for governments to deny access to GIS information held by the government.

Free access can be justified when the result is a major public good enjoyed by the citizens, but this does nothing to help recover the substantial costs involved in producing the information. And, unlike other public goods, it *is* possible to implement efficient and effective charging schemes for providing information. For-profit information product and service firms such as credit bureaus do so all the time. If the city provides information at zero price while the costs to provide it are significant, the city can be accused of poor business practices for failure to recover marginal costs of production. Even worse, if the city provides information at zero cost to an individual or company for inclusion in a product or service that returns a handsome profit, the city can be accused of giving away the public trust. Therefore, pricing for access makes sense. The problem is how the price should be set.

Again, we can turn to economics for some insight. If you ask the average micro economist[5] what price to set for a product or service, you are likely to receive the advice of pricing at marginal cost of production (Ferguson 1969). This is one version of *cost-based pricing*, in which the cost of producing something is used as a major input to the determination of the price at which it should be sold. As explained above, marginal-cost pricing provides the theoretically "optimal" price for marketing in a competitive environment. Unfortunately, Arrow's observation shows that the marginal cost of producing a "unit" of information that has already been produced is very low, approaching zero. Those who "buy" the information at this price are obtaining a tremendous bargain because the original producer of the information can never hope to recover the costs of producing the information at such low sales prices. Thus, the theoretically "optimal" pricing strategy for information does not make good business sense and yields a result very close to the policy decision to provide access for free.

There are alternative methods for pricing. One is to charge "whatever the market will bear." This means finding a buyer willing to pay the highest price and selling to them. An auction would be an appropriate mechanism for such an approach. But there are two problems with this approach in this instance. One is that there really is no "market" for the kind of information found in Buena Vista's electronic databases, so it is very difficult to know in advance what price to ask to open bidding. Second, the buyer at an auction would not be interested in buying at all if the same information were going to be made available to others, probably at a lower price, later on. To make such bidding successful, the city would have to grant an exclusive license to the buyer so the buyer could then withhold or sell the information to others at his/her discretion. Such an exclusive license could be difficult to justify. In any case, a buyer will be faced with the problem of Arrow's observation as well, and will be in danger of having the asset appropriated downstream.

Another pricing method is average-cost pricing. This takes the cost of producing the product or service, divides that cost by the number of buyers who want it, and charges each buyer 1/N of the total cost. This is "fair" in that everyone pays the same price for access, but it is unrealistic and unworkable in this instance because no one knows how many buyers there are at any given moment, and the number of buyers is likely to change over time. A few buyers who buy in early at high cost will probably be very upset when many more buyers join in downstream and pay very low costs. (In fact, most new products are introduced in precisely the opposite manner, with "entry pricing" that sets an artificially low or "loss leader" price to stimulate demand.)

The pricing issue is critical for the city because any access policy, aside from denial or access at zero price, will require some kind of pricing scheme. Moreover, the particular price set for access will be politically volatile because a price set "too low" will be open to the charge of a giveaway of the public trust, while a price set "too high" will be open to the charge of unfair or inequitable denial of access, and possibly the additional charge that the city is engaging in "for-profit" activities not permitted by its charter.

Levels of Access

The need to develop a pricing scheme does not preordain the use of only one pricing or access scheme for the city. In fact, it is likely that several different access and pricing schemes will be useful because several different objectives might be served. One way to think about the access issue is via "levels" of access. Not all kinds of access are the same. Some require modest efforts to comply and do not in any sense give away the basic asset of the database itself. We draw the distinction between three levels of access, as explained below.

Access to information from the database. This means access to only a part of the information stored in the database but none of the components of the database management system. By "components" we mean not only the actual database software, which might be proprietary in any case, but also the electronic files of information loaded into the formats required for use by the database software.[6] An example of a request for information from the database would be a citizen's request for a map of the city featuring the locations of the city's parks, drawn from the GIS databases. Another example might be a request from a fire insurance company for a map showing the location of all fire hydrants and fire stations in the city. In either case, the provision of such maps might easily be seen as the rendering of a public service, and the pricing issue would be trivial. The city could choose to give such maps away for free, or it could charge a small price for each map. The price would most likely be set to cover the variable costs involved with printing each map—paper, ink, maybe a bit of labor, and something to cover the additional capital and maintenance costs incurred by producing such maps. The person or company receiving the map would be receiving a small but perhaps very useful service for a small cost; the city would be providing a small service at a small cost, and it would in no way be giving away the public trust in the process.

Access to all of certain kinds of information in the database. In the extreme sense, this means access to the full set of data in the database but not including the components of the database system. A less-extreme instance would be a request for all of a certain kind of information in a database—say, a list of all the names and addresses of landowners in the city. In either case, the implications of such requests are quite different from those for information from the database because they request a *set of data* that are valuable because they contain all of something and not just instances of that thing.

At this point, the issue of pricing becomes important because the information in the list might have cost the city a great deal to compile and load, so there is a question of whether pricing too low constitutes a give-away of the public trust. Also, such requests raise questions about the uses to which the information will be put, and these questions might or might not have pricing implications. For example, a request for a list of landowner names and addresses raises different concerns depending on who makes the request. A request made by a non-profit organization attempting to clean up the city by improving maintenance on vacant lots carries different political sensibilities than a request by a company that wants to do a blanket marketing campaign for some product landowners might buy. A low price for the former and a higher price for the latter might be appropriate, and differential pricing when issues of public policy are at stake can be justified. But such decisions are inevitably political decisions and are not easily justified by simple economic explanations. Those who are charged more are almost certain to argue that they are being discriminated against.

There are also issues for this level of access that involve propriety rather than proprietary concerns, as the discussion in the section on non-economic issues of access suggests. There are some useful experiences and precedents regarding access to this kind of information, however, because requests for certain kinds of lists (e.g., driver's license data) have been made of state agencies for many years, and states have developed policies for handling such requests. These can serve as useful guidelines.

Access to the database itself. This means access in the form of a copy of the actual database, including the information and the essential components of the database system. It really does not matter whether the requester asks for all the information in the database, or just for certain complete classes (e.g., all the names of X, all the locations of Y, all the layers of Z). The important thing is that the requester wants a pre-loaded database record that can be installed at very low cost on equipment outside the city's ownership or control, and used to generate products or services for profit, or in ways the city considers troublesome but cannot control. This is by far the most serious level of access, and the one for which the most troublesome public policy questions arise. Unlike the issues of access to information from or information in a database, requests for the databases themselves are a new frontier of access policy. Access to whole databases, with the components of the database system intact, hinges on the central question: Is access to city databases a public service, or are the databases themselves part of the public trust? The issue is cloudier for databases than for other city service-producing assets because of the problems outlined by Arrow's observation.

Let's take a case in point for illustration. A citizen justifiably expects to be given access to the products of the city's fire-fighting system if there is a fire in their house, but one could hardly expect the city to honor a citizen's request for a fire department pumper unit and crew to be stationed at their house in case a fire breaks out. And surely, no one would accept as legitimate an effort by a citizen to obtain and sell some part of the city's fire-fighting system to another city for profit to themselves. There are several reasons why such requests would be considered outlandish, but the main one is cost. There is no way a city could afford a pumper and crew for every house. But suppose that once the city bought one pumper and crew, it could clone that pumper and crew package indefinitely at nearly no cost. In that instance, the question might turn around, from "Why should the public pay to station a pumper and crew at the Jones's house?" to "Why shouldn't Jones be allowed to have a pumper and crew stationed at his house given that the cost to the city is negligible?" The question of what is "fair" has been radically transformed by the economics of the situation, and in the process, the political and business questions before the city's leadership become confused and confusing.

Interestingly, the example above sheds light on another critical issue in pricing, which is the question of just what the product consists of. In the case of the citizen requesting a pumper and crew to be stationed at their house, we would want to ask not only how much it would cost to place the pumper and crew on site, but how much it would cost to keep them there and functionally ready to fight fires. Regardless of the cost of putting the pumper and crew there in the first place, the costs to maintain and support the operation could be high. Who would pay for those costs? Again, the fairness issue arises. And this time the issue extends more directly to the case of access to the city's databases. These are not one-time-only products, like consumer durables, that the city could "sell" and then forget. Rather, these databases are like living entities: they require care and feeding to remain healthy. And for all practical purposes, the city is the only entity that can handle such care and feeding.

When someone requests access to the databases at any level, it is essential to know whether they are asking only for access at one point in time, or for continued access to the living entity of the database. For most organizations that would use city database products in ongoing work, ongoing access is essential. It is also important to ask what is really being requested when someone asks for access. Are they asking only for the rudimentary minimum to accomplish some immediate task, or are they asking for the city's ongoing *help* in accomplishing a task or set of tasks? And if they want this help, will the city be compensated for the cost of providing the help, and if so, how?

Thus, pricing can be seen as a window on a whole range of complicated questions about access to electronic databases. Naturally, pricing is not the only issue at stake here. But it is a useful tool for capturing a large fraction of concerns the city's leadership must confront in dealing with this situation. And since most sensible pricing schemes depend on an understanding of costs, we shall return to the cost issue below in our recommendations. Before doing so, however, we wish to make special note of non-economic considerations in providing access that are peculiar to the issue of databases.

Non-Economic Considerations

At various times above, we have mentioned the issue of privacy as a factor worth considering when dealing with the issue of access to the city's databases. We will expand on the general class of non-economic issues in this section to highlight the importance of these issues for the city's leadership.

It is a truism that information can be politically important. Now, more than ever, politically loaded questions arise about the ways governments deal with information. In some cases, the questions are easily answered by long-standing precedent, and even if the consequences of granting or denying access are politically controversial, the government's defense for its actions is sound. But there are many cases where the precedents are lacking or unclear, and there are no statutes or established administrative laws to cover the situation. In such cases, the actions of the city will be evaluated by the public on the grounds of the judgment shown in the decisions made. And no matter how carefully one tries to control the impression, a public sentiment that the city's leadership or employees made a "bad judgment" or demonstrated "lack of judgment" will be costly to the city.

Judgment certainly enters in on economic matters of access. The city can be criticized for bad judgment in charging an access price so high that it effectively denies access to those working for a popular cause, or for giving a copy of a city database to a profit-making firm for a price widely believed to be "too low." The discussion above deals with such issues. But the city can also be tagged with bad judgment in many instances where the economics are not at issue, and in cases where established practices would suggest no political danger. For example, a number of state attorneys general have filed suit against a major credit data reporting company claiming the company's procedures for dealing with sensitive credit information on individuals are negligent in care for those individuals. These systems have been in use for more than two decades, but this suit is causing real controversy. Suppose the case has the effect of calling into question not only this company's practices but the judgment of those who released (e.g., sold) data used by the company in these practices? And suppose, further, that the city was such a supplier. Even if the city merely supplied the credit company with data as part of a blanket access policy, under the assumption that denying the company access would be unfair discrimination against a company with legitimate requests under the terms of the policy, the public perception of the matter could be harsh. And indeed, we argue elsewhere that there is every reason to believe that GIS systems will become embroiled in serious public policy disputes that are not easily resolved on the "science" of the issue (King and Kraemer 1991).

Two other examples help illustrate the point. One example would be similar to that above of the insurance company requesting a map of fire hydrants and fire stations, but instead, assume that someone who identified himself/herself as a member of the American Nazi Party requests a map showing all the city's synagogues. What, one might ask, would they want with such a map? And from whom should a city employee receiving such a request seek guidance on what to do? The fact that the addresses of the city's synagogues are already available in such ubiquitous documents as the telephone yellow pages probably would not divert the outrage the city's provision of such a map to such a group would engender. Another example involves the city as a data requester. Suppose that the city's police department requests access to the local electrical utility's customer electricity consumption records for the past two years, in the hope that they could use the data to identify buildings being used to grow marijuana under special growing lights. Should the city make such a request? Should the utility company grant it? Both of these examples illustrate just how fuzzy the questions surrounding access can be. It is hard enough for city employees to know whether a request has problematic implications; it is much worse when existing policy gives employees no guidance on what to *do* when they do recognize such problematic requests.[7]

Buena Vista's city's leadership must think broadly about the access issues and go beyond the merely economic concerns discussed above. Naturally, it is false economy to try to anticipate all possible downside occurrences and make policies to cover them. And the world is a risky place, in any event. Nevertheless, examples like those above illustrate the kinds of political trouble that can emerge around the issue of access, and attention to these political ramifications is needed.

THINKING ABOUT OPTIONS

The discussion above uses the concept of pricing to illustrate the complexities of the access issue. Pricing is a powerful mechanism for establishing smooth transactions between those with something to sell and those who want to buy it. But use or establishment of a market is not the only way parties can interact successfully. In fact, there are instances where competitive markets do not work well. A good example is with so-called "natural monopolies" such as occur in the provision of certain public utilities. It makes no sense to run multiple water pipes to each person's home so they can "choose" which water company they will buy from. Rather, all the homes in a given area are served by a single water provider, which is organized as a regulated public utility that provides water service as a monopoly. Electric, gas, telephone, and cable TV utilities are usually organized the same way. In exchange for the power of monopoly status, these organizations are closely regulated by governmental or quasi-governmental agencies. They are regulated in two senses: 1. they must ask for and receive permission to do anything that affects the prices or quality of their products (e.g., increase/decrease service levels, build new facilities that raise their costs, or raise their rates for whatever reason); and 2. they are restricted to only those activities that are directly pertinent to the services for which they were granted monopoly rights.

In other instances, there can be good reason to relax expectations of competition in order to achieve a higher good for all the parties involved. Thus, the federal government has relaxed provisions of anti-trust legislation to permit competing electronics companies to cooperate in joint research and development enterprises. The Microelectronics and Computer Corporation is an example of such a cooperative partnership. Organizations can also make individual donations to specially created non-profit entities that carry out work beneficial to the organizations overall; the Electric Power Research Institute is an example from the world of electric power utilities. Government agencies often engage in special partnerships and joint-powers agreements with other government agencies to accomplish missions of general benefit. And in some cases, they can join together and create non-profit firms to carry out special tasks that each agency could not do as well. An example of this is the San Diego Data Processing Authority, which is owned by, and provides data processing services to, the city of San Diego, San Diego County, and a number of other governmental jurisdictions and special districts.

The partnership or cooperative approach can be a particularly sensible route to pursue in cases where all parties share common goals and need common resources from one another. For example, both the city and local public utilities have a common desire to improve aspects of their services to citizens of the city. If the city and the utilities can cooperate in research and development projects that build capabilities all around while keeping the costs of building those capabilities low for each organization involved, the citizens are likely to benefit through more efficient service delivery and lower costs. Obviously, there are important legal considerations to be investigated in any such partnership scheme. And it is likely that such partnerships will be much easier to design and implement when the agreements are between governments, non-profits, and/or regulated public utilities. Bringing in for-profit, unregulated firms to such a partnership raises serious questions that require intense scrutiny. In some cases, such partnerships might not be appropriate no matter how much benefit the city receives.

The advantages of a partnership generally come in three forms. First, in situations where the real value of an enterprise is not clear, a partnership provides a good way of exploring the options and sharing costs in a manner that is fair over time. This could be the case with the city's databases, which surely have a value, but no one is quite clear what that value is. If an outside organization's objective in using the city's databases is to build capability that will benefit the citizens of the city, it makes sense for the city to join in the effort rather than to "sell" the databases it owns.

Second, a partnership can be very important where one party has particular assets and skills that it cannot "sell" to the partnership but that it can "contribute" to the partnership through in-kind compensation or other agreement. This could be the case in Buena Vista, where all the expertise with the city's databases resides in the city itself.

Finally, partnerships provide a forum for ongoing negotiation and readjustment of the relationship between the parties. Well-constructed partnerships contain the mechanisms necessary for conflict resolution and consensus building. Markets, on the other hand, depend mainly on the expectation that dissatisfied buyers or sellers will simply stop exchanging with one another and do business with someone else. This works when there are multiple buyers and multiple suppliers because the aggregate forces of the market level things out. When the market lacks these features, it is said to be "imperfect," and other forms of interaction, such as partnerships, become attractive.

The point here is that partnership arrangements can be successful. Ways have been found to make partnerships work between intensely competitive firms in the private sector. Surely, there are ways to make them work among governments, non-profits, and regulated utilities. Such opportunities should be explored as alternatives to a strict market system of interactions, in part because the market approach might not work well, and in part because the partnership approach might work very well indeed. Finally, partnerships can have added political advantage when the shared objectives of the partners truly involve improvement of public welfare.

SOME RECOMMENDATIONS

Local governments face a genuine challenge, and it is not clear how well they will cope with the problem. However, in the spirit of the good Samaritan, we offer six recommendations based on the discussion above.

1. *Begin to develop sound policies for public access, but focus access policy attention first on the most glaring problems.* The GIS database in our scenario city of Buena Vista has been the subject of requests for access at all three of the levels discussed above. The city needs to pay special attention to those requests for access to the database itself, including the components of the database system. This is the first priority. However, the city also needs to address questions of access at the level of information from the databases and information in the databases.

2. *Avoid giving in to demands and pressures for access that would produce premature decisions.* In the long run, the city and the users of its databases will benefit from a sound policy that makes sense politically and economically and that is stable under the law. Premature decisions are likely to make it more difficult to attain such policies and are prone to cause ill will and bad relations between the city and the potential users of its data. This situation should be avoided at all costs. Moreover, the current legal situation appears to be sufficiently ambiguous to make it unclear what the city's obligations and options are in this regard. Taking the matter to court might "settle" it for the moment, but it is hard to tell which side would win—and in any case, the victory would probably be Pyrrhic for whoever did because the real issues are still murky and the political fallout from such a fight would be unpredictable.

3. *Get ready to "go to market" with the city's databases, but not with the expectation of actually doing so.* Buena Vista's leadership should know better than anyone else the cost basis of its databases, the probable market value of those databases, and the liabilities and exposures it incurs if it grants access to the databases at different levels. Only through such mastery of the market characteristics of its databases will the city be in a strong position to argue its case in negotiations with potential users of the databases and in the crucial courts of public opinion and law. Getting ready to go to market means conducting serious analyses of cost and value

related to the databases. Such analyses might be done with city staff, by outside consultants, or by a mix of the two. We recommend, however, that the city not simply contract out this task. Rather, the city needs to build its own internal expertise in this area to be able to maintain its advantage in negotiations and public discussions.

4. *Consider "partnership" approaches to dealing with special cases of data sharing.* There are instances where it does not make sense to pretend that a market-like solution will work well, either because no market exists or because the market is imperfect in significant ways. In such cases, a more cooperative and interactive approach can yield major benefits to all concerned. This is especially true in cases where an individual or organization wishes to make use of Buena Vista's expertise and information resources to implement projects that will benefit the citizens of Buena Vista directly, but through non-city channels. Naturally, all such partnerships must be undertaken in a businesslike manner, with full attention to the rights, responsibilities, and sources of recourse for all parties.

5. *Establish a "Committee on Public Access Policy for Electronic Information."* This committee would serve as a study group and review board for issues regarding access, as well as a key repository of ideas and skills for dealing with access policy issues. Most important, perhaps, this committee could serve the same kind of role that planning and finance commissions often serve in local governments. These commissions are usually made up of qualified political appointees who research the issues for the elected legislators. In an important sense, they are a sounding board and negotiation forum that allows serious and politically sensitive discussion to take place at some distance from the elected legislative body before the final decisions have to be made. The advantage of this arrangement is that the decorum of political give-and-take is maintained while the real work of political loss and gain is worked out. However, it is important to note that such councils must be vested with genuine political power to shape the agenda and guide discussion. A review of the failings of so-called "steering" committees for the management of computing-related activities can be found in King (1983).

6. *Make every effort to keep the elected leadership informed of the broad economic and political issues at stake in these questions of access.* The access question is an evolving set of issues; it will not be settled overnight. The issues of access are mainly administrative at this time, and with skill and good fortune, they might remain administrative issues. But as the discussion earlier suggests, these are issues that can erupt into politically serious controversies.

Notes

1. It is important to remember, however, that records often are key inputs to such "living" information systems.

2. In fact, one of the most sophisticated arguments they can make is hinted at earlier, in the discussion of Arrow's paradoxical observation: that information in the city's databases is already a public good and should be distributed at zero cost to all who want it. What do they care that the city is thus deprived of any incentive to produce such information, given that it has already been produced? Moreover, they can argue that the city had non-monetary incentives to produce this information in its mandate to serve the citizens of the city, so the issue of depriving the city of monetary incentive is moot. Anyone can hire consultants with a rudimentary understanding of information economics.

3. The free-rider problem occurs when some "good" is created that cannot be effectively controlled or regulated to exclude use by those who do not pay. A common example is streets: it would be prohibitively expensive, not to mention a terrible nuisance, to put toll gates at every intersection to capture fees from all users, so it is easier and more sensible to simply make everyone pay part of the cost up-front via taxes or other general charges, and then let everyone use the good for "free." Of course, it is possible to capture some of the costs of use through secondary means, such as taxing vehicles or their components (tires, fuel, batteries). See Hirschleifer (1980) and Mishan (1975) for further discussion of the free-rider problem in this context.

4. Pricing is an ingenious mechanism because it allows one to do three things at once. The most obvious use of a price is to recover costs and, where appropriate, raise a surplus or profit. Another valuable feature of pricing is allocation of resources. If something is in scarce supply, raising the price depresses demand. Peak-load pricing, where users of a service at periods of highest demand pay more, helps "level" demand over time. Similarly, a price can be used as an incentive for a buyer to buy more; quantity discounts are a good example. Finally, pricing provides important information for capacity planning. Through analysis of demand under different pricing conditions, it is possible to assess how "price-elastic" a product or service is. Elasticity refers to the extent to which demand rises and falls according to price changes. An inelastic product or service retains the same demand over a wide range of prices; an elastic one has demand rising and falling directly with price changes.

5. A micro economist studies economic issues related to the individual firm, including costing and pricing issues related to production and marketing. Excellent and useful discussions of these issues can be found in Dosi et al. (1988) and Rosenberg (1982). Useful discussions of the economics of computing technology

can be found in Kleijnen (1980) and Sharpe (1969), although it is curious that neither book delves deeply into the peculiarities of information, the grist for the computer's mill.

6. It is important to recognize the economic significance of the difference between information in a database and the information as *loaded* into the database. A huge stack of paper documents might contain all the information that will eventually go into an electronic database system, but the process of getting the data into that system can be difficult, time-consuming, and expensive. The experience of a major firm provides a gravid example. This firm is a Fortune 500 company, a leader in the information products and services industry. Its experiences cannot be compared directly to Buena Vista's. Nevertheless, the experience of this firm illustrates our point about the costs of loading a database. Five years ago this firm undertook a feasibility analysis of a proposed project to convert its huge on-line, flat-file record systems to on-line relational database systems. (Flat-file systems are an old, somewhat inflexible means of managing electronic data; relational database management is the "modern" way to do things.) Most people involved with the proposal believed the move to relational databases would be highly beneficial. Unfortunately, the feasibility analysis, which was very carefully done, concluded that conversion from flat-file to relational database technology would require an investment of nearly *$20 million just to load the relational database*! And this was with files that were already in machine-readable form, not paper documents. The company scrapped the proposed project and stayed with the old flat-file system.

7. Again, the issue here is not merely whether the courts will find the city's policies or actions in such cases appropriate or in error, but the political difficulties created by these kinds of situations as well.

REFERENCES

Arrow, Kenneth J. 1962. Economic welfare and the allocation of resources for invention. In Nelson, R. J., ed., *The rate and direction of inventive activity.* Princeton, NJ: Princeton University Press, pp. 609–625.

Campbell, H. 1991. Organizational issues and the utilisation of Geographic Information Systems. Sheffield, England: Regional Research Laboratory Initiative Discussion Paper Number 9.

Dansby, H. Bishop. 1992. Survey and analysis of state GIS legislation. *GIS Law* 1, 1: 7–13.

Dosi, G., C. Freeman, R. Nelson, G. Silverberg, and L. Soete, eds. 1988. *Technical change and economic theory.* New York: Pinter Publishers.

Economist. 1992. The delight of digital maps. *The Economist* (March 21): 69-70.

Ferguson, C. E. 1969. *The neoclassical theory of production and distribution.* London: Cambridge University Press.

Hirschleifer, J. 1980. *Price theory and applications.* 2d ed. Englewood Cliffs, NJ: Prentice-Hall.

Hyman, D. N. 1973. *The economics of governmental activity.* New York: Holt, Rinehart & Winston.

King, J. L. 1983. Centralized vs. decentralized computing: organizational considerations and management options. *ACM Computing Surveys* 15, 4: 319–345.

King, J. L. and K. L. Kraemer. 1988. Information resource management: is it sensible and will it work? *Information and Management* 15: 7–14.

______. 1991. Models, facts, and the policy process: the political ecology of estimated truth. Paper presented at the First International Conference and Workshop on Geographic Information Systems and Environmental Modeling, Boulder, Colorado, 1991. Irvine, CA: Center for Research on Information Technology and Organizations, University of California, Irvine.

Kleijnen, J. P. C. 1980. *Computer and profits: quantifying financial benefits of information.* Reading, MA: Addison-Wesley.

McDonough, A. M. 1963. *Information economics and management systems.* New York: McGraw-Hill.

Mishan, E. J. 1975. *Economics for social decisions: elements of cost-benefit analysis.* New York: Praeger.

Onsrud, H. J. 1992. In support of open access for publicly held geographic information. *GIS Law* 1, 1: 3–6.

OTA. 1986. *Intellectual property rights in an age of electronics and information.* U.S. Congress, Office of Technology Assessment. OTA-CIT-302. Washington, D.C.: Government Printing Office. April.

Rosenberg, N. 1982. *Inside the black box: technology and economics.* Cambridge, MA: Cambridge University Press.

Sharpe, W. F. 1969. *The economics of computers.* New York: Columbia University Press.

Richard P. Taupier

15 *Comments on the Economics of Geographic Information and Data Access in the Commonwealth of Massachusetts*

The Commonwealth of Massachusetts, like many other states in the United States, over the past several years has developed a significant base of digital geographic information (GI). Each year additional Massachusetts agencies develop Geographic Information System (GIS) capabilities and improve upon their ability to make efficient use of GI. As a result of the demand for more and better GI, data access and exchange policies are of substantial interest. Massachusetts has evolved through various stages similar to those encountered in other states and expects further evolution to occur. An understanding of this behavior and the incentives for exchange of GI requires some basic understanding of the economics of public goods, the value of GI, and the nature of efficient public organizations. This chapter explores some of the details of those issues and demonstrates that efficient public agencies that are capable of generating substantial value from GI will be inclined to enter into data-exchange agreements.

INTRODUCTION

The objective of this chapter is twofold: first, to describe the initiatives within state government in the Commonwealth of Massachusetts for the sharing of geographic information (GI) among state and regional agencies as an example of the typical evolution that such an organization might experience, and second, to examine the theory behind those initiatives by drawing upon applied economic theory concerning public goods, the value of geographic information, and the goals of public organizations.

The limited writing and presentations to date on the topic of institutional data exchange tend to focus more on the barriers to data sharing and less on the incentives. Yet, there are numerous examples where data-exchange arrangements have met with considerable success, leading to the conclusion that, at least in some situations, the perceived benefits outweigh the associated costs and risks.

ACCESS TO GEOGRAPHIC INFORMATION IN MASSACHUSETTS

The evolution of the use of Geographic Information Systems (GIS) in the Commonwealth of Massachusetts is not unlike that of many other states in the United States. A lead agency, in this case the Executive Office of Environmental Affairs (EOEA), perceived an opportunity for improving its ability to meet its goals through the development of a statewide GIS. Three related feasibility studies were funded, a plan for development was negotiated with its sub-agencies, and that plan was implemented over a five-year period. As a result, EOEA has become the major provider of digital geographic information within the commonwealth and maintains a lead position among all Massachusetts public agencies using geographic information technology.

The EOEA system, MassGIS, evolved through a number of stages relative to the status of its geographic database and its willingness and ability to share geographic information with other organizations. It began as a multi-agency system. EOEA is a cabinet-level office responsible for the coordination of five environmental and natural resource departments: Environmental Protection, Environmental Management, Food and Agriculture, Fisheries and Wildlife, and the Metropolitan District Commission. It also has a statutory responsibility to act as a clearinghouse for environmental data and to provide data-processing services to its agencies.

During the five-year implementation period, staff numbers within the EOEA agencies varied from 3,500 to 2,500 FTE (full-time-equivalent) employees, and annual expenditures varied between $500 and $275 million in combined general, capital, and federal funds. Before EOEA initiated the GIS effort, three of the five agencies expressed a desire to build single-agency GIS programs. However, given the assurance that EOEA would move forward with the development of a shared system and a multi-purpose geographic database, those agencies were agreeable to contributing to shared system resources. The combined initiative and purchasing power of those five agencies resulted in the creation of a geographic database of significantly greater extent and quality than would have been possible with each of those agencies acting separately (Taupier and Terner 1991).

The estimated cost of building the system was placed at $3 to $6 million. The EOEA decision to build a shared system was based on the premise that limiting redundant data-capture efforts and system hardware and software would contain costs and allow it to realize a higher net benefit and benefit-cost ratio. This approach has been recognized as potentially yielding the level of benefits that EOEA anticipated (Wilcox 1990). At the time the decision to proceed was made, the anticipation of economic benefits was based only upon studies conducted for other state and regional agencies.

The relationships among the five agencies evolved much as anticipated by the original planners. There are now nearly sixty users with direct access to the system software and data and as many as twenty projects in progress at any given time. The direct costs of developing the system have been calculated at $2.8 million (Taupier 1992a). Formal estimates of benefits realized through the system are only recently underway, though it is generally believed by the agencies and users that the benefits significantly outweigh the costs, and that the automated production of maps and geographic information is almost always less costly than manual methods.

EOEA has received two significant awards for its Geographic Information System, the ESRI (Environmental Systems Research Institute) President's Cup and a URISA (Urban and Regional Information Systems Association) Exemplary System in Government Award. It has spawned the development of eight other Geographic Information Systems in Massachusetts public agencies that all depend on base-line data from MassGIS. As this chapter is written, EOEA is contemplating another $6 million in expenditures for geographic information over the next three years. Whether or not the expectations of the benefits realized through this GIS effort will be supported by formal evaluation remains to be seen.

DATA DISTRIBUTION THROUGH MARGINAL COST RECOVERY

Three years into the implementation of the GIS plan EOEA began to receive frequent requests for access to its geographic information from outside its own agencies, a demand that it had not anticipated. Meeting that demand required the commitment of resources that were not available to the agency except at the expense of other programs and responsibilities. Following several unsuccessful attempts to secure additional funding, EOEA sought and received legislative authorization (MGL ch.240 s.2c,

MGL ch.653, s.138) to implement an average marginal costing scheme for geographic information in both digital and hard-copy form, an option allowed under the commonwealth's public records statute and regulations. The fee schedule divided files into three groups, with prices ranging from $50 to $150 per file.

Expenses included in the calculation of marginal cost were staff time for selecting, downloading, and handling data, computer time, consumable supplies, prorated hardware and software costs, and the staff time expended assisting users once they have the data. The value of a generic hour of combined staff and computer time plus supplies was then calculated at $50. A formula was developed that considered file size, complexity (number of attributes, detail of classification, and so on), and geographic extent as a predictor of the time necessary to accomplish all the above tasks. Files were placed into three categories corresponding to the marginal staff and computer time required for transfer of the data.

In the two-year period from January 1990 through December 1991, EOEA received fees totalling $66,450 for digital geographic data and an additional $125,755 for hard-copy maps, plots, services, and training. Of the 104 fee-based requests for digital data, fifty-one originated from private firms, ten from municipalities, eight from state agencies, and five from regional planning agencies. These numbers do not represent the true distribution of users of MassGIS data, however, since many state, regional, and federal agencies, as well as academic institutions, are granted fee waivers under a policy discussed below.

Data Exchange Agreements

Massachusetts regulations pertaining to public records also allow for the custodian of public records to waive, in whole or part, any fees that it is entitled to recover from the users of its data if it determines that it is in the public interest to do so. EOEA developed and implemented a Digital Data Distribution Policy (Commonwealth of Massachusetts 1990) that set forth the circumstances under which it would a priori agree that it was in the public interest to waive the fees it was entitled to recover. That policy has led to twelve two-party data-exchange agreements between EOEA and other state and regional agencies. In addition, the policy allows for the waiver of fees for educational purposes and unfunded research, as well as a data-grant program for matching-grant purposes. Among the terms and conditions of those agreements are requirements that set minimal data set quality standards to which the cooperators must adhere. Because data quality and documentation standards are instrumental in determining the utility and value of geographic data (Parent 1989, Taupier 1992b), it would be difficult to justify the creation of an exchange program that did not include minimum standards.

EOEA was instrumental in establishing the Massachusetts Geographic Information Committee, a council of governments, academic institutions, public utilities, and private companies intended to provide advice, policy recommendations, coordination, and education within the commonwealth to government and its agencies relative to GIS and geographic information. The Executive Committee of that organization has recently begun considering a geographic data exchange mechanism that would eliminate the need for a complex web of two-party data-exchange agreements in favor of a single statewide data exchange program such as has been established in other states (Warneke 1992).

Multi-party data-exchange agreements face a complex set of issues. Publicly held geographic information shares many of the characteristics of a public good, including the problem of free-riders. Data-exchange agreements are most easily established between substantially equal partners, both with resources of relatively equal value to share. All parties to the agreement must adhere to minimal data standards; for agencies that have yet to deal with the implementation of standards, the price of entry may be substantially higher. The decision to establish a central clearinghouse for data versus decentralized repositories or a distributed network of data libraries involves trade-offs between ease of access and the distribution of costs.

The Massachusetts EOEA is also considering data-exchange agreements with the other New England states as they share many common environmental resources. EOEA has already signed a data-exchange agreement with Rhode Island. The arrangement preferred by EOEA is a New England data exchange agreement that would pre-establish the terms and conditions for such transfers among all six states. Several geo-regional studies have already benefitted from this exchange of geographic information, and ready access to data would be of significant value.

Commonalities with Other States

Massachusetts is not alone in its approach to providing access to geographic information. All six New England states currently have authorization to make digital geographic information available on a cost-recovery basis through the application of marginal-costing criteria. All six states price data on the basis of file size and some measure of the complexity of the request (such as number of tiles) and charge for the estimated average cost of reproduction. Exactly what is included in the cost of reproduction varies from state to state.

There are three conditions that those states all apply to the use of the geographic information that they distribute, though the actual wording of those conditions varies from state to state. These are: 1. no secondary distribution; 2. use of a limit of liability clause; and 3. the requirement to cite the source of the data.

SOME ECONOMIC ASPECTS OF GEOGRAPHIC INFORMATION

The situation in Massachusetts is not atypical. Many other state and regional governments have found themselves following a similar course of action relative to developing geographic databases and facing issues of access. The issue of whether or not to provide access to geographic information through marginal pricing or some other pricing scheme is currently of substantial interest. A consideration of geographic information as a public good offers some insights on this issue.

There also exists a trend toward data-for-data exchange policies. A discussion of the value of geographic information is helpful in understanding the incentives behind this organizational behavior.

Geographic Information as a Public Good

In order to arrive at some observations concerning the economics of geographic information, it is necessary, first, to consider briefly the economics of public goods. Geographic information shares the characteristics of a public good in that consumption is non-rival and there is an absence of excludability (Rhoads 1985). And, as with public goods, there exists the problem of free-riders – those who decline to purchase the good but enjoy the benefits nonetheless. *The Utility and Value of Geographic Information* (Didier 1990) describes the three conditions of a pure public good as the impossibility of exclusion of multiple users, a global obligation to use, and the absence of a congestion effect.

It is useful to refer to Figure 15.1 in order to illustrate the rationale behind the marginal pricing of public goods. The total area (ABC) under the theoretical demand curve AC represents the total value, or society's willingness to pay for the public good in question. Assuming that marginal cost is constant and that price is equal to marginal cost (P_1 = MC), the area AED represents net social benefit (without reference to whom the benefits accrue), the difference between the price charged and the amount that all consumers would be willing to pay for the good. An agency interested in maximizing social welfare will set price equal to marginal cost and will supply an amount equal to Q_E. At any other price and quantity supplied, a loss of efficiency occurs.

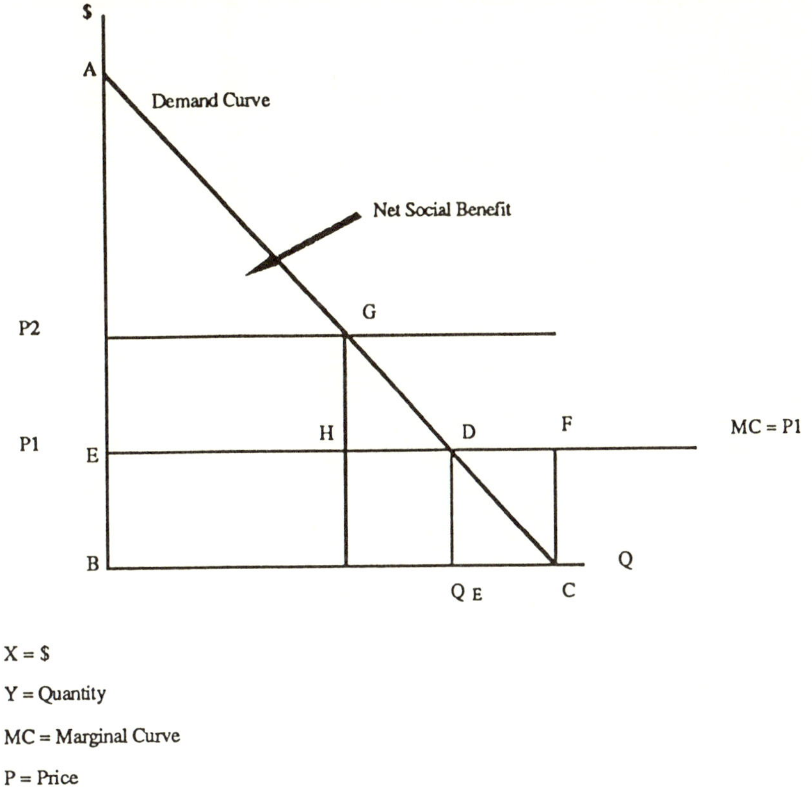

FIGURE 15.1
Demand curve for geographic information as a public good

When price is set at less than marginal cost, some users will produce products or gain utility, the value of which will be less than the marginal cost of providing the good. The difference between cost and value in this case is described as an efficiency cost. If price is set equal to 0, the efficiency costs are represented by the area CDF.

When price is set higher than marginal cost, there is also a loss of efficiency. This is because some consumers will not purchase the good and will fail to realize the value of products or utility that can be realized from the good even though the value will be greater than the marginal cost of the good. If price is set at P_2, the efficiency costs are represented by the area DGH. The public good in question could be a park, public transit, or geographic information.

There is an argument made by some users of geographic information that government should charge absolutely nothing for access to data under the assumption that public resources were used to create the data in the first place. What that argument fails to recognize is that while public resources paid for the initial cost of data capture, the marginal cost of providing access can be considerable. Such an attitude is fundamentally the same as one that would suggest that entrance to national or state parks should be free since those lands were purchased with public funds. The argument fails to recognize the loss of efficiency demonstrated above.

The prevailing attitude among public-sector decision makers appears to be that since the users of geographic information in digital form are a select group seeking benefits beyond those that accrue to the public at large, the application of fees (marginal or otherwise) prevents the public from subsidizing a relatively small group of users.

Conversely, there is an argument advanced by other policy makers that a price for geographic information closer to average cost than marginal cost is more appropriate. Such a pricing policy might allow for the recovery of costs of all or part of initial data capture. It would have several other consequences as well. The higher price would potentially exclude users from access and hence reduce total social welfare. It would transfer some of the net social benefit to the selling agency or private firm. It introduces some inefficiency into the situation by excluding benefits in excess of marginal cost that might otherwise be realized. This situation has been observed by some (Onsrud 1992) relative to Landsat data in the United States.

Arguments concerning the appropriate pricing of geographic data based on their characteristics as public goods ultimately depend upon one's view of the role of government. If one takes the view that government's role is to maximize social welfare, then marginal pricing of geographic data is the appropriate means to that end.

Some Comments on Measuring Value

It would be difficult to proceed with a discussion of incentives for the exchange of geographic information without touching briefly on issues related to the measurement of value. Efforts to measure the value of geographic information are based on the use of benefit–cost analysis as the only analytical method generally suited to the task. Initial attempts at employing benefit–cost analysis were only partially successful, and many of the intuitively suspected benefits appeared to evade quantification. Concepts such as tangible and intangible benefits applied to GIS (Dickinson and Calkins 1988) had limited value both theoretically and practically. More recent applications of benefit-cost analysis show significant progress. Some of these are discussed below.

Generally speaking, benefit-cost analysis involves the following steps:

1. Projects and alternatives to be analyzed are identified.
2. All required resources, products, and effects— both positive and negative, present and future—are identified.
3. Appropriate methods are applied to the valuation of inputs, outputs, and external effects.
4. Values are assigned to all inputs, outputs, and effects.
5. Net benefits are calculated.
6. A choice is made based on a pre-established decision rule.

These steps look simple in theory but are often very complex in practice. Placing values on the full range of outcomes and effects is rarely easy, even when considering costs and benefits that are readily measured in dollars. Benefit-cost analysis is an economic, not an accounting, procedure. Economists measure benefits and costs by a variety of direct and indirect methods employing such theoretical constructs as opportunity, cost, willingness to pay, and contingent markets. Many different assumptions go into calculating values, such as the choice of discount rate for future benefits, that can easily skew the results of the analysis. Because of these complexities and nuances, benefit-cost analysis has never been free of controversy, and its use in certain circumstances has led to heated debates. Some of its methods have been regarded with serious reservation by many economists and public policy analysts (Kelman 1981). Attempts to establish boilerplate benefit-cost techniques are ill-advised because the circumstances in which benefit-cost analysis are applied are always different, and the full range of costs and benefits is most often difficult to measure. It should not be surprising that measuring the full extent of benefits and costs associated with geographic information would prove difficult.

Environmental economists in particular have established a number of methods for measuring non-market benefits (Freeman 1979), a concept that is more useful than that of intangible benefits. Didier (1990) views all benefits of geographic information as non-market benefits, and in his *l'Utilite et Valeur of l'Information Geographique,* he applies many useful economic concepts to measuring the value of geographic information. In the application of benefit-cost analysis, Didier focuses primarily on measurement of efficiency benefits through the use of cost-avoidance calculations. Subsequent writers have expanded on this to include other types of benefits and describe methods by which theses benefits can be quantified. Antenucci

(1991) discusses five types of benefits—four quantifiable (efficiencies, expanded capabilities, unpredictable, and sales revenue) and one intangible—and demonstrates the practical applicability of these categories. In two studies conducted on behalf of the U.S. Geological Survey National Mapping Division (Gillespie 1992a, 1992b), potential benefits are classified as efficiency, effectiveness, and unanticipated benefits. This classification appears to have both a solid theoretical basis and practical applicability.

Efficiency benefits are calculated on the basis of cost avoidance, that is, by comparing different approaches to the same task with the same outcome and representing the difference in cost between the approaches as the benefit of using GIS. This is the area of benefit measurement most easily understood and calculated, and attempts are often made to justify GIS development on the basis of efficiency benefits alone, since they are the most easily measured.

Effectiveness benefits are representative of the types of benefits, often measurable in quantifiable economic terms, that could not be achieved without the use of GIS. They include difficult-to-measure benefits such as reduction of risk, better decisions, and improved outcomes. These are benefits for which it is often necessary to consider calculations based on measures of willingness to pay or other methods from which benefits can be assigned indirectly.

Unanticipated benefits can obviously be measured only after the fact but could fall into either category. It may be possible, however, to develop methods by which the frequency and magnitude of unanticipated benefits can be predicted and hence incorporated into planning and feasibility studies (Gillespie 1992a, 1992b).

It is quite possible that, even with the progress being made in the application of benefit–cost analysis to geographic information, we are being successful primarily in establishing the lower bounds of value. The cost-avoidance approach to efficiency benefits, for example, considers only those benefits that accrue directly to the user. There are many instances where the direct benefit to the user may represent only a small fraction of the total value of benefits.

Consider a situation in which the benefits can best be described as a combination of efficiency and effectiveness benefits as well as a reduction of risk—a benefit frequently beyond the reach of quantification. Underground storage tanks (USTs) represent a significant risk to environmental resources and human health. In Massachusetts, for example, there are in excess of 50,000 USTs, and it is estimated that as many as 10 percent of those USTs are leaking or in danger of leaking. Suppose that the average cost of cleanup of a leaking UST is $50,000, determined on the basis of the cost of removal of the tank and cleanup or removal and replacement of the contaminated soil and water. The best manual method for locating old USTs before they begin to leak can succeed in preventing only one hundred cases of contamination each year at a cost of $100,000. By using GIS, at a net cost above the manual method of $1 million, analysts are able to prevent 500 cases of contamination per year, representing an additional savings of $20 million per year in environmental cleanup costs. The benefit–cost ratio of that additional $1 million (effectiveness benefit) is 20 to 1. The efficiency benefit in the first one hundred cases is –$120,000. Based only on the efficiency benefits, the decision would be not to proceed with GIS implementation; based upon the effectiveness benefit, a decision to proceed would be in order.

Beyond the benefits indicated above, there is a category of benefit created as a result of the reduction of illness and premature death. Assume further that a study has been conducted that shows that contamination from USTs is expected to cause 200 to 250 cases of serious illness and mortality over the next ten years, and that the discounted medical cost alone associated with those cases is $50–$75 million. The removal of 500 USTs before they can cause contamination is expected to reduce risk of illness and death by 50 percent. The benefit of this outcome is thus placed at $25 – $37.5 million. The total range of social benefits (total societal cost avoidance) through use of GIS can be placed at $45–$57.5 million, and the greater portion of those benefits is associated with the reduction of risk of illness and mortality and does not accrue to the agency applying the GIS solution but to the community at large.

In theory such an assignment of benefits is not unreasonable. Society already demonstrates a willingness to pay for the cleanup of hazardous waste and has implemented laws to that effect. The direct benefit of avoiding the cleanup of contamination caused by 500 leaking USTs has thus been established at $25 million. It is justifiable to assign $25 million as the lower bound in this case; it is an indication of willingness to pay. The upper bound is some amount more difficult to establish, but it could well be in excess of the $57.5 million asserted above.

In practice, there is a tendency to include only those benefits of GIS that accrue to the particular agency involved in making the decision. But public policy makers should not exclude from their analysis the type of social benefits that have been discussed above. To do so is to set aside extremely valuable gains in public welfare, precisely the kind of gains that should be expected from government institutions. As greater skill, insight, and creativity are applied through GIS to problems for which there are currently no acceptable or only partial solutions, we should expect to derive greater value from geographic information.

The value of geographic information cannot easily be separated from its use nor can its benefits be arrived at separately from the value of the final decisions and social outcomes to which it contributes. It is those agencies and organizations that are capable of extracting significant value from the use of geographic information that should most willingly incur its costs.

Incentives for Exchange of Geographic Information

It is hoped that the preceding material has effectively demonstrated several key points: that the value of geographic information is not intrinsic but lies in the uses made of the information and the products derived from it; that successful efforts have been made to measure the benefits of GIS and geographic information; that for agencies committed to maximization of social welfare, marginal costing of geographic information is most efficient.

In the public sector, efficient organizations are cost minimizers; they produce a required level of output at the lowest possible cost. A theoretical goal of government from an economic point of view is to maximize social welfare. And, at times when fiscal constraints are greatest, there are increased expectations that dollars spent by government agencies will be expended in those areas where the social welfare/cost ratio is greatest (setting aside equity considerations for the purpose of this argument).

Agencies involved with GIS may be users of geographic information, suppliers, or both. Each of these three types of agencies will in all likelihood respond to a somewhat different set of dynamics. An agency that supplies geographic information will be motivated to do so in one of two different ways: it will have a mandate to produce such information as an input into the decisions of other agencies, or it will produce geographic information for its own use. If an agency produces information as an input into the decisions of other agencies, then as a cost minimizer/welfare maximizer it will produce such information as efficiently as possible. An agency that produces information for its own use will, as a cost minimizer, do so as efficiently as possible and will not pay more for the production of that information than the value of the final products it can produce.

Agencies that are users of geographic information will, as cost minimizers/welfare maximizers, prefer first to obtain information at marginal cost as the least expensive choice. Secondly, if data are not available at marginal cost, there is an incentive to engage in cost sharing with other user agencies (average cost). Finally, an agency will incur the full cost of production but only to the extent that the cost of production is less than the value of the final products and outcomes and, from a more global point of view, if the benefits are valued more than the benefits of other projects that the agency might undertake.

The majority of agencies using GIS are both users and producers of geographic information (but not all producers are necessarily suppliers). User/producer agencies will, as cost minimizers, wish to obtain data at marginal cost; as welfare maximizers, they will supply at marginal cost. Because effective (in applying geographic information) user agencies can, by definition, produce greater value per input of geographic information, they will be inclined to pay more than less effective users for that information. They are restrained by the need for cost minimization.

Effective user/producer agencies want data because it is the means by which they produce benefits and maximize social welfare. Given the choice between providing their own data at marginal cost (zero net gain to the agency) or exchanging their data for new data (from which additional benefits can be derived), the choice from both cost minimization and welfare maximization perspectives will be to make the exchange. A user-only agency will be limited to using dollars to pay marginal costs. A supplier-only agency will be inclined to accept only dollars, unless through the exchange agreement it is allowed to supply new users and hence minimize internal agency costs in supplying new data.

These various scenarios concerning incentives and behavior hold true only to the extent that certain assumptions are made about the goals of public agencies. In reality, agencies are not single-minded in their objectives, nor do they operate on the basis of clear, consistent principles. Yet, to the extent that many agencies do operate in part on such principles, it can be expected that they will often behave in the fashion described herein. It is also possible to modify or relax the various assumptions applied here and evolve different or more complex scenarios concerning the behavior of agencies relative to the exchange of geographic information.

It is also possible to look at related issues and consider how agencies might behave. The problem of free-riders (in the context of public goods) is a case in point. Agencies attempting to minimize costs will prefer to obtain data at marginal cost. Unwillingness on the part of some agencies to participate at average cost (through cost sharing) may mean that for a single agency the cost of some potentially valuable data sets will be in excess of the benefits that single agency can derive. From the perspective of society at large, the more users there are the greater the value of the benefits that will be generated. Yet, some potentially valuable data sets will not be produced because of single-agency cost-minimization goals. Hence, administrative rules need to be created to mitigate against such a loss of potential social welfare. The use of no-secondary-distribution clauses is an effort to get at this problem, but such clauses also introduce some potential inefficiencies.

CONCLUSIONS

This chapter has presented an example of the evolution of access to geographic information within a representative institution with a significant commitment to GIS. In addition, it has demonstrated that economic theory has a contribution to make relative to understanding the behavior of such institutions. The fact that agencies are not always clear concerning their operating principles, that principles may change, or that some actions may be inconsistent with general principles of operation should not be construed as reasons to assume that decisions to exchange geographic information are made without consideration of greater social goals.

While agencies behave differently in practice than they might in theory, empirical observations have been made that suggest that some public agencies are indeed cost minimizers and social welfare maximizers, or at least that they aspire in those directions. Clearly, public agencies are not single-minded in their behavior, just as not all private firms are effective profit maximizers (assuming that is their intention).

As a greater number of public agencies become effective users of geographic information, they will place greater value on access to data. Exchange agreements and vehicles may take many different forms, but it is not unlikely that they will adhere to the principles discussed here. The dual objectives of generating valuable social outcomes and the need to minimize costs will act as incentives for agencies to enter into data-for-data exchanges.

REFERENCES

Antenucci, J. C. et al. 1991. *Geographic Information Systems: a guide to the technology*. New York: Van Nostrand Reinhold.

Dickinson, H., and H. W. Calkins. 1988. The economic evaluation of implementing a GIS. *International Journal of Geographical Information Systems* 2, 4: 307–327.

Didier, M. 1990. *The utility and value of geographic information*. An edited rendering into English by J. Shepherd, Birkbeck College, University of London, England.

_____. 1990. *l'Utilite et Valeur de l'information geographique*. Paris: Editions Economica.

Executive Office of Environmental Affairs. 1990. *Digital data distribution policies.* Boston, MA: Commonwealth of Massachusetts.

Freeman, A. M. 1979. *The benefits of environmental improvement.* Resources for the Future. Baltimore: John Hopkins University Press.

Gillespie, S. 1992a. *Test of digital benefits model.* National Mapping Division, U.S. Geological Survey, Reston, Virginia.

_____. 1992b. *Digital benefits test.* National Mapping Division, U.S. Geological Survey, Reston, Virginia.

Kelman, S. 1981. Cost-benefit analysis: an ethical critique. *AEI Journal on Government and Society* (January/February).

Onsrud, H. J. 1992. In support of open access for publicly held geographic infomation. *GIS Law* 1, 1: 3–6.

Parent, P. 1989. Geographic Information Systems: issues arising from the proliferation of information. *Journal of the Urban and Regional Information Systems Association* 1, 1: 17–26.

Rhoads, S. E. 1985. *The economist's view of the world: government, markets, and public policy.* New York: Cambridge University Press.

Taupier, R. P., and M. Terner. 1991. MassGIS: a case study of the Massachusetts Environmental/Geographic Information System. Urban and Regional Information Systems Association 1991 Conference, San Francisco, California, August, *Proceedings* 2: 87-98.

Taupier, R. P. 1992a. *Calculating the benefits of GIS: the view from Massachusetts.* Unpublished paper presented at the National Governors' Association Conference on Making Information Work. Washington, D.C. January.

Taupier, R. P. 1992b. State distribution of digital geographic information: the how, what and why of it all. National Governors' Association Conference on Making Information Work, Washington, D.C., January, *Proceedings,* pp. 219–226.

Warneke, L.; J. M. Johnson, K. Marshall, and R. S. Brown. 1992. *State geographic information activities compendium.* Lexington, KY: Council of State Governments.

Wilcox, D. L. 1990. Concerning the economic evaluation of implementing a GIS. *International Journal of Geographical Information Systems* 4, 2: 203-210.

Harlan J. Onsrud

16 The Role of Law in Impeding and Facilitating the Sharing of Geographic Information

The widespread use and accessibility of geographic data sets in combination with the capabilities provided by other communication technologies are raising numerous public concerns. Among these are the effects of such technologies on personal privacy, access to the information used by government, work product protection laws, legal liability for errors and inadequacies in Geographic Information System (GIS) products and services, and concerns that information infrastructure arrangements will contribute to the widening of socioeconomic gaps among members of the social system. As developers and purveyors of a powerful social resource, the GIS discipline needs to identify the consequences of use of the technology and consider those consequences in the light of their general social effects. The policy implications of different geographic information sharing arrangements need to be explored, and the legal conditions and constraints that will affect the ability to share geographic data widely among divergent groups of potential users of GIS need to be fully discussed.

INTRODUCTION

In considering the social effects of increased accessibility to geographic information, the GIS discipline's goal should be to expose and consider those policies, practices, and technical capabilities that increase the beneficial consequences and decrease the detrimental consequences of geographic information sharing arrangements. Determining what constitutes a beneficial versus a detrimental social effect, however, is a value-laden judgment.

Increased sharing and dissemination of geographic data will be highly valuable in better dealing with a broad range of critical social problems: the adverse effects of overpopulation, the use and stewardship of increasingly scarce resources, and environmental problems from local level to global scales. However, it is also widely accepted that information and the knowledge it brings are a source of power. Those with the most efficient access to accurate information are in a better position to take economic advantage of that information or to promote their political stance on issues than those without such access. Wealth creation in modern society often has more to do with access to and use of information about capital, land resources, and labor than it does with, for instance, the current possession of such resources. Thus, economic power and political power are often viewed as the ultimate stakes in the race to capitalize on efficient information systems.

Because geographic information has potential value to those with effective access to it, this realization gives rise to the desire to exercise ownership rights over the information. Thus, the power that information provides is antipathetic to sharing. In addition, this desire to control information is in direct tension with recent technological realities that make the copying, dissemination, and sharing of information very inexpensive (Brand 1987). As technology improves, this tension will only increase. Further, as databases become more accessible, comprehensive, and useful, tensions will increase between the desire to control geographic information for private gain versus the desire to use the information for public benefit.

Sharing of geographic information necessarily presupposes the existence of relationships among individuals, organizations and/or governmental units. As spatial databases are developed that are accessible to others than those who created them, relationships among the involved parties and societal control over the databases will be adjusted through legal mechanisms. That is, the value judgment on whether a social effect caused by a geographic information sharing arrangement is beneficial or detrimental to society will be made within the framework of our legal system and within the context of the functions and limitations of law.

THE ROLE OF LAW

The law has many functions. These include settling disputes, maintaining order, and providing a ". . . framework within which certain common expectations about the transactions, relationships, planned happenings, and accidents of daily life can be met" (Mermin 1973, 6). The law also " . . . [secures] efficiency, harmony, and balance in the functioning of government machinery, . . . [protects citizens] against excessive or unfair government power, . . . [protects people] against excessive or unfair private power, . . . [and ensures people] an opportunity to enjoy the minimum decencies of life" (Mermin 1973, 7–8).

The law also has limits. "[Because] basic socioeconomic conditions and interests have a shaping influence on the law, one cannot expect the law to turn around and completely transcend those conditions and interests" (Mermin 1973, 11). The law also " . . . dare not get too far away from popular attitudes, habits, and ideals. . . ." (Mermin 1973, 13). In other instances, " . . . difficulties and risks in the administration of an otherwise desirable right may make the granting of the right more harmful than beneficial" (Mermin 1973, 42).

These general legal principles seem so basic that perhaps they do not warrant repeating. Yet their restatement helps point out that the GIS community needs to resist altering laws that "stand in our way" until the proposed changes have been evaluated with respect to the law's general functions and limits and the larger social implications of such changes have been exposed and debated. Written laws, legislated or judge-made, reflect policies that are currently supported, or at least condoned, by society. Seldom are laws founded on a single underlying policy. The social policies supported are often complex and intertwined in a delicate balance.

Changing a law to achieve a specific purpose desired by some members of the GIS community may upset the balance among the current social policies supported by the law. The result may be a string of consequences or ramifications wholly unintended and highly undesirable. Yet, it can also be argued that the rapid advancement and diffusion of GIS and other communication technologies have already changed the social condition. Because of these technological changes, a rational balance fails to currently exist, and laws must be altered in order to accommodate the new social condition. Regardless, the legal and social ramifications of proposed organizational and technical alternatives for widespread sharing of geographic information should be thoroughly evaluated prior to advocating any change in law to support one or more of those alternatives.

INFORMATION IS DIFFERENT

Information has become a dominant resource in society. "Information (organized data, the raw material for specialized knowledge and generalist wisdom) is now our most important, and pervasive, resource. Information workers are now more than half the U.S. labor force" (Cleveland 1985, 185). Through the application of new information and knowledge we have learned we can gain more for less: more products, productivity, and efficiency for less labor, capital, and physical resources.

However, information as a resource possesses characteristics that are inherently very different from the traditional resources of land, labor, and capital. As such, laws developed for controlling the exchange of these other resources may not work well (or at all) in attempting to control the flow of information. Some of the unique characteristics of information described by Harlan Cleveland (1985) are paraphrased as follows:

- *Information is expandable.* It expands as it is used. It is consumed yet not depleted. Its growth is bounded only by the time and capacity " . . . available to human minds for reflecting, analyzing and integrating the information. . . ."
- *Information is not resource hungry.* When compared to other products of value in the marketplace, few physical and biological resources are needed to produce and transmit information.
- *Information is substitutable.* It can and often does replace capital, labor, and physical materials.
- *Information is transportable.* Unlike other resources, it can travel at the speed of light. Its transport is far less affected by geography than other resources.
- *Information is diffusive.* "It tends to leak—and the more it leaks the more we have." Business and organization approaches depending on the exclusive control or possession of information are likely to succeed " . . . only in more and more specialized fields, for shorter and shorter periods of time."
- *Information is shareable.* "Things" are exchanged but information is shared. Even after you give it away or sell it, you still possess it.

These unique characteristics of information, along with its sudden emergence as a valued marketplace resource, have caused many of the assumptions and concepts that underlie our current laws for regulating resource-exchange relationships to be thrown into question. Cleveland (1985, 187) goes on to argue that:

> [T]he inherent characteristics of physical resources ("natural" and manmade) made possible the development of hierarchies of power based on *control* . . . , hierarchies of influence based on *secrecy*, hierarchies of class based on *ownership*, hierarchies of privilege based on *early access* to valuable resources, and hierarchies of politics based on *geography*. (emphasis added)

> Today, because of the dominance of the information resource ". . . the old means of control are of dwindling efficacy, secrets are harder and harder to keep, and ownership, early arrival, and geography are of dwindling significance in getting access to the knowledge and wisdom which are the really valuable legal tender of our time.

Our legal system is adapting on a daily basis to the changes that the information age is bringing about. For those involved in developing GIS databases and sharing information in them, social issues likely to loom large in the near future include concerns for personal privacy, access to the databases used by government, work product protection (e.g., copyright, patents, licensing, contracts), legal liability for errors and inadequacies in GIS products and services, and concerns that information infrastructure arrangements will contribute to the widening of socioeconomic gaps among members of the social system. In each of these areas, existing laws and restrictions were formulated to uphold policies that society either supports or has chosen to live with. In enabling or promoting more efficient handling and widespread sharing of information we are inviting conflict with the existing framework of laws and the underlying power balance that those laws support.

LEGAL INCENTIVES AND BARRIERS TO THE SHARING OF GEOGRAPHIC INFORMATION

It can be argued that the existence of a particular law has never been an incentive to invent or develop any innovation. One does not develop a GIS database or develop a data-sharing arrangement *because* legal protection for that database or data-sharing arrangement is available. The incentives for individuals and organizations to create and utilize an innovation are typically profit and recognition. In addition to monetary gain, the incentive of profit encompasses such objectives as increasing efficiencies, decreasing costs, and increasing personal satisfaction or pleasure. Moral conviction that implementing an innovation is the "right thing to do" may also be a driving incentive in some instances. There is no doubt that many innovations are pursued with little regard for personal reward or recognition simply because individual innovators believe that a proposed change will benefit the general public. Specific laws that prevent us from taking some action in order to gain that profit, recognition, or moral objective are typically viewed as barriers rather than incentives to innovation.

However, the United States legal system when viewed as a whole, while not itself an incentive to innovate, provides an environment that allows the incentives of profit, recognition, and moral conviction to ripen. In the early stages of developing a technological innovation, the goal of U.S.

law generally is protection of innovators and entrepreneurs. Familiar intellectual property law protective devices used at this stage include patent, trade secret, and copyright laws. As an innovation is exploited (i.e., with products being introduced to the marketplace and modified over time), balancing rights in the competitive marketplace gains greater importance. In this later stage, regulatory, commercial, and tax laws take on greater roles. As a general rule, innovations that have a marketplace value and are adequately financed and marketed typically have little difficulty in the United States in overcoming any inconveniences imposed by these laws.

Difficulties arise primarily when actions are proposed that attempt to alter or circumvent rights protected by existing laws. Certainly, laws must change in response to current social and technological realities. Changes in the balance of protected rights are inevitable. However, changes in the law should not alter the balance of protected rights so drastically that the legal system's ability to foster incentives for innovation or support social equity is significantly reduced.

It should also be mentioned that if your market for providing digital data to others is secure, you are unlikely to attach much importance to legal protection of your database. If others find it difficult to compete with your database services, with or without the protection, this means you probably don't need more than minimal legal protection. Some of the reasons why others may find it difficult to compete with you include the following: you have a highly efficient operation; you possess "know-how" or an established infrastructure that others do not have; you have access to information that others do not; your database services are necessary for purposes internal to your organization anyway and can be made available to others at a very small additional cost; your system is subsidized by taxpayers; or one of many other possible reasons. Perhaps the value of the data you are providing is primarily due to its timeliness or its guarantees of reliability. In such instances, even if others copy the data sets you possess, they still need to be able to compete with your updating capabilities or your ability to guarantee the data.

If the market for your data products and services is insecure, you are much more likely to want to maximize any protection you can gain from the intellectual property laws. Because of the inherent characteristics of information noted previously (including its "leakiness"), maintaining market security in a database—the value of which is not dependent on its timeliness—may be very difficult. Lack of market security and heavy dependence on intellectual property laws in order to gain economic advantages from a GIS database may foretell preoccupations and disputes over rights to the data in the future.

CONSTRUING OPEN RECORDS LAWS AS SUPPORTIVE OF WIDESPREAD SHARING OF GEOGRAPHIC INFORMATION

The current debate surrounding open access to government-housed geographic databases provides a ready forum for illustrating several of the issues raised above. Open records laws are often looked upon by local government administrators as barriers to the creation of useful GIS operations. Yet, the argument can also be made that such laws, when viewed from a broader context, are critical in providing a legal environment that supports the diffusion of GIS innovations and GIS data sharing generally.

To explore the status quo of open records laws as an incentive to the widespread use and sharing of geographic information and to demonstrate the effects of a change in one area of the law on other areas of the law, consider the following hypothetical scenario:

> *Local government administrators in several cities in state X are strongly advocating alteration of the state's open records law in order to exempt GIS data sets from the provisions of that law. The reason for doing this, of course, is that the local government administrators see a potential means of paying for the creation and maintenance of improved land information systems other than through their general tax revenues. In these times of tough budgets, they see an opportunity to generate income through the sale of products and services resulting from their GIS operations. They also believe their income will be considerably reduced in providing these products and services if private firms and private individuals are allowed to copy these publicly financed spatial data sets at the cost of duplication.*

Restricting access to public records goes against the plain letter language of most existing state open records law. Those seeking to change such laws must overcome the underlying policy arguments on which the laws are based, foremost of which is that open records laws keep government accountable. In addition, proponents of change will almost certainly encounter other credible policy arguments that favor free and unfettered access to public records.

In evaluating the legal issues that arise, we begin with the presumption that the general social policies currently supported in this and closely related areas of the law will continue to be supported in the future regardless of the impact of contemporary technology. That is, even though the letter of the law may change significantly or exceptions for special situations may be made, the underlying policies that the law was intended to support will continue to be supported, or at least will change at a much slower rate than the written law.

The following proposition and advocacy arguments in support of it are offered in the spirit of raising several public policy issues.

PROPOSITION:

Allowing wholesale duplication of publicly held land records is good for the public welfare.

ARGUMENTS:

1. *Government-sanctioned monopolies are less efficient than private enterprise.*

After a government department has gained an exclusive right to provide a specific service that generates income, there is little incentive for the government bureaucracy to improve that service. If a monopolistic government service makes a net income for the government, the operation is almost certain to be inefficient. In the land records area, the most obvious example is local registry of deeds offices. In a typical year, most of these offices in the United States have income far greater than the total expenses to run their operations (Jeffress 1991). Even in an extremely slow year for real estate transactions they seldom run at a loss. If they do, the simple solution to cover expenses is to raise recording fees rather than increase efficiencies. The captured users have no recourse but to pay. Because of their exclusionary nature and the fact that they almost always make a profit, most of these offices are at least a decade behind the proven technology that would allow them to increase the efficiency of their operations.

By analogy, if administrators of government-housed Geographic Information Systems have an exclusive right to set a price for government-gathered data, there is little incentive for them to improve the database and its administration over time or to increase the products and services provided by the GIS. However, if a government program such as a GIS operation is a burden on general tax revenues, it is likely to be more efficient over time because it needs continual justification in the budgeting process.

In the private marketplace, competition promotes efficiency and innovation. That efficiency is passed to consumers, which means that all users of the product or service gain. Thus, if twenty local engineering firms wish to copy a local government's GIS data sets en masse in order to better compete with each other in delivering services of value to the public, they should be encouraged to do so. Their need to remain competitive results in overall lower costs and more innovative products and services for the consuming public. Our federal Freedom of Information Act supports this fundamental economic principle; that is, that private enterprise and the private marketplace are more efficient than government. If this policy makes good sense at the federal level, it makes doubly good sense at the state and local levels.

Another argument sometimes made is that since local government has gathered information at taxpayer expense, government officials have a responsibility to hold this information "in trust" for the benefit of all citizens. Allowing commercial enterprises to acquire public data sets at the cost of reproduction would breach this public trust. However, the public trust doctrine is founded primarily on the principle that valuable public resources should be protected from consumption or despoliation by one generation so that those resources remain available for the enjoyment of future generations. Because information may be consumed at will and yet remains available for innumerable others as well as future generations, the argument can be made that the public trust doctrine is simply inapplicable to public information.

2. *Cost-recovery arrangements create bureaucratic overhead and legal disincentives to the sharing of geographic information.*

In performing strictly governmental functions, local governments are provided with some measure of protection against liability for their actions through sovereign immunity. However, much of this protection may be lost if the governmental unit goes beyond its traditional public service missions and moves into areas that might be classified as private commercial functions in nature. It is likely that charging "user fees" for "value-added products" (i.e., products not necessary for the performance of government) with the intent of generating income in excess of expenses would be construed by the courts as making such a GIS operation a private commercial venture in nature. If a private enterprise offers products or services for sale, basic liability principles mandate that some level of acceptable performance in the delivery of services or the quality of products is required. To make clear what its level of performance will be in order to decrease its liability exposure, the local jurisdiction will want to enter into license, contract, or alternative formal arrangements with all those seeking value-added products from the GIS operation. In addition, the jurisdiction will likely want to use contracts or licensing to ensure that persons purchasing information products or services will not distribute copies of those records free to other potential customers.

Developing written contractual relationships will create an administrative burden on the department housing the GIS. There are likely to be significant fees for attorneys and consultants in establishing the formal process for dealing with user requests as well as continuing administrative expenses over time. Once a "closed information environment" based on contracts and licensing is established, we can also predict breach-of-contract lawsuits and persistent involvement of lawyers in our GIS operations in the future. Lawyers will be necessary to deal with issues such as intellectual property rights, liability, equal treatment, antitrust, undue competition, and similar issues.

Without user fee arrangements, an "open information environment" exists. If citizens want a copy of a GIS record, whether in paper or digital form, they are typically free to have it at the total cost of duplication. The government in effect is saying: "This is a copy of a record we made in the course of doing the government's business. It may or may not have been suitable for some governmental purpose and we make no guarantees, implied or otherwise, that it may be suitable for some purpose you intend." The courts are likely to uphold this traditional position of local governments and support their sovereign immunity. In such an environment, there is little need for contracts in meeting user requests for GIS records.

Even if a local government believes it has the ability to create a restricted GIS information environment, it still needs to question whether a sufficient market really exists for the planned value-added products it intends to sell. Is the expected income sufficient to more than offset the increased administrative trouble and expense? Will the markets for products continue over time even with advances in technology such that the GIS operation will continue to produce a significant income?

GIS systems are often justified in local and state governments based on the argument that they will reduce duplication of data collection and will promote sharing among government agencies, different levels of government, the private sector, and the general public. The need to enter into contracts negates much of the willingness and practical ability to share and thus eliminates the benefits to be gained by sharing. If public GIS operations build walls made of contracts and licenses around themselves, it is difficult to see how these restrictive arrangements will foster cooperation and sharing with large numbers of citizens or other agencies.

Successful GIS operations in local governments in the United States have often been justified solely on cost savings and increased efficiencies and services in carrying out the currently defined statutory missions of that government. If a GIS project is possible and justified without pursuing cost recovery arrangements that go beyond traditional duplication charges, it can be argued that this is by far the preferred alternative. Cost recovery and user fee arrangements promote an increase in the size of government bureaucracy to produce services and products that are outside of government's role.

The underlying premise of these arguments is that local government agencies should be in the business of collecting and analyzing data only in support of their legislated public missions. If a GIS allows them to carry out these public missions more efficiently or to expand the services they can offer in support of their missions, local governments should by all means take advantage of the capabilities that GIS provide. However, if the implementation of a GIS operational environment in local government is justified only if value-added products and services are supplied and sold (i.e., those not necessary for the performance of government), perhaps the technology is not yet ripe for use by the government organization.

3. *All other rights in a democratic society extend from our ability to access information. Democracy cannot function effectively unless people have ready access to government information in order to keep government accountable.*

Geographic Information Systems are being used and will continue to be used in local government to help make permitting decisions, planning decisions, tax assessment decisions, and decisions on where to spend tax dollars. A GIS may be used to determine how a child's school bus is routed, whether and when the street in front of one's home is fixed, and whether one is served by one emergency service rather than another. All of these decisions affect people personally. The right to access information about government is very different from the right to have consumer goods such as water or electricity delivered to our homes. Access to government information involves the basic right to know on what our government is basing its decisions.

We can readily determine whether the flow of electricity or the flow of water has been cut off to our homes. But how do we determine whether the flow of information has been cut off? We don't, unless we have the right to inspect the entire body of records upon which the government is making its decisions. The potential for abuse is quite staggering if local government administrators have ulterior and inappropriate reasons for wanting to hold back information. For instance, individual citizens or citizens groups will find it virtually impossible to determine whether computerized gerrymandering has occurred in establishing the bounds of a district or in determining where tax dollars are going to be spent unless they can efficiently gain access to the same data sets. Cost-recovery arrangements and contract negotiations over providing access to government-housed GIS data sets provide a ready excuse for elected officials or government administrators to delay release of records or hold them back altogether. Because of the need to maintain confidence in our public administrators and elected officials and to avoid accusations that they may be holding back records about which citizens have a right to know, we should not restrict access to GIS data sets gathered at taxpayer expense simply because that information is commercially valuable to the government.

The preceding three arguments are but a sample of the arguments that can be raised in support of open access environments for publicly held geographic information. Note that of the arguments raised, none speaks against expanding access to government databases through on-line services and charging fees to support those services. This presumes, of course, that on-line users are free to duplicate the government-housed GIS files. Note also from the discussion above that arguments may be more aptly framed by addressing specific cost-recovery arrangements (for example, arguments relative to the creation of an "information utility" as opposed to an in-house local government GIS service).

As stated at the beginning of this section, the intent of presenting the preceding advocacy arguments is to illustrate that maintaining the status quo of open records laws may be argued as supporting the widespread use and sharing of geographic information. Thus, the intent has been to present advocacy arguments but not necessarily to take an advocacy stance on one side or the other of the "open access" versus "cost recovery" debate. Local governments attempting to cover GIS costs through some mechanism other than general tax revenues should be prepared to address the issues raised above. Counterarguments to those presented do exist (see, for instance, Onsrud 1992 and King 1995), and it is up to local jurisdictions to convince themselves of the appropriateness of the policies and practices they choose to pursue in implementing GIS in their locality.

Open access issues are only one set in a series of critical legal issues that need to be debated and reevaluated relative to their underlying social policies and in the light of recent technological advancements. Additional legal areas relevant to GIS data sharing that beg comprehensive review include personal privacy, work product protection for GIS products, services, and databases (e.g., copyright, patents, licensing, contracts), legal liability for errors and inadequacies in those same products and services, and concerns that GIS infrastructure arrangements will contribute to the widening of socioeconomic gaps among members of the social system.

CONCLUSIONS

If information and the knowledge it brings are power, the more widely information is shared the more diffused will be the power. If the diffusion of economic and political power is considered an admirable social goal, sharing of geographic information would appear inherently to be a good thing.

Yet is sharing of geographic information best promoted by advocating "free access" environments or by constructing legal means of protecting those ownership rights in information that provide economic incentives to its dissemination? Is dissemination of geographic information best promoted by allowing commercial businesses to gather and distribute any information they wish to on us as individuals, or will greater social utility and use of GIS data sets be gained by restricting the gathering or disseminating of some forms of personal information? Will greater sharing take place if we increase the reliability of data sets by making their producers and disseminators liable for damages resulting from errors and inadequacies in those data sets, or will greater societal sharing take place if *caveat emptor* policies are promulgated in the electronic information legal arena? Do work product protection laws simply need to be added to in order to cover new forms of electronic information, or is this new resource so inherently different that the assumptions upon which our work product protection laws have been built need to be reevaluated? What is the best legal environment for promoting the incentives of reward and recognition such that sharing of data will take place, diffusion of information will occur, and reasonably equitable opportunities to take advantage of the information will result?

The question remains whether the GIS discipline as a whole cares to address these broad social issues. As a discipline, we have the option of letting society react as best it can to the technological advancements being made in our field and the application of those advances to the social condition. As a discipline, we might see our primary objective as the technological advancement of GIS capabilities and leave the scholarly and practical consideration of social impacts to others than ourselves. Alternatively, we can take a proactive role—a proactive role in contemplating, discussing, and debating the social and legal ramifications of the GIS databases and processing capabilities being proposed by those inside and outside our discipline. By taking such a role, GIS applications and developments are far more likely to contribute to the elimination of bases for discrimination and unfairness in our culture than to their increase.

ACKNOWLEDGMENTS

1. This article is based on work partially supported by the National Science Foundation under Grant No. SES-88-10917. Any opinions, findings, conclusions, or recommendations expressed in the material are those of the author and do not necessarily reflect the views of the National Science Foundation.

2. This section contains extensive extracts from H. J. Onsrud (1992), "In Support of Open Access for Publicly Held Geographic Information," *GIS Law* 1, 1: 3–6. The material is reprinted here with permission of the editors of *GIS Law*.

REFERENCES AND FURTHER SELECTED READINGS

Anonymous. 1990. Access to government land information: commercialization or public benefit, issues in land information management. Paper Number 4, Belconnen ACT. Australia: Australian Land Information Council.

Anonymous. 1991. Policy issues and implications, the local government guide to Geographic Information Systems: planning and implementation. Chapter 2. Washington, D.C.: Public Technology, Inc., Urban Consortium, and International City Management Association.

Antenucci, J. C., K. Brown, P. L. Croswell, and M. J. Kevany. 1991. *Legal issues, Geographic Information Systems: a guide to the technology.* Chapter 11. New York: Van Nostrand Reinhold.

Archer, H., and P. L. Croswell. 1989. Public access to Geographic Information Systems: an emerging legal issue. *Photogrammetric Engineering and Remote Sensing* 55, 11: 1575–1581.

Brand, S. 1987. *The media lab: inventing the future at MIT.* New York: Viking.

Braverman, B. A., and W. R. Heppler. 1981. A practical review of state open records laws. *George Washington Law Review* 49, 9: 720–760.

Cleveland, H. 1985. The twilight of hierarchy: speculations on the global information society. *Public Administration Review,* 185–195.

Dansby, H. B. 1991. Recovering GIS development cost by copyright. *GIS World* 4, 2: 100–101.

Epstein, E. F. 1991. In my opinion. *Journal of the Urban and Regional Information Systems Association* 3, 1: 2-4.

Epstein, E. F., and J. D. McLaughlin. 1990. A discussion on public information. *ACSM Bulletin* 128 (October): 33–38.

5 USCS § 552. Freedom of Information Act (FOIA).

Graham, J. M. 1984. Note: fair administration of the Freedom of Information Act after the computer revolution. *Computer/Law Journal* 5: 51–76.

Henderson, F. B. 1990. Advanced satellite earth observations for environmental management during 1990-2010. Symposium on Global and Environmental Monitoring Techniques and Impacts, Victoria B.C., ISPRS, September, *Proceedings* 28, Part 7-1: 501–526.

Jeffress, G. A. 1991. Land ownership information use in real property market transactions. Doctoral dissertation, University of Maine.

King, J. L. 1995. Problems in public access policy for GIS databases: an economic perspective. In H. J. Onsrud and G. Rushton, eds., *Sharing geographic information*. New Brunswick, NJ: Center for Urban Policy Research, Rutgers University.

Lauer, D. T., J. E. Estes, J. R. Jensen, and D. D. Greenlee. 1991. Institutional issues affecting the integration and use of remotely sensed data and Geographic Information Systems. *Photogrammetric Engineering and Remote Sensing* 57, 6: 647–654.

Mermin, S. 1973. *Law and the legal system: an introduction*. Boston: Little, Brown, and Company.

Miller, R. I. 1974. *Legal aspects of technology utilization*. Lexington, MA: Lexington Books.

Onsrud, H. J. 1992. In support of cost recovery for publicly held geographic information. *GIS Law* 1, 2: 1–7.

Roitman, H. 1990. Legal issues in providing public access to an AMS: case studies and variances. *Journal of the Urban and Regional Information Systems Association* 2, 1: 38–45.

Tosta, N. 1991. Public access: right or privilege? *Geo Info Systems*, 20–62.

Westin, A. F. 1971. *Information technology in a democracy*. Cambridge, MA: Harvard University Press.

Earl F. Epstein

17 *Control of Public Information*

Public information data sharing depends upon incentives and barriers that are legal and economic as well as technological. The complex set of relations between people in regard to use of information is expressed legally and economically. These relations, constituting an information institution, determine who has what rights in the information, who controls the information, and conditions for sharing among producers and users. The legal regime includes intellectual property, freedom of information, tort, and government law. Geographic and land information relationships add environmental, land use, property, boundary, and resource law to this list. Fundamental and traditional democratic principles, especially open government concepts, and difficulties in capturing information as an economic commodity mean that distribution of public information by public agencies at dissemination costs remains good public and economic policy in the age of computers.

INTRODUCTION

Data sharing is part of the complex relations among people in regard to use of information products and services. Changes in these relations are suggested because people and organizations are encouraged to share data and information by the information management and dissemination efficiencies made possible by computers. However, technology alone is an insufficient incentive for sharing. Technology provides alternative means for achievement of an objective. Often, perspectives on the process, outcomes, and effects associated with the technology differ among people. If perspectives are negative, then barriers to use of a technology or its product are created. Therefore, data sharing is a non-technical as well as technical matter.

Cultural, legal, and institutional factors are important, sometimes crucially, in the development and use of a technology and its products (Mayo 1985). Data and information sharing is often seen as a compromise with the advantage suggested by the conventional wisdom that information is power. This compromise may be personal, organizational, economic, legal, or political. Therefore, data sharing requires incentives strong enough to overcome the barriers. The most powerful incentives may be non-technical.

INSTITUTIONS AND THE INSTITUTIONAL PROBLEM

Data are observations. Information is data in a context: data transformed by their representation in a particular medium. A map is an example of information. Representation of positional data in the specific context created by a map is unique in that the same data represented at a different scale, accuracy, or other map attribute presents different information. Knowledge is the result of the application of the human mind to information. For example, each individual carries different knowledge from examination of the same map. The development of data and information uses, including sharing, must account for these distinctions.

An institution is defined as "a custom, practice, relationship, or behavioral pattern of importance in the life of the community or society." This can be restated to define an institution as the social, economic, and legal relations between people in regard to an activity or use of an object. An information institution consists of the relations between people in regard to use of information.

Data and information are sought and used for purposes of:

- national security,
- aesthetic, scientific, and educational needs, and
- management

Reducing uncertainty in management activities (policy, decisions, and actions) is a major reason people seek information (Mackaay 1982). Information is used to reduce the uncertainty in land and management decisions to levels where people are willing to absorb the remaining uncertainty and act (Bedard 1986). Achieving the necessary levels of uncertainty is more difficult when there are barriers to the flow of information.

The benefits of investments in an information system are determined by the attributes that the technology imparts to the data and information. These benefits fall into three classes:

1. *Efficiency.* The technology changes the attributes of the data and information such that the production of products is more efficient and timely. The result is avoided costs compared with production by traditional means.
2. *Effectiveness.* The altered attributes make it possible to garner additional information from the same data. It is now practical to do operations (map overlays, analysis, and so on) that previously were imaginable but not reasonable by traditional means because of time or expense.
3. *Equity and Fairness.* The altered attributes make it feasible to expand the participation of people in the management process. Policy, decisions, and actions affect many people and their interests. If those activities are not regarded as fair or equitable, then they may not be implementable because they lack political and social support. Data and information used to support management decisions, if available, accessible, and usable by the community of affected citizens, contribute to a sense of fairness in management activities.

Equity and fairness in the management process, based on access and availability, may be the greatest, and largely unrealized, benefit of information systems. It is an expansion of democracy and an expansion of the use of data and information.

LAND MANAGEMENT AND GEOGRAPHIC INFORMATION SYSTEMS

Land and resource management is the art and science of making decisions in support of perceived objectives with respect to the inventory, allocation, and conservation of land and its resources (McLaughlin 1988). These activities are represented by the term "land tenure," the set of relations between people in regard to use of land or its product.

A basic component in land and resource management is land and geographic information, gathered in a formal, systematic process—the product of an information system. A pattern of relations between people in regard to the management and use of geographic and land information has emerged, sufficient to form an institution. There is a geographic and land information institution defined as the relations between people in regard to use of geographic and land information.

A geographic and land information institutional problem exists because people are unwilling to share data and information when they believe that they or their organization will suffer in some way if they do. Manifestations of the problem appear as legal and economic impediments to the flow of information from those who have it to others who do not but who might use it. Specific incentives and barriers appear in laws and practices that create rights in data and information that determine conditions under which sharing occurs. Economics is also a source of incentives and barriers, including the important element of transaction costs (Coase 1960).

INFORMATION CONTROL IN THE LEGAL PROCESS

The Legal System

A legal system has five elements: statutes, administrative rules, court opinions, professional standards, and commonly accepted behavior (Friedman 1984). These constitute both formal and informal elements. Portions of the legal process applicable to information of the legal process establish proprietary rights, determine access conditions, define liability for misuse, and provide a framework for the economic development of public and private information systems. Intellectual property, freedom of information, tort, government, environmental, land use, boundary, and resource law are included.

Public Information

Public information is data and information gathered, received, managed, or otherwise used by a government agency in the normal course of its mandated activities. Public information supports and records a public agency's activities.

Geographic and land information is a significant part of public information, especially at the local level where land-related information management activity is focused (Larson 1978). Much of the material assembled privately for land management becomes public because it must be submitted to an agency in order to obtain a land-use permit. This material becomes part of the public record (Daylor 1982). The result is that public agencies have large arrays of geographic and land information that they have generated or received from others in the course of their mandated activities.

Public Information Values

Four values are expressed in the information legal regime. These are public, private, commercial, and agency efficiency values.

Public values. Public values express the citizens' need to know what their governments are doing. The basis for government activity must be conveniently available to citizens in order to make this value a reality. "The database is the model of government" (Westin 1991).

These values are the basis for the U.S. federal government Freedom of Information Act and the analogous state acts that are commonly known as Open Records Laws (FOIA; Puissegur 1989). A common requirement in these laws is that records and data associated with government actions be open to inspection by citizens. In most cases inspection carries the reasonable requirement that a citizen be able to copy the document or data at the approximate cost of reproduction or dissemination. Most acts have no restriction on use of this material.

Many people believe that the purpose of these laws is to provide access to government documents and data. Instead, the goal is exposure of what governments are doing. Access is the means to achieve this goal. Information is empowerment in a democracy.

Private values. Private values are the concerns about exposure of personal and proprietary information. The same laws that establish openness of documents and data also establish specific exceptions to openness. These exceptions specify in considerable detail material that is not open because it is private and confidential.

Commercial values. Commercial values represent perceptions that public information is valuable to agency personnel, individuals, and organizations and should be held by the agencies for sale at unrestricted prices that generate revenues. These perceptions assume that agencies are (1) legally free to sell information at unrestricted prices, (2) well-positioned to respond to demand, and (3) creating an unwarranted and unreasonable public subsidy to private-sector activity when they fail to sell this material to private organizations and citizens.

Agency efficiency values. These values represent the concern that efforts to make public information available to other agencies, private organizations, and individuals will overwhelm agencies and leave them less able to execute their normal mandated functions.

PERSPECTIVES ON INFORMATION CONTROL AND DISSEMINATION

Openness and Confidentiality

Open government means that citizens must be able to determine what their governments are doing. For many years this has meant openness to records and data. In the United States, this value is as old as the republic. "[T]he generation that made the nation thought secrecy in government one of the instruments of Old World tyranny and committed itself to the principle that a democracy cannot function unless a people are permitted to know what their government is up to" (Commager 1973). The issues are political power and democracy. "Secrecy has been bureaucracy's medium. Bureaucratic power requires it" (Chernichenko 1989).

The importance of the U.S. federal Freedom of Information Act as a means to achieve government openness has been defined by the U.S. Supreme Court. "[W]ithout question the [federal Freedom of Information] Act is broadly conceived. It seeks to permit access to official information long shielded unnecessarily from public view and attempts to create a judicially enforceable public right to secure information from possibly unwilling official hands" (White 1973).

The basis of confidentiality of personal and proprietary information is established by exceptions to openness provisions of freedom of information acts. For example, the U.S. federal Freedom of Information Act lists nine exceptions (FOIA Exemptions). The ninth is concerned with geographic information, specifically "geological and geophysical information and data, including maps, concerning wells." The confidentiality of this very valuable geophysical data submitted by oil companies to the Department of Interior for oil leases has been successfully maintained (Martin 1983). Exemptions are common in state open records acts (Puissegur 1989).

Freedom of information and open records acts balance openness and confidentiality in a specific legal way. The burden is placed on the government official to show that the material an official refuses to disclose is covered by the exception, not on the requester to convince the official that the material is open. This balance favors openness. It encourages consideration of specific, carefully described, and narrow exceptions, with openness for everything else.

The development of computerized information systems raises concerns about maintenance of the traditional balance between openness and confidentiality in the electronic age (Grodsky 1990). For example, analysis or merger of data that was too expensive or took too long is now practical and efficient. This can yield information that people believe should not be generally available and leads to suggestions that there is a need for an "Electronic Bill of Rights" defining material withheld from commonly available databases and defining what constitutes use beyond a reasonable person's expectation when the data is given to an agency (Gilman 1991). The European Community has made such an effort (OECD Guidelines).

The potential problem posed by the technical efficiency is manageable by careful attention to the exceptions to public information. The challenge is to identify specific, carefully described data and information that should not be in the publicly available information files, and to segregate it from the material in files subject to unfettered flow. The result is not a license to withhold data if the exceptions are crafted in reliance on the traditional principles and practices.

Commercial and Agency Efficiency Considerations

The budget distress of federal, state, and local government agencies prompts some officials to seek greater control of public information than is available under existing laws. They appear to believe that:

- Public information is valuable, and it makes sense to sell it at prices that will sustain all agency information activities.
- Others will make lots of money with public information, and they should share.
- Systems will not develop unless legislators and managers who fund their operations are convinced that a significant proportion of the total cost of a system (hardware, software, personnel, and data) will be recovered through public information sales.
- It is unreasonable to make available at insignificant prices the computerized databases that cost millions of public dollars.

These views encourage efforts to change federal and state freedom of information laws that would allow agencies to sell public records and information at prices significantly above dissemination costs. They also encourage agencies to regulate subsequent information use by others through exercise of rights under intellectual property laws. These changes would permit agencies to significantly influence price and use of information. The agencies could establish monopolistic arrangements with private organizations to market both public information and value-added products. Agencies view this arrangement, which can be labeled an information utility, as a means of sharing not information, but proceeds.

These schemes depend on the legal and practical ability of government agencies to sell public information and control its subsequent use. The legal ability is not at all clear, given the status and relationship of freedom of information and copyright laws (Dando 1991; Kidwell 1990; Onsrud 1992). The schemes also may fail because characteristics of information make it unlike other commodities and, therefore, difficult to control practically (Epstein 1991). These characteristics include the following:

- The quantity of data and information for management purposes is difficult to measure.
- Legal regimes that protect information as property are difficult to enforce.
- Information gatekeeping is difficult when there are many electronic gates.
- Inspection is tantamount to use.
- Digitally stored information is not depleted physically by use.
- The role of information in management decisions is not fully understood.
- Public information has characteristics of a collective public good.

The legal and operational uncertainties associated with information make it difficult for agencies to identify, respond to, and sustain markets for public information. Government agencies are not designed to respond effectively to the demands for timely, appropriate value-added information even if they are legally free to do so. It is likely that collusions between agencies and others in order to maximize sales will fail.

The United States government does not generally assert copyright protection for its data and information. Instead, it distributes information for the approximate cost of dissemination. The basis for this traditional policy is found in the fundamental reason for government. The federal constitution states that the purpose of the government is to "promote the general welfare" (U.S. Constitution). This concept applies as well to state governments and their constitutions. Public information disseminated at cost supports and ensures general welfare.

American history has many examples of the dissemination of public goods, such as public information, to stimulate economic development. Two geographic information examples are useful illustrations. One is the 150-year dissemination of nautical charts by the U.S. Department of Commerce at prices such that the sum of sales revenues is a fraction of production costs. This is done to promote and sustain commerce. The result is a large, dependent constituency of boaters, as well as pilots and shipping companies, that lobby legislators when chart budgets are threatened. A second example has a negative character but supports the argument. Publicly supported, remotely sensed data are sold by a company established by specific law to recover as much of the total cost of production as possible. Prices for the data are high. The constituency willing to pay is small. The result is that, as in the nautical chart example, the sum of sales revenues is a fraction of production costs. However, a huge constituency of local governments and organizations that would routinely use the data cannot afford to. The lost opportunity costs are high.

Agencies derive benefits and revenues from sources other than the sale and control of public information. These are:

- *General and dedicated revenues, cooperative and subscription arrangements with other public and private organizations, and data input (filing) fees.* These are the traditional and, by far, the largest source of sustenance for public information systems.
- *Better land and resource management.* The great potential for geographic and land information is its application to land and resource management problems. These value-added uses are barely explored. The more people who can afford and have the freedom to use this material as a result of active public information dissemination, the more creative and timely value-added uses there will be. This development will be hindered in its early stages by restricted access and limited use based on price and control. Agencies will lose a dependent constituency that supports general and earmarked revenues for information system development.
- *Agency efficiency.* Agency concerns about being overrun by those who seek public information are misplaced. These concerns were expressed when the U.S. federal Freedom of Information Act was passed in 1966 but were not realized (Feinberg 1986). The same technology that frightens agencies with the specter of more data requests also makes dissemination easier. Active efforts to disseminate public documents and data electronically can increase the productivity of agencies, allow them to reduce the number of in-person data requests, permit them to shift personnel to more complex activities, avoid the need for additional personnel, and limit the increasing cost of agency activity (Frank 1986).
- *Sale of specially crafted value-added information.* Government agencies can, despite the difficulties, respond to the demand for this material and do so within traditional freedom of information laws. A special product, not normally created or used in a government activity but derived from a public database, can be sold at unrestricted prices. Here, the agency stands on an equal entrepreneurial footing with the private organization that can, as well, take the database and respond to the demand. They both benefit from the competition.

- *Confidence in government.* Active efforts to distribute public information at dissemination costs enhance public confidence in government. This may be the greatest benefit to agencies.

Efforts to sell and control the use of public information will undermine these benefits. Control or sale of public information is not in the best interests of the geographic and land information community.

CONCLUSIONS

Traditional balancing principles apply to public, private, commercial, and agency efficiency values as the electronic information age develops. Open government values are so basic to a democracy that they will be sustained in the transition from a paper to an electronic records and information system. In the balance between openness and confidentiality, the exemption mechanism provides the basis for maintenance of confidentiality.

Significant revenues from the sale of public information at prices significantly above dissemination prices are an unlikely prospect and may not be possible legally and practically. Agency efficiency and public support are better served by active data sharing and dissemination efforts.

Agencies should be active in the dissemination of public information as the electronic age develops. "Technology does not have its own democratic imperative" (Nader 1991). Information policy, practice, and law should proceed from a civic, non-commercial perspective when public information is involved. These principles ensure the relatively unfettered flow of public information, a condition that is both good public policy and economics.

REFERENCES

Bedard, Y. 1986. Doctoral dissertation, Department of Surveying Engineering, University of Maine, Orono, Maine.

Chernichenko, Y. D. 1989. *New York Times Magazine.* April 4: 62.

Coase, R. 1960. The problem of social cost. *Journal of Law and Economics* 3, 1: 44.

Commager, H. S. 1973. Quoted by Mr. Justice Douglas in *E.P.A. v. Mink,* 410 U.S. 73, at 106.

Dando, L. P. 1991. Open records laws, GIS, and copyright protection: life after *feist.* Urban and Regional Information Systems Association Annual Conference, San Francisco, California, *Proceedings* IV: 1–18.

Daylor, R. F. 1982. Land information in the permit process. Conference on Land Information at the Local Level, Department of Surveying Engineering, University of Maine, Orono, Maine, August, *Proceedings*, pp. 89–99.

Epstein, E. 1991. For a summary of the characteristics of information as a commodity, *see* In my opinion, *URISA Journal* 3, 1: 2–4. For an extended discussion, *see* Mackaay 1982.

Feinberg, L. E. 1986. Managing the Freedom of Information Act and federal information policy. *Public Administration Review* 46, 6: 614–621.

Frank, C. 1986. Fee or free? split views on public records. *Journal of the American Bar Association.* August.

Freedom of Information Act (FOIA), 5 U.S.C. 552, P.L. 90-23, as amended by P.L. 93-502.

Freedom of Information Act (FOIA) Exceptions, 5 U.S.C. 552(b).

Friedman, L. M. 1984. *American law.* New York: W. W. Norton and Company. Chapter 2.

Gilman, F. 1991. Presentation at the Electronic Democracy Conference, Arlington, Virginia, September 4–5. Sponsored by Government Technology (Riley Information Services, Sacramento, CA) and Computer Professionals for Social Responsibility.

Grodsky, J. A. 1990. The Freedom of Information Act in the electronic age: the statute is not user friendly. *Jurimetrics Journal* 31 (Fall): 17–51.

Kidwell, J. A. 1989. Open records law and copyright. *Wisconsin Law Review*: 1021-1031.

Larson, B., et al. 1978. *Land records: the cost to the citizen to maintain the present land information base, a case study in Wisconsin.* Wisconsin Department of Administration. 64 pp.

Mackaay, E. 1982. *The economics of information and law.* Boston: Kluwer-Nijoff Publishers. Chapter 5.

Martin, P. H. 1983. Disclosure and use of proprietary data. *Natural Resources Lawyer* XV, 4: 799–810, at 802.

Mayo, J. S. 1985. *The evolution of information technologies and social transformation.* Washington, D.C.: National Academy of Engineering, National Academy Press.

McLaughlin, J. D. 1988. Parcel-based information systems. In B. J. Niemann, Jr., and D. D. Moyer, eds., *A primer on multipurpose information systems.* Madison: University of Wisconsin–Madison. pp. 9–36.

Nader, R. 1991. Presentation at the Annual ARC/INFO User's Conference, Palm Springs, California. Redlands, CA: Environmental Systems Research Institute. April.

OECD (Organization for Economic Cooperation and Development). 1991. Guidelines. *Directives on data protection and privacy*. Paris: OECD Publications Office.

Onsrud, H. J. 1992. In support of open access for publicly held geographic information. *GIS Law* 1, 1: 3–6.

Puissegur, A. 1989. *An overview of state open records laws.* Workshop on Managing the Risks and Recovering the Costs of Geographic and Facilities Management Systems, Department of Engineering Professional Development, University of Wisconsin-Madison, April 25-27, Section 5.

U.S. Constitution. Preamble.

Westin, A. 1991. Presentation at the Electronic Democracy Conference, Arlington, Virginia, September 4–5. Sponsored by Government Technology, Riley Information Services, and Computer Professionals for Social Responsibility.

White, B. 1973. *E.P.A. v. Mink,* 410 U.S. 73, at 79.

Patrick McGlamery

18 *Libraries as Institutions for Sharing*

This chapter presents libraries as both a model and a vehicle for the effective distribution of spatial data. It begins with a discussion of the Federal Depository Library Program, a federally mandated program that "shall be to make government publications available for the free use of the general public." The chapter goes on to describe spatial data currently being distributed by the program, the potential of networking, and the library as watchdog of public domain information. The chapter then focuses on how libraries share large and complex databases among themselves. Finally, it provides a brief discussion of what is happening in libraries now vis-à-vis digital spatial data.

INTRODUCTION

In *Megatrends*, John Naisbitt proclaims, "The new source of power is not money in the hands of a few but information in the hands of many" (Naisbitt 1982, 16). If that is, in fact, the case, then libraries are in the enviable position of giving away the bank.

The library, when viewed as an institution, is both a model and an infrastructure for the sharing of spatial data. Libraries are our society's information sharers. According to the *Library and Book Trade Almanac* (Bowker 1989) for 1989–1990 there were 15,013 public libraries, 4,647 academic libraries, and 11,057 "other" libraries, or a total of 30,717 libraries in the United States. They are the repositories of great stores of our civilization's passive and active, past and current information. They provide for the controlled sharing of information in society by "rationing" information through circulation. They facilitate efficient information sharing through reference and indexing functions. But libraries also actively share among themselves, through interlibrary loan activities, cooperative purchasing, cataloging—a variety of procedures and resource sharing that have been developed over the centuries. Libraries have been building and sharing huge databases cooperatively, with shared and standardized records, for decades. An overview of libraries will not answer the problems peculiar to sharing spatial data, though it may give some clues as to the appropriate questions to ask.

FEDERAL DEPOSITORY LIBRARY PROGRAM

Libraries as institutions are the primary agents for the distribution and dissemination of public domain information in our society. The federal government has long maintained a formal network, the Federal Depository Library Program, for the distribution of federally produced, public domain, information. Section 1911, Title 44 of the United States Code, states that the basic purpose of the depository library "shall be to make government publications available for the free use of the general public" (U.S. Code 1988, 560). "The depository library system is a long-established cooperative program between the federal government and designated major libraries throughout the United States under which certain classes of government publications are supplied free of cost to those libraries for the purpose of making such publications more readily accessible to the American public" (Senate Report 87-1587, 1987, 3, House Report 8141, the Depository Library Act of 1962).

The Federal Depository Library Program illustrates the mutually beneficial relationship between the government and the library. The government provides the material; the library provides the storage, preservation, access tools, and reference of the material. Information becomes decentralized and is put in the hands of the people.

The depository library program is the fundamental reason for the growth of map libraries in the United States. Map libraries have housed spatial data in paper form. The wealth of cartographic information available to the general user in the United States is truly amazing. A moderately sized map collection in this country numbers about 150,000 sheets, rivaling many national collections. In the past two decades map librarians, realizing the importance of non-published spatial information, have begun collecting remotely sensed imagery, both as aerial photographs and satellite imagery (in paper or digital format) as well as standard maps. Boundary files derived from DIME (Dual Independent Map Encoding) files have been collected in some map libraries for at least five years. The University of California's Santa Barbara Map and Imagery Library has been actively collecting imagery in both digital and analog format since 1982.

DISTRIBUTION OF DIGITAL SPATIAL DATA

The primary source of electronic information in the Federal Depository Library Program has been the Bureau of the Census, an early user of computer technology. The Bureau has long been an active contributor to the program. Though previously the material had been published in paper in the form of printed abstracts, summary tapes were available for a relatively low cost. The cost of the tape drive and a computer of a size necessary to process them were prohibitive. Typically, tapes were acquired by only a handful of data libraries. For the 1990 decennial census, however, the Bureau produced a number of products. The distinct difference of this census is that these products are not generally paper but digital and distributed on a low-cost medium. The CD-ROM is being explored as an effective, low-cost method for distributing the types of large numeric data sets the Bureau deals in. Use of CD-ROM for disseminating information is a result of a 1987 study to investigate the equitable distribution of federal government electronic information to libraries. Now, more than 500 libraries in the depository library program have received the TIGER (Topologically Integrated Geographic Encoding and Referencing) line files on CD-ROM, and more are receiving the summary data on CD-ROM. Other digital data are being supplied to libraries as the federal government sees the CD-ROM as an effective tool for opening up their vast digital resources. The U.S. Geological Survey is currently distributing 100K Digital Line Graph (DLG), the Geographic Name Information System (GNIS), and the Aerial Photographic Survey Record System (APSRS) databases on CD-ROM. Recently, an ambitious program to publish orthophotographic images was announced.

NETWORKING

In addition to distribution by tapes and CD-ROM, networking has become a realistic option for libraries in sharing spatial data. The National Research and Education Network (NREN), a legislative initiative libraries have been very actively supporting, was signed into existence by then President Bush in November 1991. As outlined in the 1991 report, "Grand Challenges: High Performance Computing and Communications" by the

Committee on Physical, Mathematical, and Engineering Science of the Federal Coordinating Council for Science, Engineering, and Technology (FCCSET 1991, 18):

> The vision of the National Research and Education Network is of an interconnection of the nation's educational infrastructure to its knowledge and information centers. In this system, elementary schools, high schools, two- and four-year colleges, and universities will be linked with research centers and laboratories so that all may share access to libraries, databases, and diverse scientific instruments such as super computers, telescopes and particle accelerators.

Building on strong state, regional, and federal cooperative programs, libraries are poised to take full advantage of the strong potential for sharing information through networking. Presently, research libraries allow for the interactive searching of remote on-line catalogs and the subsequent borrowing through established interlibrary loan procedures. MELVYL—the on-line system of the University of California libraries, the catalog of the Library of Congress, Brown University, and Harvard University libraries all support remote querying of their catalogs through telnet. Experiments with effectively exchanging information using file transfer protocols (ftp), and studies of the possible effects of providing access to "live" satellite imagery, are ongoing.

WATCHDOG

Hearings by the Science and Technology Committee of the U.S. House of Representatives have recently discussed the issues of availability and cost of satellite imagery, privatization of the data, and global change. The library community will continue to play a "watchdog" role in this agenda.

This watchdog role is one the library community takes very seriously. In a paper calling attention to the importance of intermediaries in the dissemination and analysis of census data, Sarah Rowland (1989, 9), Statistician for the 21st Century Staff at the Bureau of the Census, says:

> The role played by the librarians in the FDLs [Federal Depository Libraries] should not be underestimated. The librarians are the most avid defenders of access to federal data by the general public; they take a dim view of any government policy that would increase the cost of access to federal information by library users.

Making TIGER a depository item has had a significant impact on the library community. It has focused library concerns by articulating software and hardware needs and user concerns to the federal data producers. The more verbal members of the library community contend that data without software does not meet the requirements of the Federal Depository Program of making information available to the people. Others take a longer view, advocating a strong, cooperative working relationship with the Government Printing Office and the Joint Committee to the Government Printer, assuring the continued supply of digital data.

SHARING AMONG LIBRARIES

Establishing and governing a federal depository program, defending the access to governmental data by the general public and, of course, providing reference to and circulation of material, are all ways in which the library acts as an agent for the sharing of information in our society. This is the "front" of the library. Behind the scenes, libraries contribute and cooperate within communities, states, regions, nationally and internationally. The library community network is an established purveyor of information throughout our society. Even small branch libraries are linked in a complex relationship to the *library* as institution. As a profession, librarianship governs and controls itself through educational standards, intra-library agreements, and international cataloging standards, among others.

Establishing and maintaining a coordinated graduate-level curriculum is one way that the profession assures a common vocabulary and experiential base, regardless of the type or size of the library. The master's degree in library and information science is a two-year degree. The degree assures that a core curriculum is taught. Licensing in many states allows for equivalent experience as a measure of excellence. Librarians in the United States—whether they work in small, medium, or large public libraries, school, business, academic or research libraries, music, map, art or data libraries—share common ethics, training, and ideals.

Cataloging is an important part of library work. It is a complex procedure for describing an item, assigning subject access and an appropriate classification number for shelving. In 1980, the Anglo-American library community formalized the Machine Readable Cataloging record (MARC) by recodifying the fundamental cataloging rules. *The Anglo-American Cataloging Rules,* 2nd edition (AACR2; Gorman and Winkler 1988) established standard procedures for the description and subject cataloging of any item. MARC exists for a variety of formats: books, serials, maps, sound recordings, musical scores, computer files, and others. These have allowed libraries to share their cataloging work through a database available on international bibliographic utilities. For the past twenty years, what the library user has seen as a paper card in the card catalog has been printed from a database MARC record. These records hold descriptive, subject, and holdings information.

There are two major bibliographic utilities: Online Computer Library Center (OCLC) and the Research Libraries Information Network (RLIN). I will discuss OCLC in this chapter. Cataloging operations in libraries have used OCLC since the early 1970s. There are now more than 25 million MARC records in OCLC, and the system is shared among 13,000 libraries in forty-six countries. In 1990, OCLC reported adding 252,989 MARC map records, about half provided by contributors and half by the Library of Congress. Library of Congress map cataloging is done by a team of six catalogers who work in one location, hold monthly meetings, and share a common sense of work community. The contributing catalogers are dispersed around the world. Through rules published by the Library of Congress, periodically disseminated rule interpretations, regional workshops, and meetings, both types of catalogers are working toward a standardized MARC record.

MARC: THE ELEMENTALS OF A BIBLIOGRAPHIC DATABASE

The MARC record is a complex database record that has emerged as a standard for bibliographic databases (On-line Systems 1991, 2):

> In the late 1960s, the Library of Congress developed the MARC formats for communication of bibliographic information in machine-readable form. These MARC formats identify bibliographic data for computer recognition and manipulation. In the mid-1970s, as variations on these LC-developed MARC formats began to be used by other libraries and library services (OCLC among them), the formats used by the Library of Congress became known as LC-MARC formats. Since the early 1980s, however, LC-MARC formats have come to be referred to as USMARC formats because they are the standard for MARC records in the United States.

Because the MARC record is the most basic unit of the shared database, a minimal description is in order for the purpose of this chapter. The MARC map format will be used as an example. The MARC record contains its information in two types of fields, the fixed and the variable.

The fixed field carries information in coded form. In general, fixed-field information describes the work cataloged and the record itself, and is usually information that does not appear on a typical printed card. Since the field is fixed and the information is coded, libraries can take advantage of the computer's ability to manipulate and retrieve data.

The variable field contains information that varies in scope from map to map. Each field is identified by a three-digit code called a tag. Tags are used to identify aspects derived through the cataloging procedures using AACR2.

Tag groups:

0xx	bibliographic control numbers
1xx	main-entry headings (i.e., corporate or personal authors)
2xx	title, title paragraph
3xx	physical description
4xx	series statement
5xx	notes
6xx	subject-added entries
7xx	added entries and linking entries
8xx	series-added entries and variant forms of entry
9xx	local-use fields

These relate in form and content to the rules established in AACR2. The MARC record is a carrier of the descriptive information derived through the cataloging process using AACR2. A look at some of the spatially relevant fields in the MARC record can give a taste of the extent to which catalogers go to describe maps.

Fixed fields:

- *Relief.* Relief codes such as:
  Relief: a (contours)
  Relief: b (shading)
  Relief: e (bathymetry, soundings)
  Relief: f (bathymetry, isolines)
- *Ctry.* Country of publication codes such as:
  Ctry: nz (New Zealand)
- *Base.* Base map element codes such as:
  Base: aa (Aitoff Azimuthal Projection)
  Base: ba (Gall Cylindrical Projection)

Variable fields:

- *034.* Coded Mathematical Data such as:
  034 1 a $b 253440 $d E0790000 $e E0860000 $f N0200000
  $g N0120000

(where subfield delimiter $b is the constant ratio linear vertical scale, $d is the HDDDMMSS code for the westernmost longitude, $e the easternmost, and so on.)

- *255.* Mathematical Data Area such as:
  255 Scale 1:250,000; $b Transverse Mercator proj. $c (W 74° 40'–W 74° 40'/N 45° 05'–N 45° 00')

- *651.* Subject-Added Entry, Geographic, such as:
  651 0 Mansfield (CT : Town) $x Maps, topographic

The relevance of the MARC record to the GIS (Geographic Information Systems) community is twofold. First, there are the creation and maintenance of the MARC format, which was devised to automate the card catalog. The catalog serves a primary library function—that of providing access to material. The automated catalog exists in an on-line environment, aspects of which are the individual library's on-line public access catalog and the OCLC database. Second, there is the library community's mission to catalog by describing the item and the subject of the item. The infrastructure that supports that mission extends into most communities in this country and, in fact, around the world. Not only established telecommunications, but also shared cataloging, record and maintenance standards, common vocabularies, and training curricula come together to provide a strong economic incentive for close cooperation between the library and GIS communities. Most recently we have seen the MARC tags as parallel components of spatial metadata standards.

INTERLIBRARY LOAN

Libraries ship an enormous amount of material through the mail and in-house transport in order to provide for the information needs of society. The interlibrary loan (ILL) system is automated, linking libraries around the nation and the world. In 1990, OCLC reported 4,909,035 on-line interlibrary loan transactions using 365 million locations' holdings, linking library materials in the OCLC database. The OCLC ILL module allows libraries to query regional holding libraries first, querying geographically distant collections when the need arises. A system of credits and rights to borrow controls the flow of material. In the ILL system we see libraries using a shared database to facilitate the effective sharing of materials for society's needs.

DEALING WITH SPATIAL INFORMATION

Recently, urged primarily by the distribution of census and TIGER data, libraries have begun to mobilize for the acquisition and control of, and access to, spatial data. During the XIth Annual North American

Cartographic Information Society (NACIS) Meeting in Milwaukee in October 1991, map librarians and users met to discuss establishing a shared agenda for the emergence of digital cartographic information in libraries. The following agenda items were offered as a point of departure for the discussion and involvement for the library community in general, and the map library community in particular. Five critical areas of attention were articulated. The roles of the library institution in the emerging field of spatially referenced information are collecting/archiving, cataloging/indexing, networking, distributing, and education.

Collecting and Archiving

The problems of the collection and long-term storage and preservation of digital information, especially digital cartographic and database files with spatially referenced fields, need to be addressed. In the evolving arena of digital information, hardware, software, data formats, and standards are critical. The problems of the nonstandard "publishing" of electronic information, the resultant problems of identification and procurement, as well as adherence to standard formats (as they are developed), demand attention. Libraries, especially research libraries, store large quantities of material, and they have for centuries. Problems of space, prioritized use, varieties of media format, and preservation of information are continuing concerns. For example, libraries have confronted the problems of preservation and conservation of information materials with national and international strategies. These strategies have provided initiatives such as insisting on materials published on acid-free paper and nationally coordinated comprehensive microfilming projects of brittle books.

Cataloging and Indexing

Navigating through electronic information is confounded by the sheer volume of information and its dynamic and unpublished nature. As more sophisticated descriptive cataloging techniques are developed and standardized and made available, item-level cataloging needs to focus less on the distribution media and more on the file level, and field level. The multidimensional layering process of GIS analysis demands increased refinement of the descriptive art of the library cataloger.

Libraries have developed, over the past centuries, sophisticated processes of bibliographic (and cartographic) control and access through the development of the descriptive and subject catalog records. These processes have culminated in the past few years in the establishment of an international database of MARC records. MARC formats exist for monographs, maps, serials and, among others, computer files. Currently, computer files are seen primarily as flat text files and software. The complex nature of GIS-related data must be identified.

Networking

Large numeric databases, which are the basis of digital cartographic formats, reside most comfortably, and are processed most effectively, on mainframe computers. The usefulness of databases, however, is felt most keenly on the personal computer and work station. This presents a dilemma that is, in part, solved by networking. Remote storage of information, ready distribution of data, and appropriate linkages that connect available information need to be investigated and addressed. Libraries have a long history of aggressively utilizing computer networks. Bibliographic utilities, such as OCLC, have existed and flourished for decades, distributing bibliographic citations and library holdings nationally and internationally. These systems have allowed for a high level of information sharing among the national and international library communities. More recently, the Internet has allowed for interactive perusal of research library bibliographic databases like MELVYL (the system of the University of California) from remote sites throughout the world.

Distributing

Data compilation continues to be the most difficult aspect of modern map production. GIS analyses' voracious appetite for information and the large and complex nature of numeric data files and databases are evolving. Distributing electronic information in appropriate file formats, with documentation, will streamline geographic analysis.

Libraries have developed extensive and complex distributions systems. Copying machines, microform copying, scanning, document delivery, and faxing are as much a part of those systems as are local, state, regional, and national interlibrary loan networks.

Education

User education is a considerable hurdle to the effective use of GIS and digital cartographic information. GIS is a whole new way of dealing with information, database management, image processing, and spatial analysis. College-level, secondary, and elementary school curricula will be necessary. Staff education on the use, storage, extraction, and retrieval of multi-dimensional files is an issue that will have to be dealt with in library schools.

Libraries have offered bibliographic instruction, beginning with the use of the card catalog in elementary schools, to workshops on navigating Dow Jones Online and BRS Afterdark. Map librarians are accustomed to the user who needs help with the graphics format of maps and readily supply interpretive legends and advice on scale and projections. Standardization of the master's in library science curricula and conferrence of the degree have established a consistent professional foundation. This foundation has been the underpinning to shared cataloging, interlibrary loan, and collection development.

ARL GIS/LITERACY PROJECT

Beginning in January of 1992, the Association of Research Libraries (ARL) began its GIS Literacy Project. The project seeks to introduce, educate, and equip libraries with the needed skill to provide access to spatially referenced data. The goals are to:

- develop a core of GIS professionals in the library community willing to lend time and expertise to new applications, user training, and education programs related to GIS;
- stimulate and encourage the establishment of "centers of excellence" for GIS;
- promote research, education, information, and the public right to know through improved access to government information;
- initiate library projects to explore new applications of spatially referenced data; and
- introduce GIS to public, academic, research, state, rural, and large urban libraries to address the diverse information needs of disparate user communities.

Initially thirty libraries were selected to participate in the project. The representatives of those libraries attended a training session sponsored by the Environmental Systems Research Institute (ESRI) prior to the ESRI Annual User Group meeting in Palm Springs, California, in June 1992. The training included a half day of a general overview of GIS, followed by in-depth training of ESRI's ArcView. Thirty more libraries followed in the fall with training at ESRI's facility in Redlands, California.

In addition to the education of the GIS "core," the project will produce an evaluation of the year's experiences in addition to a "What next?" that will take the form of recommendations to further courses of action. This project has the potential for quickly developing a relevant niche for spatial data in libraries. Issues of data storage, preservation, and resource sharing and access will be discussed. Reference service and levels of instruction will be developed and the information will, in time, be integrated into the normal workflow of the library.

CONCLUSIONS

Libraries are institutions for sharing information in our society. They have evolved over the past several hundred years to acquire and store our information for the long term, to catalog and control information by describing it, and to access and loan information. The Federal Depository Library Program is indicative of the role of the library as it relates to public domain information. This role is well-established for paper information. Electronic information is simply the latest in a series of information challenges that libraries have dealt with.

Probably the most critical aspect of electronic information from the libraries' point of view is the dynamic nature of electronic information. The lack of standards for the data, data formats, data descriptions, and data storage present problems, probably because libraries are so standardized. In order to take advantage of the library infrastructure, the GIS community will need to develop ways of accommodating existing library standards and aiding in their adaptation to the needs of geographic information users.

REFERENCES

Bowker. 1989. *Library and book trade almanac 1989–1990,* 34th ed. New York: Bowker.

Federal Coordinating Council for Science, Engineering, and Technology (FCCSET). Committee on Physical, Mathematical, and Engineering Sciences. 1991. Grand challenges: high-performance computing and communications. Washington, D.C.: The Council.

Gorman, M., and P. Winkler. 1988. *Anglo-American cataloging rules,* 2d. ed. Chicago: American Library Association.

Naisbitt, J. 1987. *Megatrends 2000: ten new directions for the 1990s.* New York: Morrow.

On-line Systems (rev. 1991). OCLC, Inc., Maps Format. Dublin, Ohio.

Rowland, S. 1989. The role of intermediaries in the interpretation and dissemination of census data, now and in the future. Internal document.

United States Code: containing the general and permanent laws of the United States, in force on January 3, 1989 (1988).

U.S. Congress. Senate. 1962. Revising the laws relating to depository libraries. Senate Report 87-1587. Washington, D.C.

Jack Dangermond

19 *Public Data Access: Another Side of GIS Data Sharing*

There have been many discussions of, and experiences with, the sharing of digital geographic data between and among various public and private organizations. Most of these discussions have been associated with the topics of joint sponsorship, cost sharing, technical interchange, agency conflicts, and legal as well as administrative conflicts. A completely separate set of issues arises when considering the sharing of data with the public and the distribution of data to the public. The issues deal not only with the needed technology but also with the more fundamental question of how governments should relate to citizens. A number of the behavioral, organizational, and institutional issues are covered here; these include a review of the models and methods that public agencies at the federal, state, and local levels of government have used for distribution of their data (including spatial data) to the public; a description of some of the public issues associated with both the distribution of data and the connection of the public to the databases that governments use in making policy and conducting their operations; an outline of some of the key geographic data-related issues that should be addressed or are of interest to citizens organizations; a review of some of the technical means that could be employed to support government data sharing with citizens; a consideration of some of the institutional mechanisms and barriers that must be taken into account in creating a successful information link between government and the public; and, finally, some indications of the benefits that such citizen participation in government might be expected to produce.

INTRODUCTION

While the idea of sharing Geographic Information Systems (GIS) data among organizations has been discussed over the years, most of these discussions have focused on sharing among levels of government, among government agencies, and among government and private organizations. The issues discussed included the advantages of data sharing in cost reduction; the nature of the agreements that had to be reached among the sharing organizations; the mechanisms by which shared data collection, automation, storage, and use could be effected; and similar practical matters.

A rather separate issue, but one with potentially profound implications for the GIS field as well as for the ways in which government functions, is the idea that government, and perhaps private industry as well, ought to do much more in sharing GIS data (and other kinds of data as well) with citizens. This chapter outlines some of the issues that arise out of a consideration of this idea.

SOME EFFECTIVE MEANS BY WHICH PUBLIC AGENCIES NOW SHARE THEIR DATA WITH CITIZENS

In thinking about public access to GIS data, it is useful to recall the means that governments now use to provide public access to the information they possess. Governments have in place many means for data sharing, and spatial data are shared, in one form or another, through many of these. While the sharing of maps with the public is perhaps the most obvious of these means, spatial data are also extracted and used as parts of many other data uses; perhaps not surprisingly, the sharing of raw data, distribution of electronically stored digital spatial data, and direct access to government databases are not common.

Public agencies share some of their information with the public by means of various books, publications, reports, notices, budget documents, printed records of legislative activities (proceedings, hearings, investigations, and the like), various printed documents provided by executive, legislative, and judicial entities (registers of proposed regulations, judicial opinions, regulations, executive orders, administrative actions, and so on), and other documents prepared for distribution. Governments also supply information to the public by means of press releases and interviews with the press or other media. The government also produces its own films, recordings, television programs, and similar materials for public release or with the understanding that the public can have access to these materials. In many circumstances, government opens to the public its various functions, such as trials, legislative sessions, legislative hearings, some administrative hearings, and the like. Government also often provides responses to public inquiries in the form of answers to questions over the telephone, letters in reply to questions, and face-to-face meetings with public officials or employees. Spatial data play some role in all these types of government information release.

The public also already has direct access to some government-maintained information bases or databases, usually those containing matters of public record (e.g., some census data, property data, building permit data, birth and death records, property transfers, some tax records, and so on). Citizens usually may obtain copies or extracts of these records, though direct electronic access is not usually provided. Information in some of these databases is spatially referenced (e.g., taxation, property, census, and other data), though the reference systems are usually rather inefficient from a spatial perspective.

In addition to direct citizen access to the government's own records, some government agencies provide public access by means of document depositories (libraries and the like) where printed copies of government documents can be examined and often can be borrowed.

In the case of GIS information, it is clear that most of the information to which the typical users of GIS have access comes from government sources in one way or another. Printed maps, electronically stored data, digital data of many kinds, imagery, census data, and a variety of other data and information forms usually come directly from government sources, though they are also increasingly being processed and repackaged by commercial organizations along the way to GIS users. Usually these data are provided to citizens in printed form (maps, reports, tabular data), as contrasted with the electronic form in which they are commonly shared with other levels of government or private businesses. This distinction between the means of citizen access and the means used by organizations may, however, have more to do with demand and past custom than with policy.

A number of government agencies, such as the U.S. Census Bureau, the National Oceanic and Atmospheric Administration (NOAA), and NASA have created special facilities by means of which individual users of their GIS-related data and information can obtain access to, or copies of, information of interest. Federal agencies, such as the Defense Mapping Agency, NASA, and others, are also moving toward providing geographic information in even more convenient electronic forms, such as CD-ROM. These latter forms are beginning to be available at costs that individual citizens can afford.

Despite these various means of public access, there are obviously many kinds of government geographic information that either have not been readily available to the public or that have not been available to the public under any circumstances. These are geographic data that are classified for national security reasons or restricted in use or circulation for other reasons. In the United States, at the federal level, for example, the Freedom of Information Act was intended to make even more information available to the public, especially the kinds of information that government, for whatever reasons, had been reluctant to release. Some of the releases of such information have been controversial, have involved litigation, and have gained national attention. It is believed that important geographic information is part of the government information to which access has been restricted, but the Freedom of Information Act means seem not to have been used very often (if at all) to acquire such information.

It has been suggested that with the rapid thaw in relations between the West and the Eastern Bloc, a great deal of geographic information acquired for intelligence purposes might be made available for public use. As this chapter is written, this has not yet occurred.

It is interesting to note, in this context, that in Europe these same kinds of issues are being raised about information collected by private businesses. The argument there is that citizens want to know what information these businesses have collected about environmental, toxic chemical, product hazard, and other kinds of issues; businesses, interestingly, are now beginning to release their data (at least selectively) in order to assuage citizen concerns, demonstrate the quality of their products, show what efforts they are making toward creating a cleaner environment, and so on. Few if any stirrings of this kind seem to be occurring in the United States.

Even this brief review indicates that providing public access to government information is already a major enterprise, one to which modern technologies are being applied, and one that is already surrounded by some controversies.

ISSUES ASSOCIATED WITH DISTRIBUTION OF DATA TO CITIZENS, OR DIRECT CONNECTION OF CITIZENS TO GOVERNMENT DATABASES

Assuming that citizens already have access to many forms of information in printed forms (including maps) and have access to a good deal of electronic data in extracted form (extracts and reports on census data, for example), the chief remaining issue in obtaining citizen access to geographic data would appear to be the means by which citizens can obtain direct access to government GIS-related data, either by means of connection to government databases or by obtaining complete copies of the data in these databases.

Some of the same issues arise in considering this issue as arise when considering data sharing with other levels of government, sharing among agencies, or sharing among government and private organizations. These issues include data security (e.g., accidental or deliberate destruction, loss, alteration, or degradation of data), confidentiality and secrecy, privacy, means of access, and costs.

While the issues may be similar, the means of resolving them are likely to be quite different when the "public" desiring access may range from a national or international business down to an individual person (as opposed to sharing data with a well-recognized government entity or a private organization with whom a specific contractual relationship has been established). It is one thing when an individual citizen wants to satisfy his/her curiosity about something; it is another when a business wants to obtain a copy of a complete database so that it can commercialize the database as a product. (The question of commercialization of government data and questions relating to how government should go about recovering its costs or a portion of the profits of such activities are beyond the scope of this chapter.)

Some new issues also arise when public access is considered, issues that are of much less importance when discussing data sharing among organizations with professional staffs. These new issues include, for example, the need for education and training of the public about geographic data sharing (e.g., what data exist, what do they mean, and how can they be used; what technical means of access exist, how are they acquired, how maintained, how is training in their use obtained; and the like); and problems associated with providing equal access to a public facility (e.g., access for the unskilled, the poor, the handicapped, non-English-speaking, the computer-illiterate, and so on).

The issue of costs also takes on new meaning when the public is directly and individually involved. There is the need to keep access costs low enough so that individual citizens really can afford them. How the costs are to be distributed, whether and to what extent access fees are to be subsidized by public monies, what priority public access should be given in the government budgeting process, how business and educational users are to be treated, and similar issues are also raised. These issues will almost certainly require that government devise some new ways of doing their business: new policies, new procedures— probably new ways of thinking about how governments should serve the people.

Finally, when the results of all these required decisions are summed up, to what do they lead? Assuming several million possible users of GIS data within a decade or so, what arrangements will be required? Do all these considerations lead to the creation of a public information utility, one akin to electric utilities or telephone companies, or can new private businesses be created to supply the information needs of these citizens?

DATA SHARING ISSUES THAT SHOULD BE ADDRESSED OR THAT HAVE BEEN RAISED BY CITIZENS ORGANIZATIONS

The idea of sharing GIS data with the public is an important one, and it obviously raises many issues. Some of these issues already have been raised by citizens organizations. Their views and the views of citizens who would be partners with government in data sharing need to be taken into account. The kinds of issues that have been recognized include those already mentioned above, as well as some others outlined below.

First, and perhaps most important, is the view that the public has a fundamental right to virtually any data that has been collected using their taxes. Closely allied with this view is the notion that governments, in general, should be far more open and responsive to citizens; making geographic data available to the public is, in this view, a part of that openness. It is also proposed that such geographic data sharing should receive a much higher priority in government budgeting than it now receives. It is further argued that geographic data should be made available at no more than the cost of production of copies (comparable to the prices for publications from the Government Printing Office, for example).

SOME TECHNICAL MEANS FOR PROVIDING CITIZEN ACCESS TO GOVERNMENTS' GEOGRAPHIC INFORMATION

Added to all the above considerations must be at least two technical considerations: What characteristics should public-access GIS technology possess, and are such technical means available?

Public-access GIS technology, first, should be inexpensive. Probably it should be about as expensive as personal computer technology for word processing is today; ideally it would be significantly less expensive. It should also be as easy to use as word-processing software, preferably easier.

In principle, public-access GIS devices should more closely resemble the telephone and the television set than they resemble the personal computer. The technology should cost no more, either for initial purchase or for monthly connection fees, than a service like cable television. Like cable TV or television sets, public-access GIS technology should be widely available; ultimately, it should be available in every home.

The access system should be independent of the user's natural language; it should be graphic, interactive, provide a high resolution color image, and provide hard-copy capability, at least as an option.

The cost targets for such a system probably cannot be met at present but, given the steady pace of computer development, the targets are within reach, probably sometime in the next decade.

The technological capability for providing public access to geographic data probably exists now, but it is not yet as user-friendly as it needs to be. Certainly a GIS as easy to use as a telephone or television set does not yet exist.

Based on present capabilities, there appear to be two routes to widespread delivery of geographic data to the public: access by networking and access via mass-produced, inexpensive, high-capacity storage media (CD-ROM is probably the present prototype medium).

Networking will likely become widely available in the United States in the next few years, probably through fiber-optic cables bringing cable television, information services, telephone services, and the like. GIS services might then be provided through the same system.

CD-ROM or successor technologies are already capable of providing mass storage and distribution of geographic data, but the price for individual copies of databases is still in the hundreds or thousands of dollars rather than the dollars or tens of dollars that are likely to be required to make the data readily available to the public.

INSTITUTIONAL AND OTHER BARRIERS TO BE OVERCOME

Public access to GIS data will require the cooperation of levels of government and many government agencies. In order to obtain that cooperation, some institutional barriers to cooperation may need to be overcome. Further education of staff and changes in regulations and administrative procedures may be required. The present forms and organization of GIS data and information resources may have to be changed in order to facilitate citizen access. If the perception exists among officials or staff that wider citizen access to government data represents a threat, that perception must be changed.

Before citizen access is a reality, changes in the public mind will also be required. Citizens will have to be educated to the value of geographic data and the uses to which it can be put; whereas citizens organizations may appreciate this, the public in general does not. The public will also need, at least over time, to understand the limitations of the GIS data and of the methods for using them.

CITIZEN USES OF GEOGRAPHIC INFORMATION

Despite all the above discussion, it is by no means clear what the public would do with governments' geographic information if access to the information could be provided. While some citizens organizations have ideas about this, it is not clear that non-activists will be as excited about getting this information as these organizations are.

Among the kinds of uses proposed are closer tracking of toxic chemicals locations, chemical spill locations, movement of toxic plumes through groundwater, studies of property taxation policies, analysis of road maintenance practices, examination of the quality and condition of various kinds of public infrastructures, monitoring of governments' responses to citizen complaints about local problems, correlation of public spending with the location of elected representatives' own residences, comparison of announced highway plans with actual performance, monitoring of logging and deforestation, monitoring of environmental quality, examination of the spatial components of various economic activities and government activities (e.g., where do most military bases get sited, where do most government grants go in relation to congressional district lines, which areas within cities have the highest prices for similar goods), and so on.

Looking at any of these lists makes it clear that many of the things that the public or citizens organizations might like to do with geographic data cannot be supported because neither government, nor anyone else, has collected the data that would be required. Nevertheless, some matters of interest could be addressed but do not seem to be addressed directly by government at the present time.

PRESUMED BENEFITS OF CITIZEN ACCESS TO GOVERNMENTS' GEOGRAPHIC INFORMATION

The fundamental presumed benefit of citizen (public) access to geographic information is that such access could play an important role in raising the consciousness of citizens about the many spatial aspects of government activity and the spatial aspects of the world within which the citizen lives. It is further presumed that this will lead, in turn, to greater participation by citizens in their government, greater support for government by citizens, and an improvement in conditions in the world.

The underlying arguments are, in part, that at least some portion of citizen apathy about government is due to the seemingly inevitable distancing of government from citizens as government has become ever larger and more complex. Citizens are mistrustful of what they do not understand and cannot grasp. They are less willing to pay taxes in support of activities that they cannot see and usually know little or nothing about. By making more information available about what government does, and providing it in the graphic form of spatial information, citizens and governments both can benefit.

Moreover, as citizens know more, they are likely to become more inquiring; this, in turn, is likely to induce governments to become more responsive to citizen interest and concerns. This process may well feed on itself, creating still more interest and ever greater responsiveness.

While these kinds of arguments have long been used to justify education of the public about government, information release to the public, and similar activities, the introduction of GIS technology may bring new significance to the whole notion of public access by providing information to individual citizens in a much more thoroughly individualized and tailored form. Moreover, the inherently graphic nature of geographic information may be a much more efficient medium than have been the text forms in which information has often been provided by government in the past.

CONCLUSIONS

Some of the considerations in providing public access to governments' GIS data have been outlined. From this it seems clear that public access to GIS data will be technically feasible within the foreseeable future, and that, if various non-technical problems can be overcome, such access will be feasible. *But many issues remain.*

In comparison with other government activities, including government-to-government geographic data sharing, what priority should be assigned to public access to geographic data? Realistically, what is the timeframe within which such access will become a reality? Is access to geographic data a useful model for citizen access to other kinds of government-held information? Who should spearhead efforts to make this access real: non-governmental organizations, citizens organizations, government, or vendors of GIS technology?

What is the significance of public access to the GIS field as a whole? Until it becomes a reality, this will be a difficult question to answer. But rapid as the growth in the use of GIS technology is, the number of users is still in the tens of thousands, and the number of kinds of applications is relatively small.

Yet, we all live in a spatial world, and we deal with spatially related problems every moment of every day. If GIS technology becomes available to millions of people, their application of it may well produce revolutionary changes in our daily lives, changes that even seasoned GIS experts cannot now foresee.

There is some reason to believe, then, that geographic data sharing with the public may be among the most important issues now before the GIS community.

Section IV

Creating and Maintaining Geographic Data-Sharing Environments

Creating and Maintaining Geographic Data-Sharing Environments

The fourth section addresses the questions that arise when spatial data-sharing activities among organizations have commenced and issues of their further development and improvement are raised. The first two chapters, by Bossler and Frederick, respectively, address national issues that are largely overlooked in earlier sections of the book. Bossler's focus on the importance of standards and upon the need for research in the standard-setting process is followed by Frederick's description of the recently organized Federal Geographic Data Committee (FGDC). Readers of Section I will recognize the "implementation of rules and procedures method" for ensuring cooperation in spatial data-sharing activities. In contrast, Cooke and Sperling, in the third and fourth chapters in this section, describe how work on one basic spatial database is being coordinated in order to enhance its ability to be shared. The database addressed, the U.S. Census Bureau's TIGER (Topologically Integrated Geographic Encoding and Referencing) files, includes the digital description of features, such as roads; their attributes, such as address ranges for the structures along those roads; and their relationship to census geography. This effort is being coordinated at federal, state, and local levels in conjunction with important maintenance and data enhancement activities of the private sector.

The next chapter by Alfelor describes the efforts of state departments of transportation to add attribute detail to the TIGER databases. A major motivation for these efforts, he asserts, is the benefits expected to occur as the emerging Integrated Highway Information System (IHIS) becomes the basis for improved management and operation of federal, state, and local transportation infrastructures. The improvement of spatial databases for the transportation infrastructure is a goal specified in the 1991 U.S. Intermodal Surface Transportation Efficiency Act, thus giving rise to a related superordinate goal as described in Section I by Pinto and Onsrud. Highway management is an area in which decision support is needed, and the ability to recall all transportation infrastructure information by a common geographic coordinate reference represents a major improvement in capability over traditional systems. Considerable progress has been made in distinguishing among problem domains (see Azad and Wiggins, ch. 2) and determining different spatial data requirements for different domains. Spatial information requirements for the national-level highway system, for example, are different from the requirements at the state level. Alfelor concludes by assessing the organizational and institutional factors that may impinge on the success in developing IHISs.

The next two chapters, by Goodchild and Stage, deal with issues relevant to the development of standards. The development of useful standards is seen by many as being absolutely critical in increasing the ability and willingness to share. Although neither the Goodchild nor the Stage chapter is wholly technical or wholly institutional in nature, Goodchild focuses primarily on technical issues that still need to be addressed in the standards arena, whereas Stage focuses primarily on developing appropriate institutional frameworks for arriving at and maintaining useful standards that are suited to the needs of the involved parties.

Goodchild discusses the difficulty of sharing data when, as is usually the case, the recipient of data from another source needs information describing the quality of the information received. He discusses typical sources of error in spatial data, including errors that arise during processing. Goodchild proposes the concept of "metadata" as an effective means of transmitting information about quality. Metadata is attached to the data themselves and is updated as the data are processed and modified by Geographic Information Systems (GIS) operations. He describes some emerging standards for describing data quality, such as those in the recently adopted FIPS (Federal Information Processing System) Spatial Data Transfer Standard, as well as recently developed methods for describing the lineage of spatial data. Goodchild concludes by noting that lack of information on data quality is a major impediment to data sharing, and that basic research is needed and software development is required if metadata is to be used effectively.

Stage describes the problems experienced by the Florida Growth Management Data Network Coordinating Council (GMDNCC) in developing a GIS. The GMDNCC worked on developing ways of facilitating the sharing of spatial data. The state of Florida Geographic Information Network is a federation of independently held databases linked together by standards and management structure. Technical advisory committees develop standards and recommend them to upper management. These, following further review, are presented to GMDNCC and, if adopted, are officially promulgated to other state agencies. Stage describes in some detail this mechanism for setting, discussing, and approving standards, as well as the benefits and drawbacks of the approach. He then describes the tools developed and implemented to document the data, including its quality and accuracy.

The next chapter focuses explicitly on the overlap between organizational and technical sharing issues. Evans and Ferreira discuss the need for technology to be adapted to typical "messy" organizational realities of data sharing and the need for organizations to adapt to the ever-changing technological conditions. They illustrate their contributions in these areas in three application areas: GIS in transit operations and management; pipelines for spatial data management; and federated spatial databases. They conclude by identifying research needs that distinguish the particularly geographic nature of spatial data that impose additional requirements in data sharing.

Most of the chapters in this volume begin with the assumption that at least one of the parties to the sharing arrangement desires increased access to geographic information for the purpose of making better decisions, such as better resource allocation or policy decisions. Scientific investigators are also heavy users of geographic information, yet their goal in gaining better access to geographic information through sharing is to advance scientific understanding. The new knowledge they expose may ultimately be very useful to decision makers, but scientists themselves tend not to be as closely involved in resource allocation and policy decision making. The last two chapters provide overviews on the current status of sharing geographic information among scientists. The chapter by Rushton and Frank provides examples of the types of sharing currently occurring among social scientists and describes mechanisms for sharing likely to be used in the future. The chapter by Goodchild that follows accomplishes the same for sharing among physical scientists.

Section IV makes clear that much remains to be learned about spatial data sharing. The number of actors, the great variation in the requirements of different users of spatial data (even for data on a single theme), and the absence of standards on so many relevant matters ensure that complexity and incompleteness will be distinguishing features of the data-sharing environment for many years to come.

John D. Bossler

20 *Facilitating the Sharing of Spatial Data: Perspectives from the Mapping Science Committee*

Within the National Academy of Sciences/National Research Council (NAS/NRC), under the general guidance of the Board on Earth Sciences, the Mapping Science Committee (MSC) attempts to "serve as a focus for external advice to the federal agencies on scientific and technical matters related to spatial data handling and analysis. The purpose of the committee is to provide advice on the development of a robust national spatial data infrastructure (NSDI) for making informed decisions at all levels of government and throughout society in general." This chapter presents some personal perspectives on issues discussed and knowledge gained by the MSC over the past several years.

INTRODUCTION

The National Academy of Sciences/National Research Council (NAS/NRC) frequently acts as an external review or advisory board for individual or groups of agencies. The Mapping Science Committee (MSC) was formed in 1989 by the NAS/NRC to provide advice to federal agencies involved in spatial data handling.

The mission statement of the committee begs for a definition of "national spatial data infrastructure." The committee has had a number of discussions concerning the definition of the NSDI, particularly with regard to limiting its scope, but we have considered the NSDI as:

> The total ensemble of geographic information at a nation's disposal describing the arrangement and selected attributes of features and phenomena on the Earth as well as materials, technology, and people

> necessary to acquire, process, store, and distribute such information to meet a wide variety of needs. In its broadest sense the infrastructure also includes the cultural, environmental, economic, political, legal, and educational values and institutions that support, facilitate, and shape its character including the forms in which spatial data are represented and utilized throughout society.

In order to work with this broad definition, it is essential that one accepts the idea that an infrastructure exists today and has existed for many years. A goal of the MSC is to help strengthen this infrastructure and make it function more effectively. One of the improvements that the MSC believes can be made is to enhance the sharing of data among federal agencies, state and federal agencies, and local and state agencies.

NEED FOR STANDARDS: A REQUIREMENT FOR SHARING

For the purposs of this chapter, "standard" means an accepted set of documents or processes related to spatial or geographic digital data. Specifically, I am discussing standards necessary for sharing data, *especially transfer standards.*

It seems axiomatic to say that some type of standard is necessary in order for people to share data and information easily. This thought has emerged as a central point of the MSC and has occupied a great deal of the committee's time. The need for standards to reduce the costs of sharing data has been reinforced recently (Moyer and Niemann 1993) and has been the focus of textbooks for years (Sullivan 1983). The need is amplified now due to a number of factors such as increased use of Geographic Information Systems (GIS), proliferation of spatially referenced digital databases, increased cost of converting analog data to digital data, and a never-ending demand for more from less. Of these factors, conversion costs are probably dominant. Most experts agree that these costs are from 60 to 80 percent of the total cost of an AM/FM (automated mapping/facilities management) project. Therefore, it is natural to attempt to share data to reduce these costs. It is not difficult to find support for the above ideas, but it does not seem to be easy to impress upon practitioners some of the important related ideas, such as those that follow.

Federal Responsibility

While it is arguable, there is general agreement that standards are a federal responsibility. Oddly, there still seems to be a reluctance to accept this idea by some federal, private, and professional organizations. It is important to note at the onset of this argument that the top-down (federal) development of standards has existed for many years; for example, in the mapping arena, the 7 1/2' quadrangle is a (de facto?) standard.

I believe that if we continually cut the budgets of the information-related federal agencies that the last thing they should give up is the standard-setting function. Setting standards may well be the most important thing they could do because setting standards should be done by an objective, unbiased public organization so that no one private firm is unfairly favored. Another characteristic of an organization that sets standards is that it should be inherently the organization of "ultimate" responsibility; for example, the United States Geological Survey (USGS) is the ultimate mapping authority in the nation (Petchenik 1991).

A spatial data standard should reach as many practitioners as possible, be widely available, and have the potential—at least—of being maintained, changed, and updated until it is obsolete. Federal agencies have the mandate, scope, and visibility to accomplish this development. These points confirm the assertion that spatial data standards should be developed and maintained by federal agencies. It goes without saying that this should be done in cooperation with many other organizations by soliciting their views and supplying funding.

Research

Presently, there is virtually no enthusiasm in the GIS community for the idea that universities—or for that matter, federal agencies—should perform research that fosters the development of standards. Research on standards was an explicit recommendation of the MSC in its most recent report (National Research Council 1991) to the USGS, but I am unaware of any federal research grants or contracts related to standards. Some of the questions that need research are:

- What standards should be developed and in what priority?
- How can we evaluate their efficiency?
- What are the benefit-cost ratios?
- What conceptual data model should be used for what standard?
- What primitives should be used for various data models and types?
- How are standards updated, maintained, and expanded?
- What process should be used to develop or test a standard?

Some of these questions are deceptively simple. I challenge my colleagues to consider them carefully and try to provide us with answers. I am fully aware of the fact that the USGS-developed Spatial Data Transfer

Standard (SDTS) was conceived primarily by academicians. However, at the moment I know of very little research related to this important aspect of data sharing. Moreover, the fact that SDTS will have taken thirteen years to emerge as a Federal Information Processing System (FIPS) standard reinforces the need for the above. The need for these activities is becoming clearer and has been documented (Morrison 1988).

Sustained Support

The development and promulgation of an important, complex standard require a significant outlay of resources over a long period of time. This fact seems not to be apparent a priori to many organizations that have developed standards. Important aspects of the undertaking include initial testing, user manuals and guides that include clear examples, compliance testing, a service support organization, and a maintenance function. Moreover, in order to accomplish widespread acceptance and utilization of the standard, "marketing" and "advertising" efforts must be organized and supported. Public hearings at appropriate meetings are essential. These activities cost a lot of money and require the best people in any given organization since, for example, the compliance-testing function requires a complete and thorough knowledge of the entire standard. The requirement for these types of activities needs to be acknowledged at the beginning of the standards development process.

ORGANIZATIONAL STRUCTURES AND INCENTIVES FOR SHARING DATA

In the federal arena, the newly chartered Federal Geographic Data Committee (FGDC; Office of Management and Budget 1990) plays the role of coordinating the numerous federal agencies involved in acquiring, managing, using, and distributing spatial data of all kinds. As can be seen from the annual reports of the Federal Interagency Coordinating Committee on Digital Cartography (FICCDC), the predecessor organization of the FGDC, this involves dozens of agencies and many areas of activity.

The problem is that there is no single federal agency in charge of acquiring, managing, or distributing spatial data. Therefore, duplication of effort in all these functions wastes valuable and scarce resources. Federal, state, and local institutions are collecting spatial data at a multitude of scales, levels of accuracy, detail, and categories, making the sharing of data difficult.

There is no analog to the FGDC at the state level, although there are a number of informal activities that are attempting to fill the need for coordination (National States Geographic Information Council 1991). And, of course, at the local level the situation is even worse. Nonetheless, it is surprising how many informal groups have been formed to discuss GIS/LIS problems. One needs only to browse through the proceedings of any AM/FM conference to be convinced.

Organizational Enhancements

Sharing data presents difficult problems, and an enhanced organizational structure would serve the nation well. Therefore, consideration should be given to strengthening the organizational structure at the federal level. When one views that consideration in the context of the information age, our lack of competitiveness in certain critical technologies, and an already nearly unworkable bureaucracy, the need for a high-level (departmental) government-wide reorganization looms over us. However, cabinet reorganizations have been attempted at periodic intervals and have failed primarily because of turf issues. I believe it is just as impractical today as it was when it was last attempted.

Other options include an independent agency—like NASA—that for the lack of a more imaginative name could be called the National Information Administration (NIA). While this idea has great appeal to me, it too is probably out of the question. Next in the chain of hierarchical organizational possibilities is probably a government-owned corporation like the Tennessee Valley Authority (TVA) or perhaps a council on which could sit the Secretaries of the "information" departments (e.g., Interior, Commerce, Transportation, and so on).

Another alternative is to strengthen the FGDC in its present home in the Department of Interior by providing a significantly enhanced budget and increasing legislative authority to implement coordination activities.

At the state level, a number of agencies have created groups within existing organizations, councils, and even societies to deal with data sharing and other coordination activities (e.g., Wisconsin Land Records Committee, Ohio Geographically Referenced Information Program Advisory Committee). The effectiveness of such organizations ranges from good to poor, but on the whole I suspect it is less than desirable. Locally, only a few groups have formed coordinating organizations.

In all cases it appears that one factor causes a certain degree of success. That factor is the implementation of incentives for cooperation.

Incentives

A number of incentives for data sharing and other forms of cooperation appear to have worked—and in some cases very well. For example, Richard Yorczyk of the Horizontal Network Branch of the National

Geodetic Survey (NGS) reports that they have received data from other organizations for 65,000 horizontal geodetic control points since 1980 (telephone conversation 1992). This data donor program works because the donors (state, county, and private organizations) want to assure the accuracy of the points they observed and earn a stamp of approval from the nation's highest authority on geodetic control. It was essential that the NGS develop a standard in order to implement this mechanism. This standard for submitting geodetic information to NGS is known in the vernacular as the "Blue Book" (U.S. Department of Commerce 1980; Federal Geodetic Coordinating Committee 1989). The cost of these data can be conservatively estimated at 65,000 x $1,000/point = $65,000,000. The *value* of such data was the subject of a study by Epstein and Duchesneau (1984), and while cost and value are not always equal, they are generally positively correlated.

One of the best examples of a successful incentive program at the state level can be found in North Carolina (Holloway 1986). More than a decade ago, North Carolina began providing seed money to counties that followed certain mapping and other standards developed by the state. These standards, in turn, comply with federal standards. As far as I know, every county in North Carolina has taken advantage of this program. Since the amount of funding given to each county is small, it appears that money may not be the only motivating factor, and that other less tangible factors like those mentioned in the preceding paragraph are important.

Nonetheless, direct monetary incentives could be easily constructed. For example, OMB could augment the budget of agencies that share data, eliminate duplication, coordinate, and so on. Money could be budgeted for data-sharing programs.

BARTERING—A NEW/OLD PARADIGM

An old method of acquiring goods and services—bartering—is becoming popular in the information age. One of the reasons that bartering decreased in usage was that it became very difficult to transport and deal with large quantities of goods physically. Otherwise, it was a rather effective way to satisfy a need. Computer-readable data and information are easily ported from agency to agency or individual to individual. This can be done by mailing a disk or, more likely as time goes on, by transmitting electronically. These data can comprise the entire database itself. Therefore, the previous problem associated with bartering has been eliminated. Moreover, in many cases, annoying procurement problems can be bypassed!

Most of us have had first-hand experience with this mode of operation. Recently, I participated in a deal between two companies to share data because the data could be sent easily and the sale of these data by either party benefited both parties.

Hence, data sharing might be fostered *if* we can identify data that is available and needed by both parties. This type of deal—bartering on an ad hoc basis—implies that an unusual amount of effort will have to go into making each of these unique arrangements. This was also a nemesis of the "old" bartering system. I suggest that, despite this drawback, the number of these arrangements will grow in the future because of the ease of transporting these data and the fact that the need for cash has been eliminated. The key for sharing data in this fashion is determining what data is *available* and what is needed.

CONCLUSIONS

The MSC has found that significant improvement in the NSDI can be made if data sharing is facilitated (National Research Council, Mapping Science Committee 1993). Aspects of the improvements have been outlined above. We have discovered, I believe, three important facts:

1. Data sharing is inexorably linked to usable standards and incentives.
2. Incentives do not necessarily need to be monetary.
3. Bartering may be an effective way to share data.

If these statements are true, the problem has been defined and we need to turn our attention to building standards, incentives, and facilitating ideas like bartering. While economic benefits are clearly an incentive, other factors such as quality of product, prestige, and pride in the parent organization may play an extremely important role in motivating organizations to adhere to standards. Perhaps defining and determining incentives is a legitimate research task. Moreover, I suggest that standards research and development be placed on the list of important and fruitful research topics. Lastly, we need to develop completely new paradigms for sharing data.

ACKNOWLEDGMENTS

I am grateful to serve with my colleagues on the MSC, and I have enjoyed their vision and clear thinking as we attempt to understand and improve our national information infrastructure. I have learned a great deal from them, and many of their thoughts are recorded in this chapter. My thinking has evolved considerably over the past few years because of the MSC discussions. Having said that, this chapter is a reflection of my present ideas and should not be considered to be the views of the MSC.

REFERENCES

Epstein, Earl, and Tom Duchesneau. 1984. *The use and value of a geodetic reference system.* U.S. Department of Commerce, National Oceanic and Atmospheric Administration, Rockville, Maryland.

Federal Geodetic Coordinating Committee. 1989. *Input formats and specifications of the National Geodetic Survey Database. Vol. 1: Horizontal control data.* Rockville, MD: Federal Geodetic Coordinating Committee, National Geodetic Survey.

Holloway, D. 1986. Land records management in North Carolina. *Popular Government.* Chapel Hill: Institute of Government, University of North Carolina.

Morrison, Joel. 1988. Digital cartographic data standards: the U.S. experience. *Cartography* 1712.

Moyer, D. D., and Bernard J. Niemann, Jr.. 1993. The why, what, and how of GIS standards: issues for discussion. *Journal of the Urban and Regional Information Systems Association* 5, 2.

National Research Council, Mapping Science Committee. 1991. *Research and development in the National Mapping Division, USGS: trends and prospects.* Washington, D.C.: National Academy Press.

_____. 1993. *Toward a coordinated spatial data infrastructure for the nation.* Washington, D.C.: National Academy Press.

National States Geographic Information Council. 1991. First meeting, Atlanta, Georgia, 27 October.

Office of Management and Budget. 1990. Coordination of surveying, mapping, and related spatial activities. Revised Circular A-16. Washington, D.C. 19 October.

Petchenik, B. B. 1991. New directions for national mapping. *Journal of the Urban and Regional Information Systems Association* 3, 1: 77–79.

Sullivan, Charles D. 1983. *Standards and standardization – basic principles and applications.* New York: Marcel Dekker, Inc.

U.S. Department of Commerce, National Oceanic and Atmospheric Administration, Federal Geodetic Control Committee. 1980. *Accuracy standards and specifications.* Rockville, MD: National Geodetic Information Center.

Yorczyk, Richard. 1992. Telephone conversation with author, 4 February.

Doyle Frederick

21 *Coordination of Surveying, Mapping, and Related Spatial Data Activities*

On October 19, 1990, the Office of Management and Budget issued the revised Circular A-16, titled "Coordination of Surveying, Mapping, and Related Spatial Data Activities." Revised Circular A-16 expands the breadth of coordination of spatial data[1] *and assigns leadership roles to federal departments for coordinating activities related to these data. Revised Circular A-16 also establishes a new interagency coordinating committee named the Federal Geographic Data Committee. The committee has established subcommittees and working groups and is coordinating different categories of data and working on issues of standards, technology, and liaison with the non-federal community.*

OPPORTUNITIES AND CHALLENGES

Today, agencies are required to respond quickly to increasingly complex problems involving a wide variety of geographically referenced data sets, such as environmental, natural resource, or socioeconomic. New and growing administrative and regulatory responsibilities assigned to agencies also have placed tremendous pressure on existing information-delivery systems. Computerized spatial data-handling technologies, such as Geographic Information Systems (GIS), have emerged as cost-effective tools for solving complex geographic problems and are assisting decision makers in finding solutions to real-world management challenges. These powerful data-handling and analysis technologies are acutely dependent on the availability and quality of digital spatial data.

Digital spatial data are a critical linkage among user organizations, hardware and software systems, and applications. Moreover, development of the necessary digital spatial databases is invariably the largest cost factor in the computer-assisted analysis of complex issues. The immediate need for, and the large cost of, high-quality digital spatial data are creating opportunities to build innovative partnerships among government institutions and the public and private sectors to avoid wasteful duplication of effort and to yield high-quality data for the mutual benefit of all interested parties.

Building innovative partnerships and improved coordination processes also will increase the nation's ability to deal with future complex societal issues and to compete more effectively in the world marketplace through the eventual development of a national spatial geographic data infrastructure, with linkages at all levels of society—federal, state, and local governments and the private sector. This national infrastructure, linked by criteria and standards, will facilitate sharing and efficient transfer of digital spatial data between producers and users, which will increase the availability and timeliness of information such as new street and road networks, demographic and cultural patterns, and land-use and land-cover changes.

The Office of Management and Budget (OMB) realized that greater efficiency and effectiveness are offered by the use of spatial data and related technologies but was concerned about the potential for waste caused by duplication of effort and a lack of coordination. In 1989, OMB asked the Federal Interagency Coordinating Committee on Digital Cartography (FICCDC) to analyze the need to expand the coordination of federal spatial data use and to review and recommend potential revisions to Circular A-16. The FICCDC recommended that: (1) the breadth of coordination carried out by the committee be increased by the addition of other types of spatial data, such as geologic, resource (including soils, wetlands, and vegetation), cultural and demographic, and ground transportation; (2) the name of the committee be changed to the Federal Geographic Data Committee (FGDC) to reflect this broader coordination responsibility; and (3) the new committee and its responsibilities be incorporated within a revised and expanded OMB Circular A-16. The FICCDC recommendations of a revised Circular A-16 were reviewed and discussed by federal agencies and the non-federal spatial data community. On October 19, 1990, revised Circular A-16 was signed by OMB Director Richard Darman.

THE FEDERAL GEOGRAPHIC DATA COMMITTEE

The FGDC (Office of Management and Budget 1990):

supports surveying and mapping activities, aids geographic information systems use, and assists land managers, technical support organizations, and other users in meeting their program objectives through

- promoting the development, maintenance, and management of distributed database systems that are national in scope for surveying, mapping, and related spatial data;
- encouraging the development and implementation of standards, exchange formats, specifications, procedures, and guidelines;
- promoting technology development, transfer, and exchange;
- promoting interaction with other existing federal coordinating activities that have interest in the generation, collection, use, and transfer of spatial data;
- publishing periodic technical and management articles and reports;
- performing special studies and providing special reports and briefings to OMB on major initiatives to facilitate understanding of the relationship of spatial data technologies with agency programs; and
- ensuring that activities related to Circular A-16 support national security, national defense, and emergency preparedness programs.

Committee Goals and Objectives

The committee's goals include increasing the nation's ability to compete more effectively in the world marketplace and to resolve complex issues, and improving the efficiency and effectiveness of federal programs. These goals recognize that the United States needs timely and accurate data to compete with the information-based economies of developed nations. Geographic data allow information to be used in new and powerful ways. Geographic data and related technologies are important tools for understanding complex national and global environmental, resources, and societal issues and for proposing and evaluating possible solutions. Customers of federal programs have increasingly high expectations of service from federal agencies and are pressing to have these services delivered more quickly. Geographic data offer opportunities to deliver current services at lower costs and to provide new services.

The committee has four major objectives to accomplish these goals: (1) promote the development of a national digital spatial data infrastructure; (2) reduce duplication and waste; (3) promote sharing of digital spatial data; and (4) promote wise use of spatial data technologies. The development of a national digital spatial data infrastructure provides a cooperative framework for sharing data among federal, state, and local government agencies and the private sector. Reducing duplication and waste includes avoiding the duplication of work already complete or in progress and identifying high-priority areas where scarce resources should be applied. Sharing of digital spatial data will be promoted by providing information on data development and availability, by encouraging data exchange through the development and implementation of standards, and by understanding the application of data in different fields. The committee will provide forums for information exchange on technologies and their use and will encourage the development of procedures to assess the benefits and effectiveness of technologies to promote the wise use of spatial data technologies.

Examples of activities of the FICCDC to be continued and enhanced by the FGDC include developing a National Geographic Data System (a system of independently held and maintained federal digital spatial databases, linked by standards and criteria; updating the FICCDC technical report "A Process for Evaluating Geographic Information Systems"; and continuing the inventory and producing revisions to the "Summary of GIS Use in the Federal Government."

COMMITTEE MEMBERSHIP AND STRUCTURE

Circular A-16 names the departments of Agriculture, Commerce, Defense, Energy, Housing and Urban Development, Interior, State, and Transportation; the Federal Emergency Management Agency; the Environmental Protection Agency; the National Aeronautics and Space Administration; and the National Archives and Records Administration as members of the FGDC. The committee subsequently admitted the Tennessee Valley Authority and the Library of Congress. The Department of the Interior chairs the FGDC. Other federal departments and independent agencies with activities or interest in surveying, mapping, or related spatial data should be represented and can request membership by writing to the Secretary of the Interior.

The FGDC will establish, in consultation with other federal agencies and appropriate organizations, the standards, procedures, interagency agreements, and other mechanisms to carry out its coordinating responsibilities. The committee may recommend to OMB additions to, revisions of, or deletions from Circular A-16 and supporting documents. Subcommittees, working groups, and advisory committees may be convened. The new subcommittees and working groups build on coordinating mechanisms established under the previous version of the circular and the FICCDC working groups.

The FGDC established a series of subcommittees, working groups, and a coordination group to undertake the responsibilities assigned by the

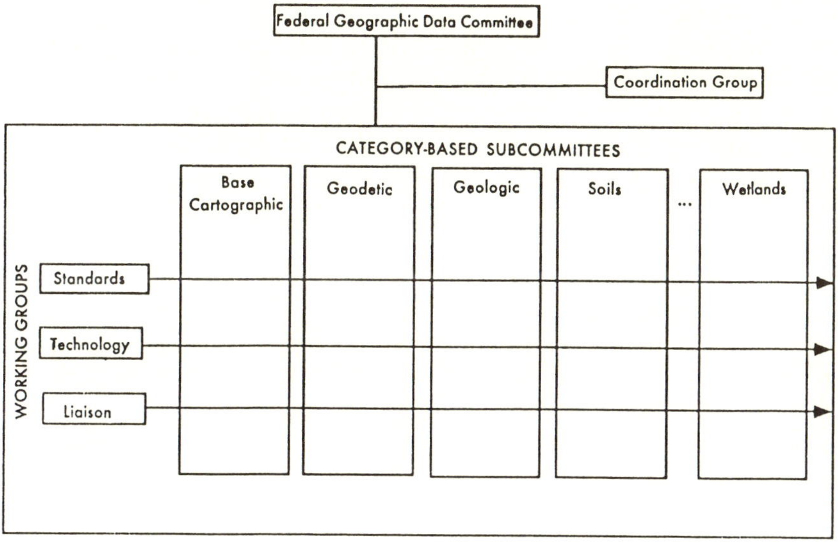

FIGURE 21.1
Structure of the Federal Geographic Data Committee

circular (Figure 21.1). Subcommittees coordinate activities and develop standards for the categories of data to be coordinated under the circular. Each subcommittee is chaired by a lead agency identified in the circular for the category. Members include agencies that have program needs involving that category. There may be further subdivisions of these groups to handle specific issues. As additional spatial data categories are identified, lead agencies will be designated and subcommittees will be established to develop standards and coordinate activities.

Working groups deal with issues common to all spatial data categories and have been established for three such issues: standards, technology, and liaison with state and local governments, academia, and the private sector. Members include federal agencies interested in these issues. The coordination group provides a means for the subcommittees and working groups to interact and coordinate their activities. The coordination group is composed of the chairs of the subcommittees and working groups, the committee secretariat, and representatives of FGDC member departments and independent agencies and complementary coordination groups.

TABLE 21.1

Spatial Data Categories and Lead Agencies under the Revised OMB Circular A-16

Category	*Lead Agency*
BASE CARTOGRAPHIC	U.S. Geological Survey
CADASTRAL	Bureau of Land Management
CULTURAL AND DEMOGRAPHIC	Bureau of the Census
GEODETIC	Coast and Geodetic Survey
GEOLOGIC	U.S. Geological Survey
GROUND TRANSPORTATION	Federal Highway Administration
INTERNATIONAL BOUNDARIES	Department of State
SOILS	Soil Conversation Service
VEGETATION	U.S. Forest Service
WETLANDS	U.S. Fish and Wildlife Service

SPATIAL DATA CATEGORIES AND LEAD AGENCIES

Surveying, mapping, and related spatial data categories are characterized by spatial data that are a part of, or frequently analyzed with a reference to, the base topographic data and geodetic network of the nation. These national spatial data categories are of interest to many agencies and represent a substantial part of the national spatial geographic information resource.

Revised Circular A-16 expands the coordination to ten categories of spatial data: base cartographic, cadastral, cultural and demographic, geodetic, geologic, ground transportation, certain international boundaries, soils, vegetation, and wetlands data. Most of these spatial data categories are collected, maintained, and disseminated under national programs established by law. The circular also initiates coordination for other national spatial data categories for which there are broad interest and resources spent by FGDC member departments and independent agencies.

Circular A-16 assigns to the departments of Agriculture, Commerce, Interior, State, and Transportation the responsibility to lead the coordination of these national spatial data categories. This responsibility includes facilitating information exchange and data transfer; establishing and implementing standards of quality, content, and transfer capability; and coordinating the collection of spatial data to minimize duplication of effort. Each department has authority described in its mission or implied as part of its program responsibilities. The departments delegated this responsibility to the organizations are listed in Table 21.1.

This leadership is carried out under the policy guidance and oversight of the new FGDC. All lead agencies except the Department of State will carry out this responsibility by chairing FGDC subcommittees. Because of the unique nature of its coordination responsibilities for the portrayal of international boundaries, the Department of State will use other means to coordinate these activities.

The Federal Geographic Data Committee guides and oversees the activities of the subcommittees and working groups. The subcommittees coordinate activities related to each spatial data category. The working groups deal with issues common to all spatial data categories and promote consistency among the subcommittees on these issues. The Coordination Group provides a means for the subcommittees and working groups to interact and coordinate their activities.

INTERACTION WITH STATE AND LOCAL GOVERNMENTS AND THE PRIVATE SECTOR

Another important focus of the FGDC will be to provide guidance and promote cooperation among federal, state, and local government agencies and between the public and private sectors in the collecting, producing, and sharing of spatial data. The liaison working group is developing strategies to assess opportunities for coordination with non-federal spatial data users. The objective of this activity is a partnership among federal, state, and local government agencies and the private sector in the development of a national spatial geographic data infrastructure.

INTERACTION WITH OTHER COORDINATION GROUPS

Revised Circular A-16 builds on the FICCDC and earlier versions of the circular. The authors of the revised circular recognized the existence of other coordination activities related to spatial data and avoided duplicating these efforts. The FGDC's coordination with these complementary coordination groups will ensure the compatibility of activities. Discussions with representatives of other coordinating groups have resulted in suggestions of joint technical meetings, close coordination among the committees' secretariats, and involvement of representatives of the coordinating groups at meetings.

CONCLUSIONS

Revised OMB Circular A-16 greatly expands the breadth of coordination of federal spatial data activities and encourages the cooperation of federal, state, and local government agencies and the private sector in the development of the nation's spatial information resources. The successful development of this resource depends on all of these groups working together.

The FGDC has established a structure for coordinating spatial data activities. The subcommittees and working groups are coordinating agencies' activities, developing standards, facilitating data exchange, and investigating means of closer cooperation among government institutions and between the public and private sector.

Note

1. Spatial data are geographically referenced features that are described by geographic positions and attributes in an analog and/or computer-readable (digital) form.

ACKNOWLEDGMENTS

This article originally appeared as D. G. Frederick (1991), "Coordination of Surveying, Mapping, and Related Spatial Data Activities," GIS/LIS 1991 *Proceedings* 1: 374–380.

REFERENCE

Office of Management and Budget. 1990. Coordination of Surveying, Mapping, and Related Spatial Data Activities (Circular A-16). Washington, D.C.: Office of Management and Budget.

DONALD F. COOKE

22 *Sharing Street Centerline Spatial Databases*

Street centerline spatial databases (SCSDs) are good examples for the study of data sharing, as there is a long SCSD history to examine involving private, quasi-private, and all levels of government organizations. The history provides examples of sharing both of data and the labor of data preparation and maintenance. SCSDs have such widespread applicability and potential value in public and commercial applications that it is suggested that the national SCSD and a few other data sets be afforded a special category of "sharing": that of spatial data infrastructure. Establishment of an effective spatial data infrastructure will require many changes from the present government operations. This chapter offers suggestions to this end.

INTRODUCTION

This chapter is based on the premise that data sharing, or the lack thereof, is a complex phenomenon, the study of which may benefit from experiences with a specific example. The street centerline spatial database (SCSD) is chosen specifically to focus discussion on a small and coherent body of experience.

I pick the SCSD for two reasons, the first being personal involvement (Cooke and Maxfield 1967) in the development of SCSDs in the United States. Second, it is becoming apparent that the United States needs a "spatial data infrastructure" consisting of certain basic spatial databases and their support mechanisms, which serve large numbers of varied constituents and applications. It is generally accepted that a national SCSD is one of the basic databases because of its broad applicability (Mapping Science Committee 1993).

While the history that follows provides examples of spatial data sharing, in a broad view it reveals a series of largely uncoordinated though well-intentioned federal departmental efforts yielding surprisingly robust products that have serendipitously spawned a variety of businesses that promise to promote thoughtful and efficient economic development in the 1990s.

There is a special case of spatial data sharing: where there are so many potential sharers and so much public benefit delivered by certain databases that they deserve to be institutionalized as a core data set of the National Spatial Data Infrastructure and made broadly available for the public good. The SCSD epitomizes this case.

DEFINITIONS AND BACKGROUND

Spatial Data and Spatially Referenced Data

There is not yet a general agreement on what constitutes spatial data. This point requires clarification. It is useful to draw a distinction between "spatial data" and "spatially referenced data."

Spatially referenced data include practically all data, whether they are census tabulations, customer lists, or the tax assessor's files. Though all are spatially *referenced* by a street address, municipality, zip code, or census tract, they remain fundamentally thematic data. It is hard to come up with any thematic data set that is not spatially referenced in some way.

"Spatial data," on the other hand, means information describing the relative positions of people, things, or events. Spatial data are the information that we're accustomed to storing primarily on maps. Spatial data are the data that are not thematic. Spatial data take the form of a structured representation of coordinate measurements, topological relationships, and common georeferencing "names," like street addresses, zip codes, and census tract numbers. The primary output of a Geographic Information System (GIS) digitizing function is spatial data, which the GIS links to thematic "spatially referenced" data.

Spatial Data Infrastructure

Infrastructure is "an underlying base or foundation; the basic facilities needed for the functioning of a system." "Infrastructure" evokes an image of roads and bridges for most people, and the analogy between spatial data and transportation infrastructures is instructive.

For example, our highway infrastructure delivers no direct societal benefit by itself. In fact, highways remove land from other use, mar the landscape, and disturb wildlife habitat. We tolerate them because of the benefits that *vehicles* provide: moving people and goods from place to place. Note that most vehicles are useless without roads. Similarly, spatial data sets deliver no direct benefits and consume considerable resources for their construction and maintenance. But we need spatial data sets because they are the foundation that application software packages need—and these applications generate benefits in planning, site analysis, demographic studies, and routing and dispatching.

Both spatial data and transportation infrastructure are multi-modal. Just as our transportation infrastructure has components supporting land, air, and water transport, our spatial data infrastructure has components supporting various kinds of spatial analysis and operations. We need harbors for ships and airports for planes; we need parcel-level spatial data sets for land-information applications, and street centerline data sets for demographic and logistical applications. Just as highways are not good facilities for landing airplanes, even a detailed and highly accurate parcel database is not useful for applications requiring a more modest (smaller-scale) level of detail.

Finally, the infrastructure has its own infrastructure! Our highways have a huge and largely hidden support infrastructure of funding mechanisms, maintenance procedures, and construction standards. Airports need air traffic control systems, linkages to other transport modes, and standards for construction and operation. When we agree on which national data sets comprise our most basic spatial data infrastructure, these data sets will require a large investment in labor and capital supporting maintenance, standard setting, and enforcement and distribution.

Just as the task of maintaining all levels of the transportation infrastructure is shared by private contractors and government at all levels working together, the task of building a national spatial data infrastructure is likely to require similar commitment to sharing and involvement by the same parties.

INFRASTRUCTURE VERSUS SHARING

Why does the issue of "infrastructure" intrude on a study of "sharing"? The transportation analogy developed above helps explain this.

As the country was settled, farmers built crude roads from their barns to their fields. As more people settled in an area, adjacent settlers cooperated in sharing the labor of establishing roads from farm to farm and to local markets. As the markets evolved into incorporated municipalities, governmental units took over maintaining portions of the road system "in the public good." This has evolved into our present system of interstate, federal, state, county, and town roads.

Similarly, some organizations, such as a water company, digitize a spatial database to manage their network of pipes, pumps and valves. Such a spatial database—like the farmer's road from barn to field—is probably of use only to the originating organization.

Other organizations may create a spatial database useful to several other agencies (e.g., a large-scale topographic survey of a town). This is the typical case where "data sharing" is appropriate since such efforts are costly but, if done with consideration of needs and standards of other agencies, present an opportunity to save money and time for all participants through cooperation. Most studies of data sharing concentrate on this case and the obstacles that often prevent sharing data or the effort of creating spatial data.

Some spatial data sets, however, are so broadly useful that they deserve not just to be "shared" but to be recognized at all levels of government as "the foundation of the economy in the 1990s," as one GIS industry leader puts it (Dangermond 1992). Such recognition should extend beyond "sharing" considerations to making a set of critical spatial data sets as accessible and easy to use by analysts as our present highway system is for motorists.

Which spatial databases require "infrastructure" treatment? My proposals follow.

1. *A basic topographical layer to serve the same general functions provided by the USGS "quad" maps for the past 120 years.* Note that digitizing quad maps does not satisfy this requirement! Due to technology shifts and the need for current information, digital orthophotoquads have replaced quad maps as the country's logical topographic base.
2. *A land-use/cover layer.* A land-use/cover layer is crucial to conserving agriculture and wildlife and to promote thoughtful development in contrast to mindless growth. It is also essential to being able to measure the success or failure of a policy of "no net loss of wetlands."
3. *A cadastral layer.* The United States is a century or more behind other developed countries in managing this crucial information. The federal government owns about a third of the nation's land outright (and becomes the custodian of even more each time a savings and loan institution fails!) We need to reject the conceit that the United States can continue to operate in a twenty-first

century land market with eighteenth century land recordation procedures. Federal leadership on standards and practices and economic incentives for land record modernization are long overdue.

4. *The street centerline spatial database.* Due to efforts by the Census Bureau and USGS (below), this is the most mature layer. A current and affordable SCSD holds tremendous promise in improving productivity in business, in improving effectiveness of facility siting, deploying field resources (sales and repair people), and managing logistical operations such as deliveries. It is on this portion of the infrastructure that this chapter focuses.

THE STREET CENTERLINE SPATIAL DATABASE

Although a more contemporary object-oriented view of SCSD is evolving through the Spatial Data Transfer Standard (STDS) process (Moellering 1992), street centerline data sets developed in the United States have historically had the following characteristics.

- As its name suggests, the subject of the SCSD is the centerline of streets. In practice, it has proven useful to include boundaries of water features, railroads, political borders, and other such lines in SCSDs.
- Applying topological theory and standard vector database practices, each logical record in an SCSD represents a line segment between two nodes, separating two areal units. In practice, census block numbers have served as area designators in U.S. SCSDs, although technically these are an application-specific overlay to the two-cell structure defined by line segments.
- Nodes are required at all line intersections, at ends of cul-de-sacs, and places where an attribute like the name of a street changes. Node location is measured latitude and longitude; no serious effort has yet been made to encode altitude (nor is altitude required for many applications).
- Commonly required attributes are recorded for each line segment: street name, address ranges on each side, and zip codes. Endless other attributes, like speed limits and turn restrictions, can be added for specific applications, but the need to link to data files spatially referenced by street address is practically universal, elevating address attributes to a status of fundamental importance in this layer of the infrastructure.

- If the street bends between nodes, the internodal path is represented by one or more "shape" point coordinates.

A street centerline data set maintained to these specifications is adequate—in fact, optimal—for many classes of spatial applications such as site analysis, demographic studies, vehicle navigation and route guidance, dispatching, and routing and scheduling. The SCSD does not contain detail at the parcel level. Another facet of the spatial data infrastructure— the cadastral data layer—must support parcel operations. Also, the emphasis on man-made artifacts (streets) limits usefulness of an SCSD in natural resource applications; the land-use/land-cover component of the infrastructure is more useful in these cases.

HISTORY

The first street centerline data set built in the United States was the SACS (Street Address Conversion System) database described by Robert Dial (Dial 1964). Dial proposed that all addresses in urban information systems be augmented with a geographic coordinate reading. This would allow analysts to digitize arbitrary study area polygons and retrieve the geocoded addresses by point-in-polygon computations. Dial pointed out that this permitted any amount of redefinition of retrieval polygons at the whim of the analyst, while the coordinates would stay constant on the data records.

Three years later, researchers at the New Haven Census Use Study made the first DIME (Dual Independent Map Encoding) files for West Haven and New Haven, Connecticut (Cooke and Maxfield 1967). These DIME files probably were the first "topologically structured" street centerline data sets built using computerized topological edits to detect coding errors. DIME grew out of frustration researchers had experienced while trying to adapt an earlier census street-based Address Coding Guide (ACG) for use as a computer mapping base file.

Address coding guides are a technical cousin to street centerline databases. The logical unit of an ACG is a range of addresses on one side of a street instead of a line on a map. Early ACGs contained no coordinates and did not aspire to topological correctness. In the mid-1960s the Census Bureau Geography Division built ACGs for 144 metropolitan areas to geocode addresses for the 1970 census. The ACG's only function was to link addresses to 1970 census block codes. When the added value of geometrical and cartographic features was demonstrated by the Census Use Study, the Bureau converted to DIME.

Between 1970 and the present, DIME metamorphosed from "ACG/

DIME" through the "CUE" (Correction/Update/Extension) program to "GBF/DIME" (Geographic Base File/Dual Independent Map Encoding), and finally became "TIGER" (Topologically Integrated Geographic Encoding and Referencing), with nationwide coverage achieved in 1990.

Two parallel efforts bear mention: the Digital Line Graph (DLG) files made by the U.S. Geological Survey and the ZIP+4 database maintained by the U.S. Postal Service.

The USGS DLG series is a digital version of the paper "quad" series maps published at various scales by the Survey since the nineteenth century (USGS 1985). Although the largest-scale quad maps use double-line representation of streets, this is but a cartographic artifact; the width of the street casings on quad maps is not related to actual street width. Consequently, the USGS chose a single-line street representation for its DLGs.

Two of the DLGs are interesting to students of SCSD. The 1:100K series is complete nationwide and forms the geometric basis for rural portions of TIGER. The USGS and Census are justifiably proud of the interagency cooperative arrangement for sharing the cost and effort of producing both the new 1:100K maps and DLGs and rural TIGER coverage.

The 1:24K DLG series is especially tantalizing to SCSD aficionados, since (unlike urban portions of TIGER) its coordinate quality is good enough for virtually any SCSD application. Unfortunately, 1:24K coverage is meager, with less than one-tenth of the transportation coverage completed. In addition, neither the 1:100K nor the 1:24K DLG series carries street name and address attributes demanded by most SCSD applications.

In contrast with DLGs, the U. S. Postal Service's Zip+4 database is a cornucopia of address information (USPS 1990). This file defines the universe of mailable addresses nationwide and, unlike other files described here (which are practically static), Zip+4 is updated monthly. Unfortunately, Zip+4 has an ACG format, contains no coordinates, and is not topologically structured. Broad use of Zip+4 is further compromised by copyright claims made by the quasi-private Postal Service.

DATA SHARING IN SCSD BUILDING

Many different types of data sharing are evident in the history of GBF building. In the mid 1960s, the Census Bureau realized the value of local knowledge and map resources in constructing detailed spatial databases and induced more than 100 municipal and regional agencies to participate in ACG production. The Census Bureau provided instructions, maps and worksheets; local agencies provided manpower and access to local maps. This "sharing" exercise was greatly facilitated by applicability of HUD 701 local planning funds to the ACG effort.

This exercise was repeated in the late 1970s in the CUE program but with subtle differences: initially, the bureau provided maps and worksheets but no funding. Some local agencies supplied manpower and local map resources but had to look to the Comprehensive Employment and Training Act (CETA) program to pay CUE workers. A large number of local agencies balked at this arrangement and failed to respond to the bureau's enticements. As the 1980 census drew near and the Bureau's operational budget grew, Census changed its funding incentives and paid first one-half, and finally three-quarters, of the local agencies' costs to draw the final holdouts into the CUE program. This infuriated some early (unfunded) adopters of CUE and soured prospects for future cooperative efforts with them.

Between 1986 and 1988 the Census Bureau contracted with four private companies to build metropolitan TIGER coverage from 1980 GBF/DIME files and map resources recently updated by the Census Bureau Field Division. Data sharing was limited to the purchase by the Bureau of fifteen GBF/DIME files that had been significantly updated by municipal or regional public agencies during the 1980s.

In 1990 and 1991, the Census Bureau held four day-long conferences aimed at forging TIGER data-sharing alliances with state governments (Census 1991). Though state-level agencies do not have direct access to local geographic knowledge and materials, the Bureau appears reluctant to set up communication with potentially hundreds or thousands of local agencies interested in ongoing TIGER updating. The Bureau continues to pursue cooperative arrangements, but so far only with public agencies.

Public access to Census geographic products has always been open and outstandingly generous. DIME and TIGER are interim work products of the Bureau's main (statistical) mission; distribution of digital maps is not a mandated Bureau function. The Bureau went so far as to establish a permanent Data User Services Division to educate users and distribute data, including secondary spatial data products. This practice of distributing work products is called "goodies over the fence" by some, to make the point that benefits from the "goodies" so distributed are accidental dividends to taxpayers and not the primary result of deliberate governmental missions.

Product Pricing

During the first two decades of ACG and DIME, the Bureau sold copies of geographic files for "cost of copying," just as it sold its data products. The cost rose from $70 per reel of tape to $140 during this period. When high-density 9-track tape (6,250 bpi) became available in the early 1980s, one could buy a complete copy of GBF/DIME from the Bureau for less than $3,000.

It is significant that the TIGER/1:100K DLG data production partnership between the USGS and Census included a "joint marketing" agreement for TIGER. This resulted in a new data-pricing policy influenced by the USGS, which had a history of pricing DLG products above cost of copying. When the pre-census edition of TIGER was released in 1989, the artificial prices ($175 for the first county in a state plus $25 for each additional county) added up to $87,000 for a nationwide copy. Though lower than some previous price estimates for TIGER, this departure from previous access costs jeopardized broad dissemination of the new TIGER resource.

However (as predicted by Census officials who were ambivalent about the new pricing policy), a private company bought one copy of TIGER and contracted for private production and distribution of eighteen other copies, resulting in nineteen organizations sharing access to nationwide TIGER by splitting the acquisition cost and benefiting from a commercially competitive cost of copying. Under this arrangement, a nationwide copy of TIGER cost each organization $10,000 instead of $87,000.

By 1991, new data publication technology—CD-ROM—helped lower economic barriers to sharing the "final" TIGER release. The Bureau now distributes TIGER primarily on CD-ROM for $250 per disk. Although this remains a somewhat arbitrary price, the nationwide total of $11,000 (forty-four disks) no longer constitutes a barrier to widespread TIGER use.

But as soon as barriers of price and lack of nationwide coverage are lowered, others appear: the three "C's" of completeness, currency and coordinates. Briefly: Census's TIGER—a work in process—has incomplete address attributes, tolerable coordinate accuracy at best, and no facility for ongoing update.

Demand for a street centerline data set is such that several private companies currently compete with the Census Bureau to provide improved SCSD products. In aggregate, their budgets are greater than that of the Bureau's Geography Division. Given hypothetical operational efficiencies often attributed to the private sector vis-à-vis government, it is entirely possible that effective custody of this layer of the National Spatial Data Infrastructure has already slipped out of governmental hands into the private sector—a possibility deemed not in the best public interest by a National Academy of Sciences Committee (MSC 1990, 45).

LEGAL, ORGANIZATIONAL AND ECONOMIC ISSUES

Andrew Frank and Peter Crossfield describe technical problems in data exchange (Frank 1992; Croswell 1991), most of which are being solved as the U.S. SCSD approaches maturity. Despite a quarter century of work, the goal of SCSD development has not yet been explicitly stated in the form of content and performance specifications. However, most issues like object-class definition are approaching resolution. What remain largely unresolved are legal, organizational and economic issues.

Legal Issues

The U.S. government's contention that taxpayers own the data produced by its agents (governmental organizations) is practically unique in the world and is probably the fundamental influencing factor in data sharing in the United States. On one hand, it drives the "goodies over the fence" phenomenon: widespread dissemination of spatial databases like TIGER and incentives for the private sector to provide added-value secondary distribution of spatial data. Unfortunately, this generous policy also relieves governmental agencies of the necessity of looking outside their individual missions and wrestling with the much harder issues of anticipating future needs of the country as a whole.

In contrast, Great Britain's Ordnance Survey (OS) has total control through Crown Copyright over primary and secondary distribution of its paper and digital map products. The OS consequently has recognized the need for a strong marketing function to listen to external needs and prepare a product line anticipating future developments like in-car route guidance which require specialized spatial data sets. As one OS Director General put it, "Being market-oriented means that an organization takes its principal guidance from the present and future needs of users and not from what it has always done in the past, what it likes doing, or even what it does best" (Ordnance Survey 1991, 161).

During the 1980s a debate raged over "cost-of-copying" for information products versus "cost recovery." This appears to be resolved in favor of "cost-of-copying," but in the meantime agencies like the USGS and the Census Bureau suffered as a consequence of charging more than copying costs for data sets. Our legal mandates—at least for federal data sets—now appear to be clear, and future departures from cost-of-copying guidelines should diminish.

A final legal consideration that will certainly affect spatial database activities is the recent *Feist v. Rural Telephone Service Company* ruling of the Supreme Court (Supreme Court 1991). Feist probably puts the information (though not the representation) in most commercial street maps and atlases into the public domain. This will probably hurt publishers of paper maps, which depend on copyright for protection of map products they sell. Ironically, Feist may benefit private spatial database vendors if it is interpreted to permit copying of "facts" from paper maps into electronic databases. Since computer databases are normally licensed rather than sold, vendors are able to incorporate intellectual property protection (beyond copyright) into license agreements with buyers. This may enhance the ability of private spatial database vendors to compete with government data sets like TIGER.

Organizational Issues

The United States currently lacks an effective organizational structure for dissemination of federal spatial data sets. Despite the 1-800-USA-MAPS service offered by the USGS, a minority of spatial data transactions are initiated or consummated through this channel. This is entirely understandable considering that federal spatial databases are maintained throughout many different agencies.

In the case of the SCSD, the consequence will probably be a robust, competitive market served primarily by private vendors. If competition succeeds in moderating prices, then the public is likely to be well served by this structure—perhaps better than it would be under government control.

Other parts of the spatial data infrastructure will not fare as well, however; the citizenry as a whole would benefit from stronger federal leadership on both the land-recordation system (including the cadastral spatial data layer) and improved national information on critical land-use/land-cover themes underlying issues like agricultural productivity and management of wetlands and wildlife habitats.

Economic Issues

Most economic issues are touched on above and evidence themselves in examples from the history of DIME and TIGER. In summary:

1. Pricing of government databases should cover cost of copying only, and organizations should strive to take advantage of economical publication media like CD-ROM. High fees like those attempted by USGS for DLGs and Census for tape distribution of TIGER will fail because entrepreneurs will inexorably provide inexpensive copies of useful uncopyrighted data sets.
2. Government agencies must charge something to cover costs; some minimum user fee will deflect unintended use. For example, the USPS used to distribute ZIP+4 free, sending out about a dozen reels of tape each month to many subscribers. USPS recently started charging a modest fee for tapes to deter the ZIP+4 program from becoming a source of free magnetic media.

3. Funding ultimately comes from a broad tax, a user fee, from private resources, or some combination of these. Dependence on a user fee alone will jeopardize building spatial data infrastructure; there may not be a critical mass of users to support data-set construction. Similarly, private funding alone is unlikely to be sufficient to build even a portion of spatial data infrastructure; almost all data vendors piggyback their efforts on publicly funded files like TIGER.

   In the absence of federal initiative, the public at large might forgo some general benefit (e.g., safer traffic or cleaner air) of an Intelligent Vehicle Highway System (IVHS) database. The issue remains a fundamental philosophical definition: What is the "public good?" To what extent, and at what level, should government provide for it?
4. Privately funded databases will not be sold for cost of copying but for higher prices determined by production costs and market forces—in the final analysis, by the value delivered. Theoretically, competition will keep prices reasonable, but in practice, measuring specific demand for a spatial data set is complicated because the data set by itself is useless without an application system (software). Value of a spatial data set is measurable only as a part of the value delivered by the total application system: software, hardware and data. The price of the aggregate spatial solution relative to the value delivered is what ultimately determines demand.

CONCLUSIONS

The history of SCSD development demonstrates that even without a central spatial data agency like Britain's Ordnance Survey, remarkably valuable spatial data sets like TIGER can be created and distributed. It is not an accident that the value delivered by TIGER coincides with the aggressively positive relationship created by the Census Bureau with users of its data. Our public domain database policies and competitive private-sector activities have put the United States in the lead in many areas of spatial data use.

However, without abandoning the free-enterprise model, the United States also needs more of the kind of coordinated vision and leadership exhibited by the Ordnance Survey. We need acknowledgment at the cabinet level that spatial data use can be the engine of deliberate and effective economic development in the 1990s and beyond. We need to seize the opportunity to take advantage of the strengths of the private sector and of government at all levels.

Though some insight into sharing of spatial data may be provided by the foregoing examples, the most interesting chapters are yet to be written. As we move our spatial knowledge from paper to computer storage, we are able to benefit from great strides in hardware technology and software engineering practices driven by forces outside of the spatial data arena.

Hardware and software improvements will continue; in fact, they appear to be accelerating. Anyone who worked with computers twenty-five years ago marvels at our current computer resources. The information-processing environment of a quarter century hence will be no less remarkable.

But good spatial data is relatively ageless. While hardware power per dollar may double every twelve months, inviting users to upgrade computing machinery frequently, spatial data has a half-life measured in decades. Aside from occasional postal code changes, the information describing a street in an SCSD—if properly encoded—may be valuable for a lifetime.

Such valuable information must be shared and shared widely. A challenge for the 1990s is to determine which data sets qualify as "infrastructure"—then to build them and establish mechanisms to make them available to all who could benefit from their use.

REFERENCES

Cooke, D., and W. Maxfield. 1967. The development of a geographic base file and its uses in mapping. Garden City, New York: Urban and Regional Information Systems Association, *Proceedings,* pp. 207–218.

Croswell, P. (Plangraphics Corp., Frankfurt, KY). 1991. Remarks at GIS-LIS '91 session on spatial data standards, Atlanta, Georgia. November.

Dangermond, J. (ESRI, Redlands, CA). 1992. Address to GIS session at American Planning Association meeting, Washington, D.C. May.

Dial, R. 1964. *The Street Address Conversion System (SACS).* Doctoral dissertation, University of Washington, Seattle.

Frank, A. U. 1992. Acquiring a digital base map: a theoretical investigation into a form of sharing data. *Journal of the Urban and Regional Information Systems Association* 4, 1: 10–23.

Mapping Science Committee, National Academy of Sciences. 1990. *Spatial data needs: the future of the National Mapping Program.* Washington, D.C: National Academy Press.

_____. 1993. *Toward a coordinated spatial data infrastructure for the nation.* Washington, D.C: National Academy Press.

Moellering, H. 1992. SDTS—simplifying the transfer of spatial data. *ACSM Bulletin,* May-June: 30–34.

Ordnance Survey. 1991. *Ordnance Survey: map makers to Britain since 1791.* Southampton, England.

Supreme Court. 1991. *Feist Publications, Inc., v. Rural Telephone Service Co., Inc.,* 18 USPQ2d 1275.

U.S. Bureau of the Census. 1991. *Proposal for TIGER data exchange.* Washington, D.C.: U.S. Government Printing Office.

U.S. Geological Survey. 1985. *Digital line graphs from 1:100,000-scale maps.* Reston, VA: USGS.

U.S. Postal Service. 1990. *ZIP+4: a technical guide.* Memphis, TN: USPS.

Jonathan Sperling

23 Development and Maintenance of the TIGER Database: Experiences in Spatial Data Sharing at the U.S. Bureau of the Census

The U.S. Census Bureau has played, and will continue to play, a vital role in the development, maintenance, and sharing of spatial and attribute data for Geographic Information Systems (GIS) on the local, regional, national, and international levels. The Census Bureau's development of shareable geographic data files, the GBF/DIME (Geographic Base File/Dual Independent Map Encoding) files for the 1970 and 1980 censuses, and the TIGER (Topologically Integrated Geographic Encoding and Referencing) database for the 1990 Census, have provided a major impetus to the rapid growth and diffusion of GIS technology. This chapter discusses the Census Bureau's experiences in spatial data sharing during these two file-building projects as well as from ongoing experiences in developing Memoranda of Understanding with federal and state agencies to update and improve the spatial and attribute data in TIGER. On the basis of these experiences, preliminary generalizations are made concerning the organizational issues that may facilitate or impede the future digital interchange of spatial data.

INTRODUCTION

The entrance and persistence of the nation's lead statistical agency as a primary producer and user of both geographic and attribute data have grown out of practical but critical concerns (Tomasi 1990). The mandate to conduct a population and housing census of the entire nation and its territories every ten years, and then disseminate the resulting information quickly and accurately, has led to the need to devise innovative ways of improving the collection, processing, and tabulation of data. These improvements have aimed at four goals: increasing efficiency and timeliness, improving data quality, lowering costs, and providing new products for the data user.

Over the past half century, the Census Bureau has pursued innovations and led the government and private sector in computerization, statistical sampling and interviewing techniques, data processing, quality control, and cartographic techniques, to name but a few (Anderson 1988). The Census Bureau pioneered the first large-scale commercial use of the computer with UNIVAC 1 following the 1950 census and the development of the Film Optical Sensing Device for Input to Computers for the 1960 census. The broad use of computer mapping by local governments and private firms was strongly influenced by the Census Bureau when it developed the GBF/DIME files for the 1970 and 1980 censuses and the TIGER System for the 1990 census.

In addition to being one of the nation's largest digital geographic databases—currently sixteen gigabytes—TIGER enables the automated spatial manipulation of census data for all legal and statistical entities in the United States and its territories. Every state and local government now has the capability to create a rudimentary GIS using small-area census data, publicly available extracts of the TIGER database—TIGER/Line™ files—and the appropriate hardware and software. More than 130 private vendors currently have the capability of processing TIGER/Line files, and the number is growing.

In this respect, the development of the TIGER database may be the most important "data" file from the 1990 Census of Population and Housing. The accessibility and widespread potential application of this innovation make it qualitatively different from earlier technological developments at the Census Bureau. The TIGER database will be used not only internally to administer future censuses and surveys (see Marx 1986 for a basic rationale of the TIGER system) but externally to provide a major impetus to the development and sharing of integrated spatial information systems in the public and private sectors as well as the academic community.

A common theme in the GIS literature, generated both within and outside the Census Bureau, has been the notion that although the Census Bureau developed its GBF/DIME files and TIGER database to meet internal Census Bureau needs, their existence has facilitated applications well beyond the scope of census-taking. Perhaps analogous to the many commercial spin-offs that have been generated from wartime military innovations or from the nation's space program, the decennial census operation, conducted by the government's largest non-military assemblage of people and resources, has been responsible for a number of innovations, some already outlined, that have had a major effect on the private sector and the academic research community.

The Census Bureau's recognition of these "consumer spin-offs" traditionally has never been well articulated or coordinated because most innovations have resulted from an internal production standpoint rather than a customer-oriented design. Recent efforts toward building a national spatial data infrastructure and the Census Bureau's adoption of the principles of Total Quality Management and Strategic Planning create the external and internal incentives, respectively, to stimulate a change in that design. Paradoxically, however, the fiscal constraints of the 1990s and their impact on congressional funding, while often considered an incentive for data sharing, may also function as an impediment in the early developmental stages of new technologies.

THE DEVELOPMENT OF SHAREABLE GEOGRAPHIC DATA FILES

GBF/DIME File Development 1966–1982

Many local agencies have been introduced to GIS by the Census Bureau through its various geographic base-building projects over the past twenty-five years. The development of the Address Coding Guides (ACG) for the 1970 census and the GBF/DIME files for the final stages of the 1970 census and for large-scale use in the 1980 census were major steps toward full automation of the Census Bureau's geographic support programs. Although the ACGs provided the building blocks for the later development of the GBF/DIME files, they lacked geographic coordinates and topological structure (Marx 1986).

The design and development of these computer-readable files involved the active participation of federal, state, local, private and academic organizations. The Census Use Study, a small-area data research group sponsored by the Census Bureau from 1966 to 1969, was instrumental in creating and diffusing knowledge about a system that represented map features numerically for processing by a computer to create a geographic base file (U.S. Bureau of the Census 1973). The enhancement of this body of knowledge was of critical importance to the Census Bureau because it enabled the adoption of census-by-mail enumeration methodologies for the major urban centers of the United States.

The change in enumeration procedures was a response to the increasing costs and difficulty of conducting a traditional door-to-door canvassing of the population, a growing and increasingly urban population living within areas having mail delivery by house-number/street-name address, and the technical feasibility of linking such addresses with the geographic units used for data tabulations due to improvements in computer capabilities. For the 1980 census, the GBF/DIME files were enhanced to cover 287 of the nation's largest urban centers, representing more than 60 percent of the population but less than 2 percent of the nation's land area (Carbaugh and Marx 1990).

Data sharing during the 1970s. The development of the ACGs and GBF/DIME files involved data sharing with more than 300 local planning agencies. Sharing primarily took the form of converting analog data into a computer-readable format rather than digital exchange per se. The long-term nature of these relationships provided the Census Bureau with intimate knowledge of the resources available to local agencies, the limitations of available data, and the willingness and ability of agencies to share data. At the same time, local planning agencies and others, through their association with the Census Bureau, became increasingly aware of the potential of computer mapping, automated address matching, and spatial data analysis to meet local needs (Sobel 1978). These relationships played an important role in the later development of the TIGER system.

Monetary and other incentives to local agencies played a major role in the "data sharing" development of the ACGs and GBF/DIME files (Silver 1977). In many cases, local agencies completed work under contract with the Census Bureau or with funding provided by the Department of Housing and Urban Development (HUD), the Federal Highway Administration (FHWA), and other federal, state, and local planning sources. Sensitive to the fact that the development of the GBF/DIME files provided a practical solution to many needs of local governments, the Census Bureau offered to provide the computer programs, processing methodology, and clerical procedures for creating and updating the file, as well as a free copy of the completed file, to each participating local agency. The Census Bureau also offered to do the data keying and processing if agency resources were limited. The promise of higher quality data when these files were used for taking and tabulating the 1970 and 1980 censuses provided a further incentive for data sharing.

As a result of this process, the Census Bureau and local officials learned that the quality and currency of existing data used by planning agencies could not be taken for granted. For example, many local communities relied on the knowledge of a few people for the location of dwellings, address-numbering systems often were not systematic, tax assessor sources sometimes did not meet the bureau's quality standards, and data from different agencies often were inconsistent. Because of these situations, the initial transition to address assignment via automated processes was difficult.

The efforts of working with so many local agencies also challenged the Census Bureau. During the file-building projects, there was turnover in project personnel and key decision makers in the local agencies as well as at the Census Bureau. Some "champions" of the new methods were not reelected or moved on to other jobs. The long-term benefits of building such files were not always apparent to public administrators. Some administrators viewed the new computer technology as a threat to their role in the agency. Also, during the 1970s, there was a general lack of understanding (personal computers had not yet been invented) and/or trust in the new computer technology and the ever-present fear of "big brother" mainframe computers held by a few large government agencies. All these factors tended to hamper development activities and to result in considerable variation in the quality, time expended, and funds necessary to complete the GBF/DIME files.

Data sharing during the 1980s. Census Bureau funding of data collection and coding by local agencies to create the GBF/DIME files created precedents that would later affect the pre-1990 TIGER-building efforts. The overall reductions in federal aid to cities and states during the 1980s provided an atmosphere of "less than cooperative initial attitudes" among many local agencies struggling with smaller staffs and fewer resources than they had in the 1970s. Many agencies expected to be paid for their efforts. These situations, and the perceived difficulties inherent in working with several hundred local agencies, contributed to the Census Bureau's decision to do most of the map and address updates for the 1990 census in-house, albeit often based on materials supplied by local agencies.

Changes in the original GBF/DIME file format by local agencies exacerbated some early data-sharing efforts. Many of the larger files that had been maintained by local agencies (e.g., New York City) had been adapted and enhanced to meet local planning and administrative needs. Fiscal difficulties at the local level, time constraints on the Census Bureau, and the lack of additional programming staff in all agencies often precluded efforts to recreate the original file structure without losing the map and address updates. In these cases, the feature and address range update work had to be redone manually by the Census Bureau from digital plots and databases.

One of the general weaknesses of the GBF/DIME files was that they were limited geographically and, therefore, the ability to use the files on an ongoing basis was limited to large-scale users with on-line access to mainframe computers. Relatively few of the original GBF/DIME files given to local governments were updated and maintained during the 1980s. Even fewer of the locally updated files met Census Bureau standards for direct incorporation into the TIGER database. In practice, however, these updated GBF/DIME files provided the Census Bureau with its first challenges in digital spatial data exchange at the local level.

Digitized GBF/DIME files, attribute-rich but of mediocre coordinate accuracy, formed the cartographic base for 345 urban centers in the 1990 TIGER database. Absolute coordinate accuracy was not a primary concern in the development of the GBF/DIME files, as they were used primarily by the Census Bureau for geocoding rather than mapping purposes. Although their coordinate accuracy was well below that of the U.S. Geological Survey's (USGS's) Digital Line Graph (DLG) files derived from their 1:100,000-scale maps, these files represented features with all their respective feature names, address ranges, and 1980 geographic area codes in their correct relative location—sufficiently accurate for taking a census (Sobel 1986).

Although the Census Bureau would have liked to provide an enhanced cartographic quality, the deadline pressures of an upcoming decennial census forced management to abandon initial plans to align this information to the USGS's DLG files. Also, because of staff and time constraints, the files were sent to four private-sector contractors for digitizing of feature updates using Census Bureau-supplied updated 1980 census maps. The results from these arrangements were of mixed quality.

TIGER Database Development, 1983–1990

The institutional knowledge and experience gained from the development of the GBF/DIME files, further theoretical and conceptual advances in the field of mathematics (Corbett 1979 and White 1984), and the "enabling" availability of new and affordable technology, provided fertile ground for the Census Bureau's next challenge: the development of the TIGER database for use in the 1990 census. Whereas the GBF/DIME files covered small noncontiguous portions of the United States and were developed initially without spatial or graphic references in their design, the TIGER database covered the entire nation and its territories and was grounded in a more rigorous conceptual model of topology and space (see Boudriault 1987; Kinnear 1987; Broome and Meixler 1990).

Census Bureau/USGS cooperative agreement. In addition to data-sharing arrangements on the local level, the ability to complete the building of the TIGER database in time for the 1990 census was directly dependent on a landmark 1983 data-sharing agreement with the USGS. The USGS provided the Census Bureau with computer files of scanned versions of its 1:100,000-scale maps for the lower forty-eight states. In return, the Census Bureau assigned cartographic classification codes to the roads in these files. The resultant product formed the cartographic base for all areas outside the large urban centers covered by the GBF/DIME files, thereby enabling the Census Bureau to complete a coast-to-coast digital map base in time for the 1990 census (McKenzie and LaMacchia 1987).

Interagency cooperation with the USGS was an experience that provided valuable lessons to the Census Bureau in particular, and a model for future cooperation between federal agencies in general. Following a successful Florida pilot project, high-level management in both agencies perceived that cooperation would result in a win–win situation. Each agency would be able to accelerate its individual map production programs and, in the process, they could develop the first "large-scale" digital map file of the United States. Early negotiations ensured that neither agency would feel it was bearing an unfair burden. A schedule of meetings on a regular basis ensured communication during all phases of the cooperative agreement. The challenge of meeting the Census Bureau's decennial deadlines provided an added incentive to "get the job done" and an ongoing requirement to measure progress.

The U.S. Census Bureau/USGS cooperative agreement minimized duplication of effort in federal map automation activities and provided immediate short-term benefits to both agencies. The success of this cooperative venture demonstrated to the Department of Commerce, the Federal Office of Management and Budget (OMB), and the Congress that there were significant benefits to be derived from such activities. The success of this cooperative effort also resonated in the later development of the Federal Geographic Data Committee (FGDC) (OMB 1990), the growing impetus for further cooperative efforts in spatial data sharing, and the increasingly articulated vision of a national spatial data infrastructure (Marx 1992).

Other success factors. The success of the TIGER system and the ability of the Census Bureau to overcome organizational inertia both within and outside the Census Bureau were due to a number of factors. There was a shared sense both within the Census Bureau and by the data-user public that change was needed in the geographic support process. A primary incentive for developing the TIGER database was the large number of inconsistencies between the statistical and geographic data products in the 1980 and earlier censuses, a product of the complex and clerically intensive preparation of maps, ACGs and GBF/DIME files, and geographic reference files (Marx 1986).

The enormous political and economic ramifications of the decennial census made everyone a stakeholder in the process. The growing importance of the information sector of the economy and growing public demands for more accurate, cost-efficient, timely, and accessible data products helped to promote an environment receptive to the exchange of data, expertise, and experience with other governmental agencies at all levels, the private sector, and the academic community.

Bureaucratic inertia was further overcome by staff commitment, expertise, and initiative within the Census Bureau's Geography Division. The transition from traditional to automated mapping for the 1990 census required changes in the organization and planning of the Census Bureau's mapping activities as well as in the requirements of staff in developing and incorporating new cartographic techniques and computer skills. Motivated by a decennial environment of schedules and fixed deadlines with no alternative to full automation for product delivery, and a cooperative agreement with the USGS, the Census Bureau's staff produced significant results in a relatively short time (Trainor 1990). The resources and skills gained from this experience, combined with a history of successful technical innovations at the Census Bureau, promoted a willingness to reach out and explore the potential for digital data sharing.

Some criticisms. Countering these positive aspects are criticisms of the TIGER database, such as the relatively poor coordinate accuracy of the roads in the major urban centers (the roads that came from the GBF/DIME files used in lieu of USGS DLG files), lack of address range and ZIP Code improvements or expansion beyond the 345 GBF/DIME file areas, and inconsistencies in the names and classification of streets. Public complaints about the quality of the data in the TIGER database provided the Census Bureau and its parent agency, the Department of Commerce, with further verification of the wide applicability and importance of this database beyond the internal needs of the Census Bureau.

Although valid, most of these situations were not critical for the taking of the 1990 census, the primary mission of the Census Bureau. In fact, many private consulting firms have taken advantage of these "problems" to repackage "new and improved" versions of the Bureau's publicly available extracts from the TIGER database: the TIGER/Line files. The Census Bureau is correcting many of these situations and will release future TIGER extract products with these updates and corrections.

In general, the TIGER database continues the GBF/DIME file tradition of being attribute-rich and current but with limited coordinate accuracy in the major urban centers. The USGS's DLG files, on the other hand, have high "ground truth" accuracy for the features they show but have few attributes (DLG-Enhanced Files, once released, will improve on the latter). In addition, the DLG files do not contain current information, and the USGS has not been provided with the financial resources needed to perform frequent, nationwide, and systematic updates. As stated previously, alternative methodologies for updating the files are being evaluated, including an initiative to accelerate the collection of base cartographic data using graphic or digital orthophoto quadrangles or aerial photography (FGDC 1992).

ENHANCING THE TIGER DATABASE

Since the completion of the TIGER database for the 1990 census, the Census Bureau has become increasingly aware of its vast potential as well as its current weaknesses. If viewed as a process rather than a product, the TIGER database provides an opportunity to improve statistical accuracy and data quality significantly. The updated address and Geographic Information Systems in a growing number of public and private agencies and the databases of the U.S. Postal Service (USPS) provide important means of enhancing the collection, processing, and tabulation of census data.

The planned release of a new TIGER/Line extract by the Census Bureau containing extended address ranges and ZIP Codes for all areas with city-style mail delivery is significant, not only for the inherent value of the added data (there will be an increase in address range coverage from approximately 55 percent to 85 percent of all housing units in the United States), but because it represents new capabilities that never existed before. For the first time, the Census Bureau will be releasing value-added files for data users after the census that are not directly tied to the decennial statistical data products (however, they will be used for geocoding establishments in the intervening economic censuses). This precedent also is important in that it reflects a subtle change in the traditional once-a-decade data dissemination paradigm and opens new possibilities for future digital data exchanges. Perhaps it also is indicative of the more customer-oriented approach mentioned earlier in this chapter.

Because the Census Bureau was not intended to be the nation's preeminent mapping agency, any data-sharing agreements to improve the TIGER database must be perceived by the overall organization as primarily benefiting the census-taking process. At this stage of planning for the 1992 and 1997 economic censuses and the 2000 decennial census, the Census Bureau has made an organizational commitment to the integration of the Census Bureau's related Address Control File with the intercensal update and improvement of the TIGER database. This commitment, however, is dependent on the availability of resources.

In the absence of a national updated map system, the U.S. Census Bureau has pursued a relatively high-cost map compilation strategy for the 1990 and previous censuses (Rhind 1991). Labor-intensive comparisons of reference sources, often of varying scale and quality, have been the primary means for updating census maps. In this respect, the Census Bureau has pursued and maintained close contacts and relationships with other federal agencies, state, regional, and local planning and transportation agencies, engineering firms, aerial survey companies, tax departments, utility firms, and a host of other public and private firms with current map and address reference source materials.

The development of the TIGER system and other digital spatial and attribute databases, coupled with the proliferation of more powerful computer hardware and GIS software, allows data exchange to occur in a more sophisticated, more timely, and potentially more accurate and less costly manner. From a technical perspective, there appear to be few limits to the advance of this new mode of digital data exchange. According to Cooke (1995), the technical problems of data sharing have mostly been, or are in the process of being, solved. The non-technical components of data interchange may prove more daunting.

Current Data-Sharing Plans

Current data-sharing plans at the Census Bureau are twofold. On the one hand, the Census Bureau is investigating the possibility of national sources, predominantly federal agencies, that could provide the information to keep the feature and address-range information in the TIGER database up to date. Based on a report by the Government Accounting Office (GAO 1991), which stated that federal agencies increased their planned expenditures on GIS by about 60 percent between fiscal years 1990 and 1992, the likelihood of such exchanges is promising.

At the national level, the Census Bureau has entered into or proposed data-sharing agreements with several large governmental or quasi-governmental agencies including the USGS, USPS, the Environmental Protection Agency (EPA), the Soil Conservation Service, the Federal Railroad Administration, the U.S. Army Corps of Engineers, and the Federal Emergency Management Agency (FEMA). These efforts to improve the spatial and attribute data in the TIGER database are ongoing and likely will include several other agencies by mid-decade. As the coordination of GIS activities improves in the federal sector, bilateral agreements between agencies to improve the TIGER database will likely evolve into multi-agency agreements, as has already occurred at the state and local levels (see Murakami and Greenleaf 1992).

The Census Bureau also is investigating possible mechanisms for the electronic interchange of updated geographic information with state, local, private, and academic organizations. The current preference of the Census Bureau is to coordinate these efforts and control quality at the state level rather than having to deal with conflicting data and different file formats from several thousand local governments, private agencies, universities, and other sources. Experience indicates, however, that this will not always be possible.

Role of the FGDC in Data Sharing

Increased GIS use by federal agencies as well as state and local agencies has led to renewed efforts to coordinate development, sharing, and dissemination of spatial data, primarily through the Federal Geographic Data Committee (FGDC). The FGDC, formed in late 1990 at the direction of the OMB in its Circular A-16 (OMB 1990), includes representatives of fourteen departments and independent agencies but has no direct authority, responsibility, or resources. Participation is voluntary, with decisions based on consensual agreement among its members. Given the current limits on availability of fiscal resources, several agencies have agreed to pursue cooperation in accelerating the 1:12,000-scale digital orthophoto program and the 1:24,000-scale digital quadrangle program. The Census Bureau is working on individual Memoranda of Understanding (MOU) with other federal agencies in the spirit of the FGDC and has agreed to cooperate with the USGS in devising a still more powerful data structure.

In the long run, the FGDC may provide the vehicle necessary to transform institutional relationships within the federal government as well as with state and local governments, the private sector, and the academic community. The FGDC is making an increased effort to involve the nonfederal community in its coordination work. Although the goals are lofty and the potential long-term benefits extraordinary, effective leadership at the highest levels and a concomitant commitment to the development of compatible standards will be necessary. In the absence of these developments, only significant short-term and real cost-saving benefits gained from data sharing by particular agencies will move the process forward toward a national digital spatial database.

Significantly, in July 1992 the Census Bureau and the USGS signed an amendment to the original 1981 Memorandum of Understanding that commits both agencies to merging the current DLG and TIGER databases, including information resulting from partnerships with other agencies. The development of a shared database that combines the essential geographic information needed by both agencies to carry out their respective institutional mandates will have a profound effect in further stimulating the development of a national spatial data infrastructure in the United States.

Intercensal Data-Sharing Projects, Post-1990

Current data-sharing experiences at the Census Bureau have been exploratory, and the actual mechanisms and standards for digital data interchange are still in the process of being developed. The Census Bureau is engaged in a number of activities to help promote digital data sharing during the 1990s. These activities include conferences, participation in the FGDC initiative, MOUs with other federal, state, and local agencies, pilot projects, and the planning and imminent release of the prototype version of the TIGER/SDTS (Spatial Data Transfer Standard) file.

Based on these forums and activities, the Census Bureau has begun to receive initial feedback on the non-technical impediments to and incentives for digital spatial data interchange. The Census Bureau has provided its digital geographic and statistical data sets to the public at the cost of dissemination and as a public resource (OMB 1992). Other public and private agencies, however, have different perspectives and regulations pertaining to their own data sets and on the updates they may perform. Profit is a major concern of private companies, and public agencies also are looking to their products as a source of revenue.

States and local agencies consistently have reported one or more of the following problems in trying to coordinate a GIS:

- agencies wanting proprietary control of internal data
- lack of resources in one department or institution affecting data requirements of another
- archaic systems
- managers and commissioner-level officials who know, or care, little about GIS
- staff turnover
- lack of commitment

Once a state or local GIS is operational, the lack of overt incentives to expend the additional time and expense required to feed local updates into the TIGER/Line or SDTS formats for interchange becomes more apparent. Ensuring the currency, accuracy, and quality of the TIGER database is an integral part of ensuring the accuracy and quality of the associated census data, which are used for reapportionment, redistricting, the distribution of federal funds, and innumerable planning and development programs—but this may be a very long-term and nebulous consideration for some agencies. The lack of a clearly defined formal process for data interchange by the Census Bureau could provide an even greater impediment to institutions willing to share data.

The existence of multiple geographic data file formats also has inhibited the interchange of data. Future geographic data files at the Census Bureau will be released in accordance with a recent Federal Information Processing Standard (FIPS) – the SDTS. Adoption of the SDTS involved cooperation of federal, state, and local officials, the academic community, and the private sector over an extended period of time. The Census Bureau was an active participant in the development of this new federal standard for data exchange and, to that end, released its first prototype TIGER/SDTS file (Davis et al. 1992). However, the effect of this rather complex format on data interchange is not yet clear.

Each data-sharing agreement brings forth new possibilities, new arrangements between agencies, and the potential for new products. Working with a variety of agencies and soliciting recommendations for improving TIGER, Census Bureau staff have noted similarities among the interests of many agencies. For instance, conversations with the USPS and the U.S. Department of Transportation have revealed similar interests in enhancing attributes for streets (e.g., turn and directional restrictions). In many cases, the Census Bureau would be able to use the enhancements needed by other agencies for improving the quality and cost-effectiveness of its own internal operations (e.g., routing of enumerators).

Proposed Census/USPS cooperative program. The USPS and the Census Bureau have been working together for more than thirty years in the delivery and return of questionnaires for the decennial, agriculture, and economic censuses and surveys. The increasing use of mail-out/mail-back procedures throughout the nation has made the USPS an indispensable partner in the Census Bureau's data-collection activities. The Census Bureau traditionally has paid the USPS to verify the completeness and accuracy of its decennial census address list, which was purchased from private vendors and enhanced through in-house programs, prior to the mailout of questionnaires. The Census Bureau also worked with the USPS on the development of computer algorithms to match the ZIP+4 files to GBF/DIME file records during the 1980s. ZIP+4 files, also known as the Address Management System (AMS) Files, contain potential address ranges for all areas where the USPS delivers mail.

In an effort to enhance this cooperation to the benefit of both agencies, the Census Bureau proposed a formal Memorandum of Understanding (MOU) with the USPS similar to the one it had with the USGS in the

1980s. In 1990, the Census Bureau and the USPS, with the participation of the USGS, began cooperating on a pilot project to provide a better geographic database for all three agencies. If signed, this MOU will have broad implications, not only for these three agencies, but for the GIS community as a whole during the 1990s.

The Census Bureau's proposal envisions a four-year file update/enhancement plan (fiscal years 1994-97) and subsequent ongoing cooperative efforts to update a Postal/TIGER database. As with the earlier USGS agreement, the USPS/Census Bureau MOU would formalize a seemingly well-suited alliance between two agencies. Cooperative database-building activities could reduce duplication of efforts, thereby reducing overall costs, as well as improve the geographic and attribute accuracy of the information available to each agency.

The Census Bureau's objectives for the proposed joint program during the intercensal years are (1) to obtain updated information on the location of streets with their names, address ranges, and ZIP Codes as well as the location of group quarters, office buildings, or other locations of economic activity; and (2) to improve the error-prone decennial census address list development operations. According to a 1992 GAO report, enumerator follow-up to vacant and nonexistent units alone resulted in an added expenditure of approximately $317 million to the 1990 census operation. Accomplishing the above objectives will enable the Census Bureau to reduce its critical dependence on a large, temporary clerical workforce before and during each decennial census, thereby reducing costs, and to improve the overall quality and consistency of decennial census data products.

In the overall proposal, the joint venture envisions updating the TIGER database to permit automated analysis of carrier routes and the production of carrier route maps for use by the USPS. In the pilot study, the effort to improve and maintain the positional accuracy and completeness of the TIGER database involved the use of Global Positioning System (GPS) technology with receivers mounted on USPS-supplied vehicles. In order to improve and update the geocoding capability of the TIGER system, the pilot study also updated the address ranges in the TIGER database and added ZIP+4 Codes and other USPS information. The joint venture envisioned would extend similar geocoding improvements to all parts of the United States with city-style address systems. The Census Bureau would provide the USPS with its technical and geographic expertise.

Potential impediments. Although this data-sharing scenario appears to satisfy the needs of both agencies, there are a number of organizational, behavioral, and institutional impediments that will need to be overcome. Some of these impediments are specific to these two agencies while others are generic to any data-sharing milieu. One specific impediment is that the USPS is not part of the FGDC. The USPS also is a quasi-federal agency, which means that it must justify its participation in an agreement on a benefit/cost basis including expected revenue or savings from potential products. There also are questions about what information will be shared and what will be the property of each organization.

A more general institutional impediment is the difficulty of building a single database to serve the needs of different agencies. Federal agencies have worked independently of each other for a long period of time collecting and structuring their data according to geographic units based on their own unique criteria and naming conventions. Data sharing presupposes a strong, long-term, funded commitment to reconcile what really constitutes different versions of the same reality. In order to reach agreement, a shared database initially may involve compromises that could make it less effective than two separate databases. Each agency has its own self-interest and mission that must be met first. Short-term objectives may become more pronounced and inhibit progress. The need to understand each other's terminology, organizational structure, and requirements represents another potential impediment.

Another important issue, not only for the Census Bureau but for all future data partners, is the potentially differing perception of the benefits from data sharing. One agency may believe that it is giving up more than the other agency and is bearing an unfair burden, causing a negative effect on data-sharing plans. Some groups within each agency may either be opposed to or less than committed to the idea of data sharing. While this issue was always in the background during the joint Census Bureau/USGS Cooperative Mapping Project, it never came to the forefront as a major issue.

Another potentially critical impediment common to bureaucracies is the inertia effect; that is, it is easier to stick with the tried and reasonably true methods than to try something new. Change can be intimidating and can upset a delicate balance of power in an organization.

OTHER ONGOING DATA-SHARING ACTIVITIES

In a broad sense, the Census Bureau traditionally has been in the business of sharing geographic data and has a long history of partnerships with state and local governments. The Census Bureau receives updated governmental unit boundaries from local officials on a periodic basis through its Boundary and Annexation Survey. The Census Bureau also works with local Census Statistical Areas Committees, composed of representatives of the public, private, and academic communities, in delineating the boundaries of census statistical areas such as census tracts, block groups, and census-designated places.

The Census Bureau also has worked closely with state election officials to provide the information they need for redistricting and reapportionment; with the U.S. Department of Education to incorporate the boundaries of school districts as a means to produce data for school districts; and with metropolitan planning organizations to improve the quality of the Census Bureau's address reference files, which improves the quality and usefulness of the census journey-to-work and place-of-work data, and to define traffic analysis zones in terms of census blocks to facilitate the tabulation of decennial census data for those areas. The Census Bureau has built its massive geographic database primarily from source materials acquired from state and local agencies. In a new program based on 1990 census data, data users may independently aggregate census blocks to define their own statistical areas and receive maps and data profiles of these user-defined areas from the Census Bureau on a cost-reimbursable basis.

In addition, the Census Bureau maintains and is constantly improving its huge, ongoing institutional data-sharing apparatus. Each of the Census Bureau's twelve regional offices supports information services and geographic programs that coordinate activities with state data centers and their affiliates throughout the United States. The Census Bureau's Data User Services Division combines educational functions (e.g., ongoing workshops on TIGER, census maps, and data products) with its data distribution functions. The Census Bureau participates in the kindergarten through twelfth grade (K-12) geographic literacy campaign in the United States as a means to address the important issues of access to new information technologies. Public access issues (Emergency Planning and Community Right-to-Know Act 1986) also have provided the impetus for an ongoing cooperative project with the EPA and the National Oceanic and Atmospheric Administration (NOAA) that supports the development of public domain software that links and displays environmental, socioeconomic, and demographic data using an extract of the TIGER database.

From a more global perspective, the Census Bureau shares its technological and statistical know-how with a number of participating countries through its in-house and overseas training programs. The Census Bureau also is cooperating with Statistics Canada (Haythornthwaite 1992) and pursuing talks with Mexico's statistical agency, Instituto Nacional de Estadística, Geografía e Informática, to create a North American Common

Borders Database. In summary, the Census Bureau maintains a huge ongoing apparatus that supports and complements its current efforts at spatial data interchange.

CONCLUSIONS

As the Census Bureau prepares for the 2000 census, change, once again, appears imminent. According to a recent GAO report (1992, 4), "the current approach to taking the census appears to have exhausted its potential for counting the population cost-effectively." Similar statements were made in reference to the Census Bureau's geographic support program following the 1980 census (Tomasi 1990). These statements have recurred with periodic frequency in the recent history of census-taking. In large part, the innovations that followed were a response to intense public scrutiny, a result of a process that has significant and far-reaching political (reapportionment and redistricting) and economic (distribution of government funds) consequences. Over the past fifty years, the Census Bureau has met not only the challenges brought about by massive social, demographic, and economic change, but has provided effective leadership in applying and diffusing a number of new technologies to both the governmental and non-governmental sectors.

The application and use of new technologies on a vast scale by public organizations are reshaping the internal organization and relationships within the public sector as well as among the public and private sectors and the academic and research communities. The dominant values of narrow, functionally separate governmental agencies and departments are increasingly being replaced by a system of greater complexity and interrelatedness and a growing trend toward public- and private-sector cooperation. The Census Bureau's TIGER system is an important example of this phenomenon and may be a primary motivator for such changes.

The Census Bureau's geographic and statistical products—inexpensive and ubiquitous spatial and attribute raw material—have helped transform GIS from a highly technical field dominated by large agencies, private firms, and universities, to one that is becoming increasingly accessible to many data users. Advances in our technological infrastructure thus far have been the primary factor enabling institutions to develop and share digital geographic data. Personal computers, workstations, CD-ROMs, and databases available for automation were relatively nonexistent until the 1980s. The incipient use of new telecommunication technologies, such as the Internet, may have an equally profound effect on data sharing and online services in the latter part of this decade.

As computer hardware and software continue to become less expensive and more powerful, the vast potential of GIS will depend increasingly on the behavioral, organizational, and institutional issues acting as impediments and incentives to the sharing of geographic data. The integration of various multi-media technologies and the growing capability to link a wide variety of public and private databases also raise a number of privacy issues. Deriving the full benefits of GIS and related information systems will depend, to a significant degree, on how society approaches and resolves these issues (Onsrud 1992).

Based on past experience, it will require extraordinary leadership, communication, and flexibility among agencies to facilitate the process of spatial data interchange. Data sharing will be most successful when such ventures can be justified by short-term results, verified cost reductions, improved operations, and minimal problems of data ownership. Data sharing will be enhanced to the degree that these ventures can be built around joint development projects such as the Census Bureau/USGS experience during the 1980s and the potential Census Bureau/USPS and Census Bureau/USGS experiences in the 1990s. Awareness of the need for such coordination is growing and is best reflected in the increasing number of statewide GIS committees and partnerships between the federal and state communities.

Geographic databases have been built thus far to support the mandates of single institutions or parts of an institution. All who collect and manage data for activities related to their own responsibilities will need to understand and appreciate the value of those data to others and to collect and structure their data accordingly. In order to take full advantage of the opportunities offered by these new technologies, business, government, and academia will need to develop, support, and fund data exchange on a systematic and ongoing basis as well as promote accessibility of GIS capabilities to all sectors of our society.

REFERENCES

Anderson, M. J. 1988. *The American census: a social history.* New Haven: Yale University Press.

Boudriault, G. 1987. Topology in the TIGER file. Eighth International Symposium on Computer-Assisted Cartography, Baltimore, Maryland, *Proceedings*, pp. 258–263.

Broome, F. R., and D. B. Meixler. 1990. The TIGER database structure. *Cartography and Geographic Information Systems* 17, 1: 39–47.

Carbaugh, L. W., and R. W. Marx. 1990. The TIGER system: a Census Bureau innovation serving data analysts. *Government Information Quarterly* 7, 3: 285–306.

Cooke, D. 1995. Sharing street centerline spatial databases. In H. J. Onsrud and G. Rushton, eds., *Sharing Geographic Information.* New Brunswick, NJ: Center for Urban Policy Research, Rutgers University.

Corbett, J. P. 1979. Topological principles in cartography. Technical Paper 48. U. S. Bureau of the Census, Washington, D.C.

Davis, B.A., J. R. George, and R. W. Marx. 1992. TIGER/SDTS: standardizing an innovation. *Cartography and Geographic Information Systems* 19, 5: 321–327.

Emergency Planning and Community Right-to-Know Act: Title 3 of the Superfund Amendments and Reauthorization Act of 1986. PL 99-499, 17 October 1986. United States Statutes at Large 100. pp. 1728–1758.

Federal Geographic Data Committee. 1992. *Multi-agency initiative to meet high priority requirements for base cartographic data.* FGDC Subcommittee of Base Cartographic Data. Washington, D.C.

Government Accounting Office. 1991. *Geographic Information Systems: information on federal use and coordination.* IMTEC 91-72-FS. Washington, D.C.

_____. 1992. *Decennial Census: 1990 results show need for fundamental reform.* GAO/GGD-92-94. Washington, D.C.

Haythornthwaite, T. 1992. Development of the United States–Canada Common Border Database. *The Operational Geographer* 10, 1: 28–30.

Kinnear, C. 1987. The TIGER structure. Eighth International Symposium on Computer-Assisted Cartography, Baltimore, Maryland, *Proceedings,* pp. 249–257.

Marx, R. W. 1986. The TIGER system: automating the geographic structure of the United States census. *Government Publications Review* 13: 181–201.

_____. 1992. Building the National Spatial Data Infrastructure: the data integrity challenge. Paper presented at the Regional Surveying Engineering Conference, Hartford, Connecticut.

McKenzie, B. Y., and R. A. LaMacchia. 1987. The U.S. Geological Survey–U.S. Bureau of the Census Cooperative Digital Mapping Project: a unique success story. Paper presented at American Congress on Surveying and Mapping meeting, Reno, Nevada. Fall.

Murakami, E., and K. Greenleaf. 1992. Multi-agency TIGER file updating. *URISA Proceedings* 2: 25–35.

Office of Management and Budget. 1990. *Coordination of surveying, mapping, and related spatial data activities.* OMB Circular A-16 (Revised). Washington, D.C.

_____. 1992. *Management of federal information resources.* OMB Circular A-130. Washington, D.C.

Onsrud, H. 1992. Privacy and spatial databases. Technical Program Abstracts, 27th International Geographical Congress, Washington, D.C., pp. 480–481.

Rhind, D. W. 1991. Counting the people: the role of GIS. In D. J. Maguire, M. F. Goodchild, and D. W. Rhind, eds., *Geographical information systems: principles and applications*. Longman Scientific and Technical, Essex, 2: 127–137.

Silver, J. 1977. The GBF/DIME system: development, design and use. Paper presented at the 1977 Joint Annual Meeting of the American Society of Photogrammetry and the American Congress on Surveying and Mapping. U.S. Government Printing Office, 1977-240-869/1102.

Sobel, J. 1978. GBF/DIME system—development and reference source problems. Applied Geography Conference, SUNY—University Center at Binghamton, 1: 112-121.

_____. 1986. Principal components of the Census Bureau's TIGER file. *Research in contemporary and applied geography: a discussion series*. SUNY at Binghamton, 10, 3: 1–16.

Tomasi, S. G. 1990. Why the nation needs a TIGER system. *Cartography and Geographic Information Systems* 17, 1: 21–26.

Trainor, T. F. 1990. Fully automated cartography: a major transition at the Census Bureau. *Cartography and Geographic Information Systems* 17, 1: 27–38.

U.S. Bureau of the Census. 1973. Census Use Study. 1972 International DIME Colloquium, Washington, D.C., August 27–29, *Proceedings*.

White, M. 1984. Technical requirements and standards for a multipurpose geographic data system. *The American Cartographer* 11, 1: 15–26.

ROEMER M. ALFELOR

24 *GIS and the Integrated Highway Information System*

The application of Geographic Information Systems (GIS) in Transportation (also called GIS-T) has evolved from simple graphic and map-based representation of roads and geographic features to serve specific functions in a highway agency to a more complex and powerful tool for integrating information and decision-support systems within and among agencies. This change is partly driven by the limited budgets and resources faced by highway departments that required them to find means to minimize the cost of managing and operating their transportation infrastructure including the network of highways, bridges and other facilities. An Integrated Highway Information System (IHIS) on a GIS platform avoids duplication of highway data collection effort and supports integrated decision-making at various administrative and functional levels of the organization. This chapter describes both the incentives and requirements for spatial data sharing in Integrated Highway Information Systems. Some of the technical and non-technical issues involved in implementing GIS for data sharing in Integrated Highway Information Systems are discussed, with emphasis on those arising from institutional, organizational and behavioral characteristics of highway agencies.

INTRODUCTION

The 1991 Intermodal Surface Transportation Efficiency Act (ISTEA), also called the Highway Bill, that was passed by the U.S. Congress calls for pavement, bridge, congestion, safety, and intermodal facilities management systems in state highway and transportation agencies. Much additional attention has been focused at both the national and state levels on integration of these information and management systems through initiatives of individual agencies and cooperative proposals of the states and the federal government. Some of this research concerns the development of an Integrated Highway Information System (IHIS). This system consists of linked highway databases to support analysis, reporting, and decision-making functions needed to operate and manage various highway infrastructure facilities including roads, bridges, traffic control devices, railroad crossings, and highway traffic. Management of existing transportation facilities and traffic congestion relief in urban areas are some of the major problems currently facing surface transportation and have been specifically addressed in the bill.

The Surface Transportation Legislation also has specific recommendations for research, development, and deployment of a wide variety of Intelligent Vehicle Highway System (IVHS) technologies. An IVHS aims primarily at reducing or eliminating congestion in road networks through an Advanced Traveler Information System (ATIS), Advanced Traffic Management System (ATMS), Advanced Vehicle Monitoring and Communication System (AVM/C System), Automated Vehicle Location (AVL), and Advanced Public Transportation System (APTS). Some IVHS services (e.g., ATIS, AVL, ATMS) require digitized maps and location information to be available to the vehicle operators. These maps represent the road network and other related structures to guide travelers in finding optimal routes and information about the location of road services and utilities. Real-time traffic management and control through IVHS requires information on location, nature, time and duration of construction and maintenance activities for pavement, bridges, road furniture, and other highway infrastructure. This type of information can be obtained from other management systems maintained by highway agencies. IVHS is one of the real-time data management and decision-support systems that will require integration and sharing of highway information and other geographic data (e.g., location of natural and man-made features on a map) by public and private providers of transportation services.

An IHIS can be used at the local, regional, state, or national level to organize all the data and information associated with a network of transportation facilities. Such a system provides a convenient, efficient, and cost-effective means of data storage, retrieval, reporting, and exchange. It helps integrate different specific functions within a highway agency including the design, construction, maintenance, and operational aspects of pavement management, bridge management, and traffic control. In the past, data related to each function have been stored and managed separately by individual units in the organization. This independent data management system has resulted in slow, costly, and tedious management of information and decision making at the strategic transportation network level.

One of the requirements of an integrated highway information and management system is a *reference system* that will allow data contained in each highway file to be referenced to a common point. All data pertaining to the various elements of the highway system are referenced by their locations in the road network. Location is a common attribute used by management systems to associate spatial and temporal information such as traffic volume, pavement condition, accident rate, degree of curvature, railroad crossings, and so on. Geographic Information Systems and other noncoordinate systems are currently used by many highway agencies to encode and represent spatial data such as road sections, intersections, railroad crossings, and other surface features related to transportation. In an integrated highway planning and management environment, all information and decision-support systems for bridge, pavement, and traffic management must use a common location reference system.

The benefits that can be derived from an IHIS have been recognized by many highway agencies. However, the integration process is not an easy task. Highway agencies have to deal with many implementation issues pertaining to the technology, the organization, and external factors. Efforts to integrate highway databases on the state level have been met with varying levels of difficulty due to incompatibilities in data structure of various data-management systems. Application programs and software used by different units within the highway agency differ in data requirements, computer environments, and architecture. Highway reference systems for storing, manipulating, and presenting spatial data differ for individual highway units. Although the route-milepost and link-node systems are the most common location reference systems, many states use other reference systems for identifying bridges, railroad crossings, and other infrastructure.

The same problems are encountered in integrating highway information on the national level. A national integrated highway database and decision-support system such as the Comprehensive Highway Information and Planning System (CHIPS), which the U.S. Department of Transportation through the American Association of State Highway and Transportation Officials (AASHTO) and other organizations, including the Federal Highway Administration, are developing, will identify measures to combine or establish an interface among various databases maintained by different agencies within and among states. The objective is to achieve sufficient commonality in data definition, structure, and other database features to allow data interchange, communications, and processing among various agencies. Much of the data that will be shared by highway agencies and other AASHTO member departments would pertain to the National Highway System (NHS) and other highway sections that are used in analyzing needs and evaluating the performance of the highway system. Advanced information technologies for highway data representation and referencing, including GIS and Global Positioning System (GPS), will be used. Unlike the state-level highway information system, the CHIPS database will include network-type attribute data pertaining to the major highway links and corridors (e.g., interstate highways and interregional corridors) that are required for transportation planning and programming on a national level. These types of applications do not require very detailed/disaggregate spatial data information pertaining to a highway segment, which is true for statewide applications like pavement management, bridge management, traffic monitoring, or accident analysis. Therefore, the scales for maps and other geocoded highway data that will be contained in the CHIPS databases will be different from those found within state highway departments.

This chapter addresses the fundamental issues involved in data integration *within state* agencies. Information systems technology continues to be responsive and adaptable to the many needs of the highway agency. However, the benefits of data integration cannot be fully realized unless the adoption of technology is accompanied by organizational, institutional, and behavioral changes.

THE HIGHWAY ORGANIZATION

In order to understand the implementation issues surrounding spatial data sharing for state highway organizations, it is necessary to look at their vertical and horizontal structures. Figure 24.1 shows the diagram of a typical state transportation organization. (Note: state departments of transportation have traditionally been called "highway departments" because the highway division is usually the largest functional unit.) It is not difficult to see the compartmentalized nature of the organization. Each unit or bureau exercises some degree of independence in terms of information management and decision making. This organizational setup and relationships have characterized highway agencies for many years. When computer technologies introduced new tools for managing information and analyzing data, the different units took advantage of these advancements and adopted/customized application programs to serve their needs. These applications include database management systems and other information management systems, including GIS. However, data management and

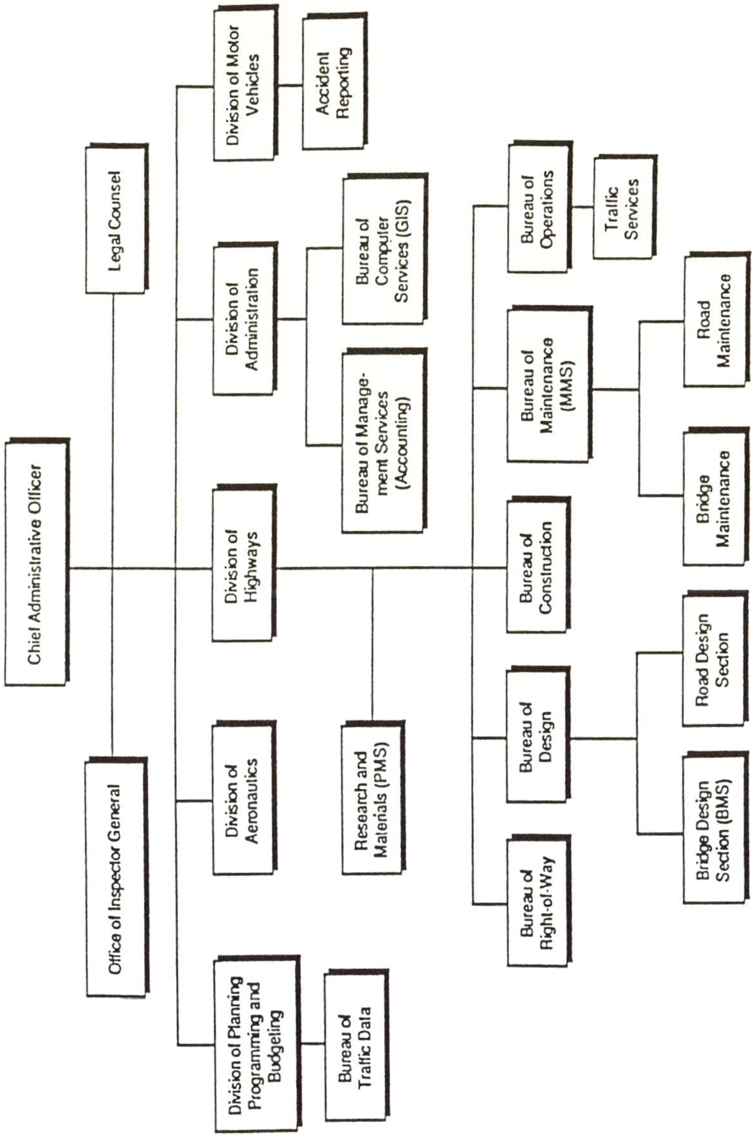

FIGURE 24.1
Typical transportation department

decision-making remained balkanized, and each unit maintained its monolithic structure. With the increased capabilities of information systems for data representation, transfer, processing, and telecommunications, data sharing and integrated decision-making can be achieved by linking together the databases maintained by these units. Highway administrators viewed this as an opportunity to link information systems and application programs to achieve coordination and avoid data redundancy. This led to the concept of the IHIS.

INTEGRATED HIGHWAY INFORMATION SYSTEM

Integrated Highway Information System (IHIS) is the term used to describe the computer-based system being developed by highway agencies on the state level to organize all the data and information associated with their network of transportation facilities. Individual units or divisions within the highway agency perform specific functions that may or may not require identical data. This resulted in a management information system (MIS) used by each unit to be customized to its needs; hence, the MIS is independent of systems used by other units. For example, road maintenance, bridge maintenance, and traffic operations exist as independent sections within the highway department. Separate bridge, pavement, and traffic management systems are used not only to store data relevant to each function but also to aid in finding short-term and long-term solutions to existing problems. From the individual unit's point of view, this may be an ideal management perspective. However, highway agencies have suffered from slow, costly, and redundant management of information and a suboptimal decision-making process at the strategic level as a result of incompatibilities among different MISs and duplication of data at the task-oriented tactical levels. This need, coupled with the existing capabilities of information technologies for system integration, is the prime motivation for adopting an integrated systems framework.

An IHIS includes a collection of databases that are grouped according to their role in transportation planning. For use in network planning, programming, and budgeting on a state level, IHIS databases contain information on the road system as a network of links with jurisdictional, functional, traffic demand, and physical characteristics and other related attributes. An IHIS also has a database manager that allows the files or records in the files to be linked, updated, searched, deleted, or accessed by other systems used by the units and bureaus that perform specific functions and operations. Finally, an IHIS has a highway reference system that allows data contained in each database to be referenced to a common point using a location tag such as geographic coordinate or route-milepost (Duffy and Wyznyckyj 1987).

There are five functional levels for which highway information is required by an agency (Paterson and Scullion 1990). They are, in hierarchical order:

1. *sectoral level* (highway system is viewed as a whole)
2. *network level* (planning, programming and budgeting of links and other components)
3. *project level* (technical and functional design)
4. *operational* (construction, maintenance, traffic and safety, specific to a particular facility or operation)
5. *research and development* (study specific, precise, and independent)

The level of detail required for the information increases from the sectoral level, which uses overall summary statistics, to benchmark-type data, which are used in detailed design and management operations. However, there are variations in levels of detail required for each functional level, again depending on the nature and dimension of the applications. Most applications including pavement management, bridge management, traffic control, and maintenance management function on the network, project, and operational levels. The level of detail can be made consistent by expanding or aggregating the information based on the functional level for which this information will be used.

The IHIS database typically consists of the following groups of data:

- roadway and roadside inventory
- traffic volume
- traffic accident
- maintenance
- structure

Roadway inventory data contains the physical and geometric characteristics of the section of the road including its location, functional classification, construction parameters like pavement type and thickness, pavement condition, alignment, and curvature. Roadside data includes traffic signs, signals, lighting, drainage, and other highway appurtenances. Traffic volume data is obtained from traffic count and weigh-in-motion surveys that collect information on road usage and composition of traffic. Traffic accident data is usually reported and recorded by motorists and the police and include detailed information on accidents such as location, number of fatalities and injuries, number and types of vehicles involved, and time and weather conditions. Maintenance data refers to types and extent of maintenance performed on the various elements of the highway. Maintenance could be in the form of snow removal, pothole repair, bridge painting, sign rehabilitation, grass cutting, and so on. Structure data includes inventory and other information related to bridges and railroad crossings that are part of the highway network (Briggs and Chatfield 1987).

Highway agencies in different states maintain different sets of databases as a result of management systems not being uniform across states. Each state is in a different stage or level of integration, and the variability in organizational structure and use of information makes it difficult to define a standard database organization that will apply to all states. Various application programs are used by units and divisions within the agency for their specific purposes. These applications include planning and design, pavement management, bridge management, maintenance management, safety management, construction and project management, budgeting, and others. While the structure and use of data vary among units in an agency, there is a global parameter with which all these data elements are referenced: location. Figure 24.2 shows the conceptual structure of IHIS.

To summarize, the benefits of data integration through IHIS include:

- integrated decision making;
- reduction of data redundancy/duplication; and
- reduction in the cost of data collection.

HIGHWAY REFERENCE SYSTEM AND GIS

Almost all functions and operations within the highway agency are referenced by location. The most common reference systems used are the route-milepost and the link-node systems. However, other reference systems are used to suit various applications. In fact, many highway agencies maintain several reference systems. For example, the Michigan DOT has identified thirty-eight reference systems classified into coordinate, area identifier, segmental location (length), and point location systems that reflect the varying degrees of detail and levels of aggregation used by different units in the department. The route-milepost system uses the route number assigned to different roads in the network as the key or primary reference, and the mileage or milepost is measured from a specific point on the route and used as secondary attribute. The modification of the route-milepost system uses an inventory number used by states to identify the road segment. The link-node system defines a link or segment of road as bounded

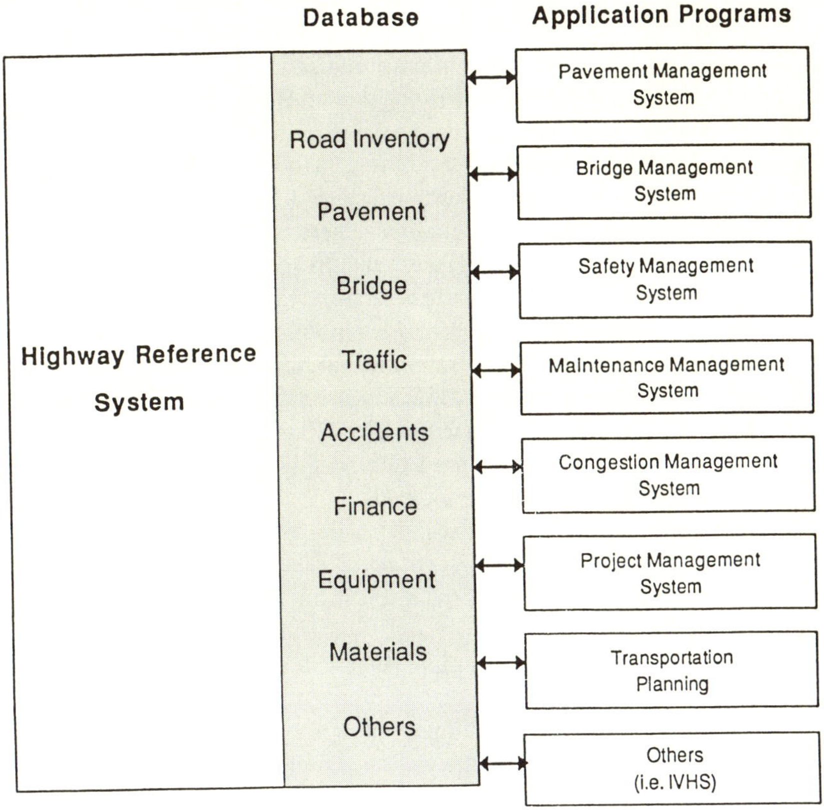

FIGURE 24.2
Ideal framework for IHIS

by two nodes in its ends. The nodes may represent intersections or points of transition from one link characteristic to another. While the route-milepost system uses points as references, the link-node system uses lines as references. More states have adopted the route-milepost system because it is easier to convert field locations in terms of mileage (Briggs and Chatfield 1987).

A reference system that underlies GIS can provide a connection between local and statewide network spatial data analyses. GIS-T has been defined as the adaptation and adoption of GIS for transportation purposes and serves as an environment or framework for managing and analyzing spatial data related to transportation (Fletcher and Lewis 1992; Lewis 1990). A recent study conducted by the University of Wisconsin for the National

Cooperative Highway Research Program (NCHRP) on the number of GIS applications within state departments of transportation (DOTs) indicated that sixty such applications were operationalized/implemented while 113 are being planned or considered (Vonderohe 1991). The applications include pavement management, bridge management, accident analysis, traffic management, highway inventory, and transportation planning, among others. The study also revealed different notions about GIS-T: for some it is simply a new computing tool, whereas others insist that it should have spatial data integration functions. Another study on statewide GIS-T implementation concluded that "there is no single optimal path to GIS-T implementation, but rather a multitude of shades and nuances" (Brown and Lewis 1991). This study analyzed the strategies that have been employed in developing and applying GIS-T in Wisconsin DOT, Florida DOT, and the Saskatchewan, Canada, DOT. Wisconsin DOT has been the leader in GIS-T implementation and has developed GIS-based pavement management, highway inventory, and accident analysis.

As stated earlier, different scales for maps and spatial data are required for different highway applications. The level of precision offered by GIS-T in locating specific points of interest on the road depends on the resolution of digital maps and their accuracy. The resolution of many digital maps is very coarse—for example, that at the 1:100,000 scale. Resolution at the USGS quad scale (1:24,000) is approximately 20 meters, but quad maps have proven to be of questionable accuracy for more detailed highway applications. Aerial photographs of higher resolution and accuracy can be digitized and converted to base maps in GIS. The creation of highly accurate cartographic base maps for GIS generally depends on the improved mapping technology, especially those maps based on the satellite Global Positioning System (GPS).

IMPLEMENTATION ISSUES

Incentives to data integration, including elimination of data redundancy, reduction in costs of data collection, and maintenance and integrated decision making, are the driving force for implementing IHIS. However, agencies are faced with many problems and usually need to overcome several barriers and constraints to achieve success. Issues surrounding data sharing are classified as those pertaining to technology and those related to the organization. These issues are mutually dependent and are all worth mentioning here. We will discuss these issues in the general context of data sharing in IHIS while specifically addressing spatial data sharing in GIS.

Technology

The difference between technology and people is that technology changes very quickly. In fact, GIS technology has been likened to a "moving target" that is difficult to contain (Vonderohe 1991). Indeed, the adoption of a particular GIS tool for data sharing remains a major issue as there are multiple technologies available. The problem is to choose from among the wide array of system options for data integration and to make sure that the system can evolve or change to respond to later developments. System design for IHIS is influenced by the following factors.

Data type and structure. Existing management information systems for representing data used by different highway units differ in data structure and storage format. Each unit in the highway department maintains its own database using its own data format, which is most convenient for its purpose, or which has been used for many years (e.g., flat files, USGS quad maps, and digital line graphs). The complexity of integrating highway information data lies in the definition and development of a database management/information system that can link incompatible data files pertaining to roadway inventory, traffic volume, traffic accident, maintenance, and bridge data. Specifically, the problems arise due to: 1. disparate data format for bridge, pavement, and other infrastructure databases (many agencies are still using standard data format of flat files); 2. variable reference systems used to identify spatial information for highways, bridges, railroad, and other structures; and 3. different programming languages and computing environments required by application programs and end users. There is the specific problem with the level of detail for representing spatial data on a digital or cartographic map (Alexander 1991). Some applications or functions require sectoral representation of road networks for which large-scale maps with low resolution can be used. However, for very detailed operation or research, the spatial information should be available in the order of a few feet or meters, making the USGS maps not suitable.

Computing environment/system architecture. The architecture of an integrated information system such as IHIS defines how it operates in the agency. This architecture should be designed based on the structure of the organization, the flow of information, and the data requirements of the functional units. Depending on the locations of the users of information, a computer network of local or wide area coverage can serve as the communication and computing environment. However, the system protocol needs to be designed to fit the style of the organization. The client–server system architecture and distributed computing model seem to be the most logical choice for any organization; but exactly how these models will be customized to each highway agency remains a critical matter.

Standardization. Whether data is to be integrated and shared by units within a highway agency or by agencies across the nation, *standards* for data interchange must be established. Standards can facilitate transfer of information from a data source to the application programs and other end users. They serve as common media or a set of rules with which communication and various transactions can take place within IHIS (Hyman and Alfelor 1991). These standards include models and procedures for data structure and representation, database languages, data transfer and communications, and computer hardware/architecture to facilitate data sharing in an integrated information environment. Spatial data standards in IHIS include models and procedures for referencing location. Complexity in defining spatial data standards stems from the different levels of detail required by different applications and, as mentioned earlier, the different types and structure of data currently used in the organization.

Organizational and Institutional

Many GIS-T researchers believe that the technological barriers to spatial data sharing in transportation agencies are not as much an impediment as the organizational and institutional transformations that must accompany the integration process. These changes have repercussions on the role each unit plays in the agency and the responsibilities associated with this role. Some of the issues and constraints associated with these changes are described below.

Staffing. People with the knowledge and technical skills to implement GIS successfully have not been easy to retain in a highway organization. They are usually in short supply and very expensive since not very many individuals have the combined background in engineering, information systems, and geography that constitute the disciplines needed to be capable of working with GIS-T in IHIS. Experience and training to gain sufficient knowledge of these disciplines usually require several years. As the level of expertise grows, the short-term demands for services of these people within and outside the organization also increase. As a result, the best technical people either develop little commitment or find other employers who offer more competitive salaries. This shortage of competent staff continues to deter the success of GIS in transportation agencies. Another source of problems is resistance to change. Most of the individuals who have worked with the agency for many years and are accustomed to traditional approaches to data processing and management will be reluctant to change and sometimes feel their job or expertise considerably threatened by new technologies, including GIS.

Cost justification. While some agencies have adopted GIS-T both as an application tool and as a spatial data integrator, its full benefits will not be realized for many years, or even decades. Initial costs of acquiring GIS and other integrated systems tools (including hardware and database development) and implementing them in the agency are usually high, and the long, uncertain period of returns to investment makes its adoption difficult to justify. Aside from the total cost of implementation, there is also the problem of allocating the costs to the different applications and users.

Data administration/location in an agency. One of the main issues facing data integration in IHIS is where the system will be located in the highway agency. This problem is the same as identifying the group or function that will be responsible for managing the system and making sure it serves its purpose (Lewis 1990). There are many possible repositories of integrated data, including the management information systems (MIS) group, the planning group, and the administrative (executive) level. The choice is influenced by the role the unit plays in managing highway information and its relationships with the other units that will be using the system. The integration of spatial data puts an added burden on the various functional groups that have to depend on the information manager for quick, complete, and reliable data. This is often viewed as a restriction of freedom and decrease in flexibility to use information.

Coordination of tasks and decision making. One characteristic of highly compartmentalized organizations is the "Don't care about others" attitude. This is true of the different functional units within the highway agencies. Turf and contention are difficult to deal with when integrated decision making is the goal. Conflict usually arises when demands by some units are not met—especially when budget is involved. Competition for highway funds often subverts any effort to achieve coordination and integrated decision making.

Support. Initial commitment to data integration usually comes from the top or upper echelons in the agency. Decision rests on these power sources (i.e., chief administrative officer) as to whether or not the agency will implement data sharing and, if so, how. However, full support is required at all levels, particularly the representatives or heads of each functional group (e.g., data processing, bridge management, maintenance management, and so on). Many highway agencies lack individuals who can serve as the "champions" for GIS and data sharing, who can communicate and interact with these different functional groups and with the powers sources on the benefits and implementation of data integration. The absence of such "prime movers" at all levels of the organization can deter any effort to implement spatial data integration.

Standardization and open systems environment. Earlier we discussed the need for standards to resolve technical problems with integrating different data formats. However, organizational barriers require that standards should be carefully chosen. For example, standards for representing highway links and nodes in the IHIS should enable all database users and application programs to navigate through the database and manipulate the spatial data in its standard format or convert it to another data structure. Many application programs are developed by private software companies and use proprietary data structures. Similarly, specialized applications may require the same generic data structure. The impacts of standards, in terms of costs and benefits, on both the users and developers of application programs should be considered. Developers of standards should realize that the various users and providers of application programs will have to change or adjust their individual data-handling functions and decision-support tools to conform to, or interface with, the open systems environment. This will require users to convert or develop a neutral interface to access the system.

MIGRATION PATH

The challenge facing highway administrators in implementing spatial data sharing in IHIS is not to allow the integrated system to disintegrate the organization. The adoption of GIS for integrating highway data should be accompanied by varying degrees and types of organizational transformation, depending on the characteristics and idiosyncrasies of the agency involved. This means that there are as many migration paths pertaining to technology adaptation, data management, system development, and organizational shifts as there are organizations. However, some general rules of thumb arising from research in organizational theory and lessons learned from the past by highway agencies themselves can provide guidelines and recommendations for successful implementation of spatial data integration in a GIS platform.

A general model of organizational transformation for utilizing geographic information technology was presented in a GIS-T symposium (Lee 1991). The model consists of four steps:

1. Recognizing the need for change
2. Creating a new vision
3. Designing the infrastructure
4. Implementing changes

In the case of spatial data sharing for IHIS, the first step requires everyone in the organization to realize that the integration process is an agency need and that integrated data should be viewed as an agency resource rather than designed for a specific purpose. When the need for spatial data sharing is realized, all the members of the organization should have a clear vision of what the integrated system will be and what adjustments must be made to implement the changes successfully. The importance and role of each unit in the process should be emphasized and clearly defined. This requires that the system be designed to meet the objective of integration with minimum adjustments from the functional units. The problem is analogous to the classical transportation network equilibrium problem where each driver's objective is to minimize his own travel time (the "Do not care about others" attitude) on a network, whereas the system's objective is to minimize the total travel time for all the drivers on the road network. In our problem, the difficulty arises when each highway unit is forced to adjust in order to satisfy the system's objective: to integrate data and decision making. The challenge is to make everyone better off as a result of the adjustment.

REFERENCES

Alexander, T. 1991. Institutional issues and technical trade-offs in implementing GPS and GIS in highway agencies. Fifth International Pavement Management/Maintenance Exposition and Conference, 4R Conference and Roadshow, *Compendium*.

Briggs, D. W., and B. V. Chatfield. 1987. Integrated highway information systems. National Cooperative Highway Research Program Report 133, October.

Brown, J., and S. Lewis. 1991. Strategic approaches to GIS-T in three state departments of transportation. Paper presented at the 1991 AASHTO Symposium on GIS-T for Transportation, Orlando, Florida, March.

Duffy, T., and L. Wyznyckyj. 1987. Review of state comprehensive computerized highway information systems. Technical Report, Federal Highway

Administration Research, Development and Technology, FHWA-TS-87-220, May.
Fletcher, D., and S. Lewis. An introduction to GIS and transportation. Pre-Conference Workshop of the Transportation Research Board 1992 Conference, Washington, D.C., January.
Hyman, W., and R. Alfelor. 1991. Standards for data interchange for bridge management systems and integrated highway information systems. Technical Paper, National Cooperative Highway Research Program Project 20-4, Task 46, October.
Lee, S. T. 1991. Geographic Information System technology and organizational transformation. 1991 Geographic Information Systems for Transportation (GIS-T) Symposium, Orlando, Florida, March, *Proceedings*.
Lewis, S. 1990. Use of Geographical Information Systems in transportation modeling. *Journal of the Institute of Transportation Engineers*. March.
Paterson, W., and T. Scullion. 1990. Information systems for road management: draft guidelines on system design and data issues. Technical Paper, The World Bank, September.
Vonderohe, A., L. Travis, and R. Smith. 1991. Adaptation of Geographic Information Systems for transportation. Interim Report of the National Cooperative Highway Research Program (NCHRP), Project 20–27.

Michael F. Goodchild

25 *Sharing Imperfect Data*

Quality is almost always an issue in working with spatial data, since almost all spatial data are of limited accuracy. Access to appropriate information on quality is essential if agencies or scientists are to share spatial data, since no user should be willing to trust data of unknown accuracy. The concept of metadata provides an effective means of ensuring such access, by integrating information on quality with the data themselves, and by updating metadata as the data are processed and modified by Geographic Information Systems (GIS) operations. Standards for the description of quality and for metadata are emerging, and procedures for quality assurance and quality control of GIS data are under development. But much more work is needed in basic research and software development if metadata is to be used effectively. The impact of effective metadata on the use and value of spatial data has yet to be evaluated.

INTRODUCTION

The theme of this chapter is that data quality is a significant hindrance to the sharing of spatial data among institutions. At the simplest level, data that meet the quality standards and needs of one agency will frequently fail to meet the different, or more exacting, needs of others. But at a more subtle level, the difficulties inherent in documenting data quality, in passing that information to the consumer of data (Beard et al. 1991), and in making it general enough to apply in a wide range of applications, in themselves form a much more problematic and challenging barrier. Not only must agencies sharing data trust each other's quality control procedures, but they must also be able to communicate knowledge of data quality—sources of error, types of imperfection, measures of accuracy—in an objective and useful way. I argue that the lack of suitable methods for formulating and communicating information on quality currently forms a significant impediment to sharing, particularly given the awkward quality problems inherent in spatial data.

Data quality is a technical issue, as the measurement and modeling of quality rely to a great extent on spatial statistics. Communication of information on quality, through metadata or techniques of visualization (Leung, Goodchild, and Lin 1992), also requires technical solutions. But the impact of the data quality issue on data sharing and decision making creates an institutional problem that can be studied, addressed, and hopefully removed, only by a combination of technical and institutional research. Social science research methods are needed to address the impact of data quality on GIS use, perhaps through a series of case studies. Yet such research would be difficult to carry out in a rigorous and objective fashion without a clear, underlying technical basis that can provide sound definitions of the various dimensions of quality. Thus the problem of data quality and its importance in data sharing can be addressed only by a combination of technical and social research methods.

The first part of the chapter introduces the data quality problem and explains its dimensions as it applies to spatial data. The next section looks at the problem of communication, through concepts of metadata and lineage, and at why these are frequently poorly developed in the spatial data-handling field. This is followed by a section on current developments in quality assurance and quality control (QA/QC). The last section examines possible courses of action that may help to remove the data quality impediment to greater sharing.

THE DATA QUALITY PROBLEM

Observational data are almost always subject to error, but spatial data seem to suffer more from imperfect quality than do other kinds of data. In part this is due to the fact that spatial data can at best only approximate the complexity of real geographic distributions, through processes of generalization, abstraction, or aggregation. It may also be due to the fact that much geographical data is the result not of precise measurement but of partly subjective interpretation: two soil scientists are unlikely to make identical soil maps of the same area. Finally, spatial data may be unusually imperfect because of the measurement errors inherent in many instruments and the simplifying assumptions about such things as the figure of the earth that are implicit in their use.

Sources of Error

The next six sections describe some of the common sources of error in spatial data. Veregin has published a bibliography and associated taxonomy (Veregin 1989a, 1989b), and reviews of many aspects of the spatial data quality problem can be found in Goodchild and Gopal (1989).

Measurement. The instruments that produce raw geographic data often have well-known accuracies. Global Positioning System (GPS), for example, provides positional coordinates with levels of accuracy that are a function of the experimental design (one receiver or two) and can be degraded through the use of Selective Availability (Leick 1990). Land-cover classes determined through aerial photography or remote sensing often have known confusion matrices, which give the probability that any one cover class may be confused with any other. Such errors enter the raw data and stay with them through the entire process of database creation and use.

Definition. In photogrammetry, remote sensing, geodesy, surveying, and other measurement-oriented spatial sciences, the variables being measured are mostly well-defined. However, in the more interpretational spatial sciences, such as soil science, geology, forestry, or geography, this is far from true, and one person's moraine can be another's kame. Measurements are not necessarily replicable between observers, and uncertainty of definition is a major source of inaccuracy in data. Although one can argue as to whether variation between observers really deserves to be called error, a system that fails to identify different parts of the map with different observers has no way of distinguishing this type of error from other types.

Lack of documentation. To extend the previous point, differences between observers become indistinguishable from other types of errors if not identified as such. For example, suppose two maps of the same area use different geodetic datums or different map projections. Without adequate documentation, the user may ascribe the differences in the two maps to error. In effect, then, lack of documentation is in itself a contributor to error. We could go so far as to define error as the unexplained differences between observations of the same phenomenon. Once the difference in datums is known, the discrepancies between the maps can be removed by a precise series of mathematical transformations.

Distortions in physical media. Much spatial data originates from maps, either printed or hand-drafted. Although some media are very stable, paper is known to stretch with changes in humidity, and folding can also introduce unwanted distortion. Digitizing and scanning of paper documents produces precise measurements based on distances from a defined origin point, so distortions of the base produce errors that accumulate with distance from the origin. Thus, although a typical digitizer can achieve precisions of fractions of a millimeter, errors on the order of several millimeters in position commonly occur due to distortion of the base. Some types of copying processes are known to introduce distortions that vary markedly with direction.

Processing. GIS is a precise technology, and GIS processing is usually carried out with precisions far in excess of the inherent accuracy of the data. Nevertheless, some GIS processes are capable of introducing substantial errors. The use of snap tolerances in digitizing, automatic sliver removal in overlay, and raster/vector conversion are all sources of uncertainty at levels comparable to the accuracy of the data. For example, if a tolerance of 2mm is used to snap lines in digitizing, it is impossible to distinguish correctly between a lake with a narrow neck, and two lakes, if the neck is 2mm or less across.

Interpretation. Very few systems for spatial data handling offer any extensive capabilities for formal documentation of data. The meaning attached to an attribute, such as class 1 on a land-cover map, is more often entirely the responsibility of the user. The label "soil" attached to a file will have different meanings to different users, each of whom may have different needs and purposes. One user may be looking for geotechnical information on soil mechanics in connection with highway construction, while another may be looking for soil fertility information in connection with forestry. Far more documentation is needed if each is to be able to make effective use of the data, but it is very unlikely that the necessary documentation will be in the database; more often it is on a sheet of paper somewhere, which may or may not have been filed with the data, may or may not be legible, and may or may not be correct. Lack of documentation in the database, and the responsibility for interpretation that this places on the user, is in itself a significant source of error in spatial data because of the different meanings that users are consequently free to attach to the same information.

The Structure of Errors

Errors of position are perhaps the more obvious type of spatial data error, although certainly not the only type, and most of the examples in the previous sections were of positional error. The error in the position of a point can be described by confidence limits in the x and y directions. However, the impact of errors in spatial data is not just a function of their magnitudes but also responds to the inherent relationships between them. For example, if the point is part of a larger object, such as a line, it may behave independently, leading to uncertainty in the shape of the object; or the object may behave as a rigid whole, every point in the object being subject to precisely the same error. Conceptually, we distinguish between cases like these by differentiating between absolute and relative positional error. Statistically, the distinction lies in the structure of correlations between errors: correlated errors (the case of the rigid object) produce very different effects from independent errors (each point behaving independently, distorting the shape of the object).

Unfortunately, few geographical objects can be described as collections of discrete points. More common are continuous curves, which may be approximated as discrete points in the database. While we can describe uncertainty in the positions of continuous curves through such devices as buffer zones and epsilon bands, it has proven much more difficult to develop appropriate theories and robust measures.

Although positional errors have been emphasized, another common source of errors in spatial data lies in the attributes of objects, rather than their positions. For example, the accuracy of digital elevation models (DEMs) is commonly described as an expected error in the elevation specified at each point rather than in the positional accuracy of those points. Structure is important here, also. For example, in the viewshed operation commonly used in GIS, the accuracy with which a visible area can be estimated is strongly affected by the correlations between errors in the digital elevation model. Independent errors at each point will create a rugged distortion in the horizon, whereas correlated errors will raise or lower the entire horizon (Fisher 1991b).

METADATA AND LINEAGE

Thus far, we have established that data quality is a significant problem in spatial data handling. There are many sources of error in the pathway from observation to database, analysis and decision, and each has its own characteristics and processes. Error is a function of information. Once differences can be described by simple mathematical functions, they can be eliminated, but differences of unknown structure – random differences –

can be treated only as error. Moreover, these mathematical functions can be lost along the way, creating additional sources of error. If the mapmaker forgets what projection was used to make a particular map, then in effect another source of error has been introduced, unless someone is clever enough to deduce the projection and its parameters.

We define metadata as information about data—knowledge of scale, date of creation, projection, legends, and the meanings of symbols, the name of the mapmaker, and any other information that may be useful to the potential user. Of course, the more widely distributed the data, the more difficult it is to anticipate uses and thus to determine what to include as metadata. The use of GIS merely makes this situation worse, by making it easier to copy, distribute and analyze data for a more and more varied set of purposes.

Two different approaches are useful in describing the accuracy of data to a potential user. One is by direct comparison between the data and the truth that it supposedly represents. Thus, we could describe the digital elevation data above as having an accuracy of 7m, because the root mean square difference between the elevation in the database at a given point, and the true elevation at that point, is 7m. Because of frequent difficulties in defining "truth," it is often acceptable to compare the data to an independent source known to have higher accuracy.

The alternative approach is to describe accuracy through the processes known to contribute to error. In the digital elevation model case, we could describe the various processes used to generate the data and include estimates of the magnitudes of errors introduced at each step. Assuming that the errors at each step are statistically independent, total error is found by summing the squares of the individual error contributions and taking the square root of the total. We will refer to these two as the "measurement" and "process" approaches to error estimation, respectively. Both are useful, depending on the context and the information available to the user of the data.

Spatial Data Transfer Standard

The federal Spatial Data Transfer Standard (SDTS) (DCDSTF 1988; NIST 1992), now adopted as a Federal Information Processing Standard, contains a lengthy section on data quality (Wortman and Morrison 1992). It establishes the principle of "truth in labeling"—that it is more useful to report all available information on data quality, and to lay out the minimal set of parameters that must be measured, than to establish minimum thresholds of quality that data sets must meet to merit different levels of quality classification. SDTS identifies five dimensions of data quality—

positional accuracy, attribute accuracy, consistency, completeness, and lineage—that can be defined for the entities in any digital cartographic database. Under each heading there are prescribed indicators of quality that must be measured or estimated and reported in the data quality statement accompanying the data set. Under the lineage heading, the distributor of the data set is expected to provide information on the stages in the creation of the data and the uncertainties introduced at each stage.

SDTS provides a comprehensive template for data quality reports. It falls short of defining precise format standards for metadata or of requiring that data quality reports become an integral part of the formatted database and inseparable from it, and as a Federal Information Processing Standard it can be enforced only for data sets handled by the federal government. Finally, as the name of the original task force suggests, its concern is primarily with cartographic data and the uses to which such data are commonly put. Traditionally, cartographers have assumed that the user will "read" the map visually, but increasingly the same maps are being used to create digital databases and support detailed and precise analysis. A standard oriented toward the quality of maps clearly cannot anticipate all of these new uses or the error descriptors relevant to each of them. For example, spatial dependence of errors, which we have argued to be relevant in viewshed estimation from DEMs and many other applications of spatial data, is not listed as a parameter to be estimated in SDTS data quality reports. Moreover, a GIS user may find it difficult to relate the measures prescribed by SDTS, which are concerned with data as input, to the measures needed to determine quality of output in particular applications.

Lineage Handling

GIS software has been very slow to adopt the idea of integrated metadata. Even such obvious parameters as the scale of the map from which the data was digitized, the date of creation, or positional accuracy must be communicated to the user through written documentation, as the system has no means of storing, handling, or reporting them. Soil maps often carry elaborate text explanations of classes, but the digitized soil map in the database simplifies these to alphabetic codes and leaves it up to the user to search out more complete descriptions of each code's meaning. Some systems will go so far as to identify class 1 with "Grimsby Loam" but will not help the user to access the one thousand words of detailed description of Grimsby Loam provided by the mapmaker, often on the back of the sheet.

There are some exceptions to this general lack of interest in metadata on the part of GIS designers. The recent version of IDRISI (IDRISI Project,

Clark University, Worcester, Massachusetts 01610) includes several standard metadata fields for each layer, and in relational systems such as ARC/INFO (Environmental Systems Research Institute, 380 New York Street, Redlands, California 92373), it is possible for the database designer to link metadata files to objects or coverages in particular implementations. However, neither of these systems provides capabilities for handling metadata, such as the automatic propagation of metadata from input to output in GIS processes. For example, a metadata handler should attempt to estimate the accuracy of the output of an operation like overlay given the accuracies of the input layers.

Lanter and Veregin (Lanter 1991; Lanter and Veregin 1992) have described a research effort at the University of California at Santa Barbara to develop a metadata handler, called Geolineus, which provides a user interface to a GIS command language with full access to appropriately defined metadata. The user interacts with information on each layer in the database through icons and a "point-and-click" interface, and the system automatically propagates metadata using appropriate rules. Systems like this replace the traditional approach in which metadata are entirely the responsibility of the user, and are often ignored, with a set of tools to inform the user of the nature of the data being processed.

QUALITY ASSURANCE AND QUALITY CONTROL

Quality assurance and quality control (QA/QC) are major concerns of agencies and activities that must deal with the potential threat of litigation. Standards such as SDTS lay out the parameters that must be measured and described in a quality report, but they fall short of providing the kinds of assurances demanded in a legal setting—that all reasonable efforts have been made to control and maintain the quality of the data. Such assurances are more likely to be provided by procedures and protocols, and evidence that these have been enforced and followed in practice, than by objective measures of quality. In other words, the legal context invariably emphasizes process rather than measurement.

In the extreme, QA/QC can place entirely unreasonable demands on process. In GIS applications, the highest level of QA/QC would require that every item of information in the database has attached to it a documented line of responsibility—that someone "signed off" on every item. While this is possible in principle, its imposition in practice would make it impossible to build the kinds of multithematic, detailed GIS databases now commonly in use. Because the data are processed by software, the same standard would apply to the code—that every line and every change have some documented line of responsibility attached to it.

The measurement approach to quality is fundamentally statistical. For example, the Circular Map Accuracy Standard (CMAS), a commonly used measure of positional accuracy of points, is defined as the radius of a circle within which the true location of the point is expected to lie 90 percent of the time. The remaining 10 percent of cases are expected to lie outside the circle, and it is possible for individual points to lie at distances of many times the CMAS. But while averages and statistical expectations are the basis of such measures, they may have little value to QA/QC procedures concerned with maintaining rigorous levels of quality throughout the database. A single mislocated point may be entirely acceptable within a statistical framework but can be very damaging to credibility in a courtroom, where maximum error is likely more relevant than average error.

Various efforts are underway to establish QA/QC procedures for GIS applications. The Environmental Protection Agency, for example, is developing a set of QA/QC procedures to cover the wide variety of spatial databases and GIS-supported decision making within the agency. These, coupled with effective reporting of data quality and handling of metadata, should help to remove some of the barriers to greater data sharing.

A USER PERSPECTIVE

Defining Metadata

Data quality is a messy issue, and the average GIS user would probably far rather ignore it than confront it. But as GIS matures, we can expect more and more to find GIS results challenged in court, and it will be less and less acceptable to rely on the overused notion that computer output speaks louder—that the output of a GIS beats a hand-drawn map every time. Increasingly, both sides in a suit will have access to the same technology, and the winner will be the one that can demonstrate higher quality, in data as well as in analysis, through rigorous procedures for QA/QC, and the use of well-defined measures of uncertainty, documented and communicated as metadata.

Standards like SDTS result from lengthy discussions but essentially reflect the perspective of the data producer who is unwilling or unable to anticipate the data's applications and therefore provides information on all possible aspects of quality. Because of its richness and complexity, it will prove difficult to automate the SDTS data-quality statement as metadata, or to handle and update it automatically during GIS processing. I would like to suggest an alternative, user-oriented research agenda, that will identify and prioritize the commonest, most primitive GIS queries and products and the metadata needed to support them. This approach would not attempt to capture everything known by the producer about the data's quality but would capture only the highest priority dimensions in forms that are processable and most appropriate to the commonest user requirement.

For example, consider the TIGER (Topologically Integrated Geographic Encoding and Referencing) database and two common uses: the determination of the geographic coordinates of a given street intersection and the estimation of the geographic coordinates of a given street address. If we assume that both queries are valid—that the street intersection and street address do indeed exist at the time of the query—then the accuracy of both responses has two components: one concerned with positional accuracy, and the other with the probability that the intersection or address exists in the database (either may be missing, most likely because of problems with currency). The metadata measures required to provide reasonable estimates of both components to the user are not complex or difficult to devise, but they may require sampling programs by the producer that go beyond current programs of QA/QC. Moreover, both components of quality—positional accuracy and currency—are known to vary spatially in TIGER, and consequently metadata may have to be coded at the feature level rather than as attributes of the entire data set. The use of GIS has vastly increased the range of applications of spatial data, but at the same time its technical nature provides a simplifying framework within which it may be possible to provide simple measures of quality for all but the commonest types of query.

The Role of Visualization

Cartography has few techniques for displaying data quality, and maps rarely try to communicate this type of information to users. Boundaries between soil classes are always drawn with the same width of line, although the transitions they represent have widely varying widths, and soil classes are shown as uniform within boundaries despite clear lack of homogeneity in the field. While such conventions may be acceptable as long as the map is seen as a way of communicating general impressions and interpretations, they become much more questionable in the precise world of a digital computer.

In GIS applications, data quality impacts both the communication of information to the user and the process by which decisions are reached based on that information. GIS output has tended to follow cartographic convention by ignoring the issue of quality. But as metadata provides increasingly useful quality information, the question of how, or whether, to present it to the user will become increasingly important. What is the impact of data quality on the decision-making process? Is that process best served by absence of quality information, or will quality become an increasingly important issue?

GIS provides us with three different approaches to the communication of data quality. Because quality varies spatially, all three are essentially visual. First, we can suppress all information on quality, as occurs in remote sensing when an image is shown colored by most-likely class—the maximum likelihood classification. Second, the user can be provided with indices of quality, by blending of colors, or the use of gray or varying acoustic tone. Third, we can present the user with realizations of an error model—in other words, with samples of possible maps (Englund 1993; Fisher 1991a), each one equally likely to be the truth given our knowledge of uncertainty. In terms of impact, the third is clearly the most dramatic, since it provides the user with a direct visual impression of uncertainty rather than an abstract measure.

The modeling and simulation of uncertainty are technical issues, but their impact on the decision-making process is clearly institutional, and we are reaching the point in the development of GIS where knowledge of the nature of that impact is becoming increasingly important. We need research not only into how to describe data quality and pass it between users, but into the nature of the impact of data quality information on the decision-making process that GIS exists to support. Who benefits from knowledge of data uncertainty, and who loses?

Tolerance to Error

The value of GIS lies in the benefits derived from its products and their relationship to the costs of product generation. As error or uncertainty increases, and products become less and less reliable, the benefits to the user fall. The effect of error on benefits is cumulative in the sense that loss of credibility in the system accumulates through time, and eventually the entire system may have to be abandoned. On the other hand, every product is tolerant of some degree of error, particularly in those aspects of the product that are not used directly in supporting decisions.

Thus, from the user's perspective, one might attempt to determine accuracy requirements by asking questions of the following form: "How much error of what forms can be tolerated in this product before there is significant loss in the benefits anticipated from it?" The well-known map of the London Underground system clearly allows large amounts of error in the geographic locations of tracks, positions of stations, and the distances between different tracks while maintaining the topological ordering of stations along tracks, and route maps often make similar distortions. The role of metadata is then in providing the information needed to compare tolerable levels of error to the levels actually anticipated in the product.

To date, efforts to address user requirements for accuracy have more often begun with the input data rather than with the product. We need good case study research to determine the feasibility of addressing user accuracy requirements from the user perspective, and of connecting these to appropriate metadata.

CONCLUSIONS

The central theme of this chapter is that lack of information on data quality (and lack of quality itself) is a major impediment to data sharing. No one should be willing to make use of data they do not trust, or whose accuracy they do not understand. But since the effective description of quality is a function of use, the task of describing quality against all possible uses is formidable, especially given the ease with which data can be disseminated and applied in novel ways in the world of GIS. In this sense, GIS is its own worst enemy: by inviting people to find new uses for data, it also invites them to be irresponsible in their use.

In this situation some degree of compromise is essential, and traditionally mapmakers have provided data-quality reports that satisfy the largest possible number of users while requiring only a reasonable level of effort. Thus, users of topographic maps must often infer quality from simple diagrams showing the level of accuracy of various parts of the map using terms such as "relatively low" and "relatively high." These can be made with little effort and are adequate for many purposes.

For better or worse, the widespread use of GIS to process spatial data has changed the parameters of this compromise. Simple quality diagrams are no longer sufficient to help the GIS user assess the accuracy of his or her output. Instead, GIS has led to an explosive increase in the applications of spatial data, and this requires a parallel increase in the effort invested in measuring and documenting data quality. Furthermore, because such documents are of no value unless they are available to the user, it is essential that quality be handled and transmitted as metadata, fully integrated with the data themselves.

At the same time, use of GIS restricts and formalizes the queries and functions performed on spatial data and allows us to analyze and prioritize measures of data quality from a user perspective. This perspective has largely been missing in debates on quality standards to date, which have necessarily emphasized the perspective of the data producer, but its use should allow metadata to be placed on a much more rigorous and user-directed footing. With better forms of metadata and better methods for tracking data quality through GIS operations, we will have the opportunity to study the impact of such information on the decision-making process itself.

ACKNOWLEDGMENTS

The National Center for Geographic Information and Analysis is supported by the National Science Foundation, Grant SES 88-10917.

REFERENCES

Beard, M. K.; B. P. Buttenfield, and S. Clapham. 1991. Visualizing the quality of spatial information: Scientific Report of the Specialist Meeting. Technical Report 91-26, Santa Barbara, CA: National Center for Geographic Information and Analysis.

DCDSTF (Digital Cartographic Data Standards Task Force). 1988. The proposed standard for digital cartographic data. *American Cartographer* 15, 1: 1–140.

Englund, E. 1993. Spatial simulation: environmental applications. In M. F. Goodchild, B. O. Parks, and L. T. Steyaert, eds., *Environmental modeling with GIS*. New York: Oxford University Press.

Fisher, P. F. 1991a. Mapping soil map-unit inclusions by Monte Carlo simulation. *International Journal of Geographical Information Systems* 5: 193–208.

_____. 1991b. First experiments in viewshed uncertainty—the accuracy of the viewshed area. *Photogrammetric Engineering and Remote Sensing* 57, 10: 1321–1327.

Goodchild, M. F., and S. Gopal, eds. 1989. *Accuracy of spatial databases*. London: Taylor and Francis.

Lanter, D. 1991. Design of a lineage-based meta-database for GIS. *Cartography and GIS* 18, 4: 255–261.

_____, and H. Veregin. 1992. A research paradigm for propagating error in layer-based GIS. *Photogrammetric Engineering and Remote Sensing* 58, 6: 825–833.

Leick, A. 1990. *GPS satellite surveying*. New York: Wiley.

Leung, Y., M. F. Goodchild, and C. C. Lin. 1992. Visualization of fuzzy scenes and probability fields. Fifth International Symposium on Spatial Data Handling. Charleston, South Carolina, August 3–7. *Proceedings* 2: 480-490.

NIST (National Institute of Standards and Technology). 1992. Federal Information Processing Standard. Publication 173 (Spatial Data Transfer Standard). Washington, D.C.: U.S. Department of Commerce.

Veregin, H. 1989a. Accuracy of spatial databases: annotated bibliography. Technical Paper 89-9. National Center for Geographic Information and Analysis, Santa Barbara, California.

_____. 1989b. A taxonomy of error in spatial databases. Technical Paper 89-12. National Center for Geographic Information and Analysis, Santa Barbara, California.

Wortman, K., and J. L. Morrison, eds. 1992. Implementing the spatial data standard. Special Issue. *Cartography and Geographic Information Systems* 19: 277–334.

DAVID STAGE

26 A Multi-Agency Management Structure To Facilitate the Sharing of Geographic Data in Florida

The Florida Growth Management Data Network Coordinating Council (GMDNCC) was created by the Florida Legislature in 1985 to facilitate the sharing of information needed for growth management planning. In 1988 the GMDNCC recognized that access to geographic information is essential to making informed decisions on this topic. With this in mind the GMDNCC stated that one of its major objectives was to promote information sharing through the creation of a "federation of independently held databases that are linked together by standards and a management structure." After three years of facilitating cooperative activities, it is the conclusion of this author that the primary impediments to information sharing are the difficulties of operating in a multi-agency environment without the appropriate support mechanism to facilitate cooperative activities. This chapter describes the author's observation of the major problems that must be addressed and the organizational structure and decision support tools that were developed to overcome these problems.

INTRODUCTION

Florida is located in the southeast portion of the United States and covers approximately 54,500 square miles. It is blessed with a semi-tropical environment and one of the longest and most accessible coast lines of any state. As a result, Florida has become a mecca for tourists, retirees, and associated industries, all of which have led to one of the highest population growth rates in the nation, growing from 5.8 million in 1965 to 13.5 million in 1992. Because these growth pressures are having a negative effect on the environment, the state passed a Growth Management Act in 1985. In conjunction with this legislation a multi-agency coordinating council was formed, the Growth Management Data Network Coordinating Council (GMDNCC), to facilitate the sharing of "growth management" information. Initially the focus of the GMDNCC was technical in nature, emphasizing hardware and software, but this proved to be of limited value. In 1988, the GMDNCC initiated a study to identify a more productive focal area: geographic information and Geographic Information Systems (GI/GIS) were identified as the key to growth management planning. For the next several years the GMDNCC worked on developing ways of facilitating the sharing of this information. During this time a Memorandum of Understanding regarding standards was signed between the GMDNCC agencies, a pilot project to create a regional coordinating council was initiated, standards committees were formed and standards developed, implementation strategies designed, policy committees created, joint funding activities initiated, and so on. As predicted by conventional wisdom, the gamut of impediments to information sharing were experienced, but it soon became apparent that these problems were only symptomatic of an interorganizational "bureaucratic void" that allows personal and organizational behavioral problems to surface and impede multi-agency cooperative efforts—hence, the sharing of geographic information. It is the intent of this chapter to describe the structure and tools that have been developed to fill this void and facilitate information sharing.

IMPEDIMENTS TO SHARING INFORMATION

The impediments to sharing geographic information are frequently described in terms of specific behavioral activities such as management framework rigidity, personal attitudes, turf battles, and so on (Onsrud and Rushton 1992), and many of these negative organizational and personal behaviors have been observed in the course of our work. It is my contention that these behaviors, which are omnipresent, are amplified as a result of the geographic information/Geographic Information Systems (GI/GIS) community being forced to operate in a multi-organizational environment. These problems are elaborated below.

- The expense of collecting spatial data is substantial, forcing agencies to seek existing data sources, frequently from external organizations.

- The specific information being collected is often outside the jurisdictional boundaries of the information collector (physical as well as the agency mandate), requiring the data to come from outside sources.
- The measurement of spatial data often requires sophisticated technology and procedures that are understood only by a small body of experts.
- The lack of a common language for geographic information (standard definitions, standard reporting formats, quality and accuracy reporting formats, and the like) impedes the development of standards and the coordination of resources.
- Upper management does not understand the issues related to geographic information and in many cases does not even know that issues exist.
- Because of the complexity of the issues and the number of organizations that need to be involved, long-term projects are difficult to sustain.

Without a multi-agency bureaucracy to ensure participation in extra-agency activities, I have observed that cooperation and compliance with multi-agency activities and standards become subject to the behavioral "whimsy" of an individual or an organization. On the other hand, it has also been my observation, and that of others (Warnecke 1992b), that most individuals in the GI/GIS business understand the need to cooperate and very much desire to do so. The recent flurry of activity to develop multi-agency coordinating bodies to provide the GI/GIS community with a mechanism to facilitate cooperation and information sharing supports this notion.

MULTI-AGENCY COORDINATING BODIES

The GMDNCC's organizational structure is but one of many that have been developed in the past ten years to address the problems associated with multi-agency activities. Warnecke has found that each state has some form of a multi-agency coordinating body (Warnecke 1992b), and she has identified three different groups.

State Groups Established to Address GI/GIS Issues

These bodies typically focus on issues related to GIS. The Wisconsin Land Information Board (WLIB), which was authorized by legislation in 1989, has one of the largest programs in the United States. It is funded by filing fees collected by the county Register of Deeds office, which provides it with the ability to induce county governments to meet WLIB's information standards and information-sharing requirements through a cost sharing program with the counties (Holland 1992). Mississippi created a central authority for GIS in 1983—the Mississippi Automated Resource Information System—which facilitates information sharing and standards by acting as a hub from which users can obtain or develop data, have it analyzed, and be directed to sources of existing data (Warnecke et al. 1992). The North Carolina Geographic Information Coordinating Council (GICC) was established in 1991 to provide technical oversight and direction to the state's overall approach to information resource management. The GICC is closely associated with the Center for Geographic Information and Analysis (Warnecke et al. 1992).

Other Groups Influencing GI/GIS

These are groups that influence GI/GIS as an element of another statewide focus that included state information technology policy, growth management, and natural resources. Florida's GMDNCC falls under this category. Another example is the Texas Department of Information Resources, a state-level information management body that was created to provide the leadership and coordination of information resources management within state government. A Geographic Information Planning Council was formed in the fall of 1991 to address statewide GIS interests and issues (Warnecke et al. 1992).

Independent State Groups

These organizations are frequently nonprofit entities, many of which are organized as geographic information councils. The State Mapping Advisory Committees (SMACs), which were originally created to focus on mapping issues, have naturally taken on the role of GIS-facilitating entities where no such organization has been developed in a state. Warnecke has identified at least five states in which the SMAC is listed as the only GIS coordinating body. The New Mexico Geographic Information Council, which includes both public and private entities, was authorized by executive order in 1987 and incorporated as a nonprofit entity to push the goals of cooperation and coordination of programs, reduce duplicity, and so on. The WLIB also began in this manner (Warnecke et al. 1992).

The Growth Management Data Network Coordinating Council

It is desirable to identify some of the forces that have shaped and/or constrained the activities of Florida's GMDNCC. The GMDNCC's management structure and the focus of its activities have been greatly affected by its legislation, staffing limitations, and the fact that it is not associated with a GIS laboratory. The enabling legislation stated that the GMDNCC's primary objective was to facilitate information sharing, giving it an executive-level council with a small staff. The chair and the staff of the GMDNCC are statutorily located in the Governor's Office of Planning and Budgeting, providing the staff with close ties to the development of state policies and budgetary issues. The staff are completely removed from any sort of applications development, allowing them to focus solely on coordinating issues. The GMDNCC does not have a regulatory function; that role is reserved for another entity in state government. Because of all of these factors, the GMDNCC's activities have tended to address multi-agency institutional management issues, procedures for standards development, and relevant policy issues.

MANAGEMENT STRUCTURE

The creation of a multi-agency management structure is the key to the success of the GMDNCC's efforts to promote standards development, reduce duplicative activity, and facilitate cooperative ventures.

Problems for Sharing GI

It has been my observation that organizations are generally structured in a vertical fashion, with the executive management defining the direction of the institution by creating policies and procedures to accomplish their goals and objectives. These guides are passed down the hierarchy to middle management and technicians who develop operational procedures and action plans to achieve specified objectives. It is at this level that measurement of past performance and the identification of new areas of concern are determined by the collection, synthesis, and analysis of information. This information is organized into an appropriate representation (executive summary, tables, maps, and the like) and periodically presented to executive management so that the institution can make any necessary adjustments. Historically, government agencies have used data that was tabular in nature, with the specific data parameters being defined by executive management (the number of cases handled, claims, errors, and so on) to measure institutional performance. As issues that governments address become more complex, the information required by management to guide its decisions has correspondingly become more complicated. Management, due to its lack of technical expertise, has had to move from describing the specific data parameters that are needed for decisions to a rather general framing of questions that require answers—for example, is there more or less mercury in the water, and is it dangerous? Deciding what information needs to be collected and how that information is acquired, measured, synthesized, and the results presented are delegated to the experts in that field.

The information being used to answer these more complex questions is typically spatial in nature, and the information collected is often best manipulated by GIS. The collection of this raw data presents some interesting problems to the information resource manager. Only the experts, naturally, truly understand this information, and subsequently they are also the only ones able to develop appropriate policies and procedures to maintain the integrity of this data, but they are not empowered to do so. Even though these managers may develop shop standards, these standards are only ad hoc in nature. If paths to executive management are open to formalize these ad hoc standards, there is no structure to promote these standards outside of their organization.

If the need is great enough, individuals will take the initiative to solicit cooperation from another agency, but due to several factors, this can have only a limited impact. Once an individual leaves one bureaucracy and begins interacting with individuals in another organization, the parties are no longer operating within any defined institution—they are functioning in a "bureaucratic void," with the total lack of structural support that this term implies. To accomplish their goals they are restricted to their personal resources, with success being dependent upon an individual's will until a formal agreement is acquired through such devices as a Memorandum of Understanding or a contract. Even when a formal agreement is acquired, it is typically limited to short-term activities. Furthermore, due to the previously mentioned reasons, agreements among more than two parties are even more difficult to acquire. Although individuals may be able to achieve some significant gains, there is nothing to perpetuate these gains when those individuals leave. The "corporate memory" of an organization—and more often than not, the initiative—leaves with them. Subsequently, many intergovernmental cooperative activities in areas of high complexity are limited to small-scale or short-term projects.

Another problem faced by GI users and managers is the inability to act on windows of opportunity. For example, in a structured meeting that we had on the Soil Conservation Service's soils database, the engineers from the Department of Transportation (DOT) were reviewing the USDA Soil Series Data Base with a Soil Conservation Service soil scientist. It was discovered that the information that was being collected to measure the corrosive properties of soils was restricted to corrosion on iron. The transportation engineers needed data on galvanized steel, stainless steel, and concrete. The DOT office had the laboratory facilities for measuring these properties, and the parties present were able to conceptually structure a way in which the USDA soils data could be sent to the DOT laboratory for analysis of these other data elements. The problem arose in finding a way to convey this concept to upper management in both institutions and thus initiating an intergovernmental cooperative activity.

Generally, we have found that the primary impediment to sharing information is the lack of formal management structure in a multi-agency environment. To address this problem we have developed tools for communication, a consensus group methodology for developing data standards, and we are currently in the process of developing a multi-agency bureaucracy to initiate standards development and implement the standards that are derived from this process. The key to success is access by the middle managers and technicians to executive management. This has been achieved by creating in the statutes in Florida an Executive Council and by signing an inter-agency agreement regarding standards.

State of Florida Geographic Information Network

The State of Florida Geographic Information Network is a federation of independently held databases that are linked together by *standards* and a *management structure*. Its primary function is to allow the experts from the various organizations to meet in technical advisory committees (TACs), develop standards, and provide a conduit through which those recommendations can be sent to upper management. The multi-agency management structure, illustrated in Figure 26.1, has three levels of bureaucracy. At level one, executive management makes decisions regarding policy and provides executive support to the staff advisory committee and TACs. At level two, a staff advisory committee identifies topics of concern, creates TACs, and identifies the experts from each of the agencies that should participate on those TACs. At level three, the TACs—consisting of experts in a specific field—come together to develop policy, procedures, and standards that are then presented to the staff advisory committee for review. These recommendations are ultimately presented to the GMDNCC and, if adopted, they can be promulgated to other state agencies.

There are six fundamental tools that are used in this structure:

1. a newsletter to provide information about ongoing activities;
2. an issue statement that documents the various activities and provides a structured methodology and project manager for TACs that can develop recommendations for standards, procedures, and policies;
3. a catalog of geographic data (data directory);

4. documentation of the "goodness" of the data (quality and accuracy report);
5. a detailed description of the data (data dictionary); and
6. formal transfer protocols using the FIPS 173 Spatial Data Transfer Standard (NIST 1992).

These methodologies and tools will be described in their appropriate sections.

Management Support and Structured Methodologies

The institutionalization of a management structure requires management support services and structured methodologies to ensure consistency and coordination. The GMDNCC has established the following: a facilitator to provide administrative and management support services; a multi-agency project manager (issue statement) that coordinates and tracks activities; and TACs (subcommittees and consensus groups) that provide an inter-agency micro-bureaucracy.

Facilitator. A facilitator is essential to the GMDNCC, the staff advisory committee, and the TACs, providing training, support services, and management guidance. This is particularly important to the TACs where a significant amount of work is required of a membership that is essentially composed of volunteers. The facilitator:

- promotes horizontal activity by providing a single point of contact, and becomes a broker of resources by developing a "corporate memory" for the multi-agency body;
- acts as an inter-agency manager by coordinating activities, ensuring the use of structured methodologies, and aiding in the development of strategic plans for the GMDNCC;
- provides an inter-agency staffing by aiding the various TACs with their administrative duties, documentation of meetings, presentations, and the development of issue statements. Each TAC, which is actually developing its own micro-bureaucracy, utilizes the facilitator to help establish the structure, identify the goals, objectives, and tasks in order that they may spend their time working on the objectives as opposed to the administrative duties of the TAC; and
- provides a management function to the GMDNCC by tracking the progress of each TAC.

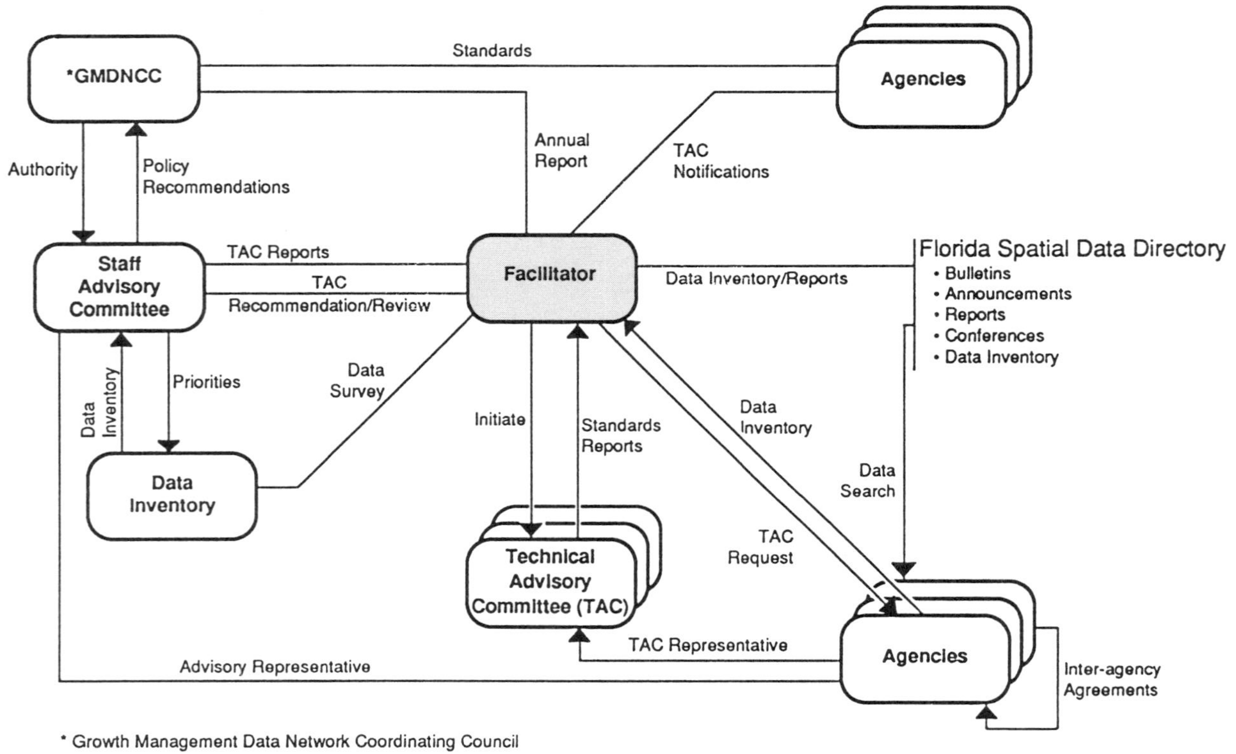

FIGURE 26.1
State of Florida geographic information network

Issue statements. The issue statement was developed as a way to focus the activities of the TACs and to function as a project manager. These amorphous bodies, which have an inherently weak leadership and no authoritative power, face a number of problems that inhibit their ability to become an effective operational organization. Turnover in state government is approximately two years, causing a constant attrition to these groups; unstructured meetings tend to produce a lot of good discussion but little planned action; members of multi-agency tasks force are essentially volunteers able to provide only the minimum amount of time outside of the meetings of the TAC; and the collective memory of such organizations is poor, at best, making long-term projects difficult. To overcome these problems, we have developed a multi-agency project manager through the means of an issue statement that is designed to alleviate these problems. (See Appendix 1. The issue statement is modified from the Environmental Protection Agency's Gulf of Mexico Program Action Plan concept.) This document:

- provides a charter and justification for participation;
- establishes an action plan with goals, objectives, tasks, and timelines;
- provides documentation of actions (a corporate memory of activities), allowing issues to be placed on hold and, when resurrected, the work can continue where it was left off even if the membership has completely changed; and
- acts as a project manager for the facilitator and co-chairs.

Technical Advisory Committees

Technical advisory committees are inclusive of both subcommittees and consensus groups and, although both groups use issue statements for their organizational structure, they are conceptually two different types of activities. Subcommittees develop policies and procedures to standardize operations; consensus groups develop standard data definitions.

Subcommittees. Subcommittees are formed to address issues that will result in the development of generic standards (see Standards section) that are related to the development of broad-based policies and procedures. For example, the Florida Public Records Law requires that state agencies make information that is collected with public funds available to the public for inspection. But the law was created before the existence of computers, and the last update to Florida's Public Records Law occurred before the development of GIS. There is a great deal of confusion in the GIS community over how the information resource community will be impacted by the Public Records Law, and as a result they are somewhat reluctant to advertise the existence of their data. A Public Records Law subcommittee was formed to address this issue and will eventually make recommendations to the GMDNCC on how to resolve these problems on a statewide basis. Prior to the establishment of this subcommittee, the only recourse for information resource managers was to consult with their legal staff to resolve a specific issue. There was no formal avenue to address this issue on a statewide basis.

Consensus group. A consensus group is a structured methodology by which standards on data can be developed (data specific standards). Conceptually, the following will take place: a dataset will be identified as having "corporate value," that is, value outside of its use to the data developer(s). A consensus group will be formed that includes the data developer(s), potential data users, and a technical person. Using the quality and accuracy report (Appendix 2) and the data dictionary (Appendix 3), the dataset will be reviewed. Suggestions will be solicited and, if possible (either by good will of the data developer or incentives by the user), enhancements can be made to the data, thus increasing the utility or "corporate value" of the data. Consensus groups accomplish two things: they act as a peer review process for the documentation of the data, and they provide an opportunity for improving the way information is being collected for the larger community. This process allows the data to be scrutinized, providing the user with the opportunity to request changes to the database. For example, let us suppose that two adjoining counties are collecting information on multi-family dwellings. If County A includes duplexes in its definition of apartments, and County B identifies duplexes as a separate entity, then it will be impossible to conduct valid bi-county studies on multi-family dwellings. By holding a consensus group before the information is collected, such problems can be avoided.

STANDARDS DEVELOPMENT AND TOOLS FOR IMPLEMENTATION

The solution most frequently presented to facilitate the coordination of data collection activities and the sharing of geographic information is the development of standards. When analyzed, one is confronted with a myriad of problems, such as identifying which elements to standardize, putting together a structure to develop standards, defining what a standard is, standardizing the standards procedure, and enforcement.

Standards Development

To facilitate the development of standards, we are incorporating a decision support tool, the Futures Planning Technique (Breto 1992) that prioritizes policies based on the projected impact of those policies on the state. We are then linking these policies to issues and data sets to be addressed. For the standards development activities, we have recognized three areas that need to be dealt with: standard formats, generic standards, and data-specific standards.

Futures techniques. A decision support methodology, the Futures Planning Technique, is being used to prioritize and focus the available resources on the most important issues. To facilitate the sharing of information, one must address issues as fundamental as which datum to use, as specific as the definition of data sets, and as broad as the development of statewide policies. Through the use of this decision support planning technique, we are attempting to identify and prioritize issues that are the most important to all of the organizations. This will link together the policies of the different agencies with the activities of the GMDNCC.

Standard formats. Standard formats are the way in which standards are reported. Currently there are several organizations, in addition to individual agencies, working on "standards" that will affect Florida. The state of Florida has the Growth Management Data Network Coordinating Council, the Tampa Bay Regional Coordinating Council, and the Florida Base Mapping Advisory Committee. Organizations developing standards outside of Florida and affecting the state are the Environmental Protection Agency's Gulf of Mexico Program, the Federal Geographic Data Committee, and three neighboring states. If geographic standards are to be universal, it is imperative that these organizations communicate and coordinate their operations. To accomplish this there must be a set of common communication formats, just as there exists within any agency. This will create a functional basis for a multi-agency/governmental bureaucracy by linking the organizations together by the creation of a common language and procedures, or standard formats. This can succeed only if these formats are useful at all levels of government, have internal value to an agency and external value to the multi-agency body, and are formally adopted by all parties. The management and documentation tools that we have developed all have these features designed into them.

Generic standards. Generic standards apply across all agencies and may be implemented through policies, the rule-making process, and legislation. For example, standard procedures for digitization define a methodology that is focused on providing a consistency in line structure across agencies.

Data-specific standards. Data-specific standards focus on one specific data set. The purpose is to define the way that data are collected in such a way that it maximizes its utility to all members of the corporation, not just the data developer. Data-specific standards require the participation of the data developer and the user community to define the data in so that all parties know and understand what the data is. In order to maximize its use, the users need to have an opportunity to define how it is collected.

Documentation Tools

Florida has now adopted the Federal Spatial Data Transfer Standard (National Institute of Standards and Technology 1992). Two elements of the SDTS were identified that we could adopt immediately: 1. the data dictionary, and 2. the quality and accuracy report. These two tools, in conjunction with the card catalog, have become the fundamental elements of our information network. The card catalog locates the information; the quality and accuracy report describes the "goodness" of the data; and the data dictionary describes the individual data elements. These formats describe the data at the most basic level and form the basis on which standards are developed.

The importance of structured documentation tools. As previously mentioned, the need for standard formats for communicating ideas is essential for coordination between organizations and governments. These standard formats include transfer protocols, operational procedures, and documentation. One of the difficulties that we have found in implementing our system is adherence to established guidelines. The documents that need to be completed are by their very nature complex, and although they have significant value to the participants within their own organizations, there is a tendency to modify the formats. This defeats one of the major objectives of documents: the creation of a standard format in which information can be compared and standards developed. The use of structured documentation tools will solve this problem by providing the user with a software package that makes the process easier and at the same time dictates the structure. We are in the process of developing such a tool for the card catalog, quality and accuracy report (Appendix 2), and the data dictionary (Appendix 3).

Data directory (card catalog of spatial data). The creation of a card catalog of spatial data is essential to the development of a multi-agency bureaucracy. We have developed a system accessible by phone modem that, if implemented within a multi-agency management structure, will provide a dynamic directory of geographic data in the state. We recognize that there are already several similar systems in existence, but we believe that they do not have all of the necessary information. Furthermore, without a multigovernmental management structure to keep such a directory updated, it will become useless in a very short period of time.

The institutionalization of such a system distinguishes it from the periodic spatial data surveys being conducted. Such a system should be able to provide the user with the necessary information to identify data sets of interest, locate the information, and provide access to a quality and accuracy report that will describe the utility of the data. This system will not be useful unless the information that is included in the directory is updated on a regular schedule. The development of a management structure to ensure the validity of this information is essential. Within each organization, a single point of contact will be assigned to keep the directory updated. Although this task may seem onerous, we have found that the need for organizations to track this information within their own institutions is just as great as the need to locate the information in different agencies. We are developing a software package that provides each agency with a tool to track their own data resources, and because all agencies will be using the same software, this information can be pyramided to a central repository.

Quality and accuracy report. The quality and accuracy report was originally conceived as a part of the Standards Development Subcommittee of the SDTS (Chrisman 1986). They recognized that requiring an organization to meet an external standard was impossible, so they designed a self-reporting format in which the provider describes a set of information about the data: "truth in labeling." We have added additional structure to these reports to overcome problems that we have found in self-reporting, such as the natural tendency to describe only the more positive aspects of the data rather than the negative. To resolve this, we took the elements that the Standards Development Committee identified and structured the report in an outline, or template, format. This produces a document that gives a more complete picture of the quality and accuracy of the data by requiring the declaration of not only what information is known but also what is unknown, and what is not applicable. This standard reporting format allows the recipients to focus on those aspects of the report that are most important to them and to evaluate the data's suitability for the recipients' use easily (see Appendix 2).

Data dictionary. The data dictionary provides two important elements: 1. documentation of the data, and 2. a basis for developing quality and control standards. The data dictionary defines each of the data elements (attributes) by describing how it is measured, its structure (if automated), and the codes used. When a database is documented, the dictionary is used as an agenda to review the data. Modifications, clarification of the document, additions, and standard levels of acceptance can be agreed upon through the review of this document. The data dictionary we have developed and tested is included in Appendix 3.

CONCLUSIONS

It is my observation that the impediments to sharing information are due to a lack of substantial multi-agency bureaucratic structure. It has been the intent of the GMDNCC to fill this bureaucratic void and provide a mechanism by which individuals with specific interests can meet, develop recommendations, and have access to executive management in order that recommended standards and policies can be adopted and implemented in a multi-agency environment. This mechanism requires a support structure, both from the standpoint of staff to provide administrative support and from an inter-agency bureaucracy that will allow these individuals to meet. Furthermore, they must also be provided the tools, techniques, and support staff to facilitate the decision-making process. Such tools include decision support techniques, structured documentation tools, consensus methodologies, and project management. Without some sort of supporting structure, the ability of inter-agency cooperation will be significantly hindered, therefore impeding the ability to share geographic information.

REFERENCES

Breto, E. 1992. A strategic plan for the Tampa Bay Regional Coordinating Council. Tampa Bay Regional Planning Council, Tampa, Florida.

Chrisman, N. 1986. Testing the Interim Proposed Standard for Digital Cartographic Data Quality. Report of the testing phase, Cycle 4, National Committee for Digital Cartographic Data Standards, Working Group II on Data Set Quality.

Holland, W. S. 1992. Wisconsin Land Information Program: modernizing Wisconsin's land records through decentralized and integrated land information systems. Wisconsin Land Information Board, Madison, Wisconsin.

NIST (National Institute of Standards and Technology). 1992. Federal Information Processing Standard Publication 173 (Spatial Data Transfer Standard). Washington, D.C.: U.S. Department of Commerce.

Onsrud, H J., and G. Rushton. 1992. National Center for Geographic Information and Analysis Research Initiative 9, Institutions Sharing Geographic Information, Scientific Report for the Specialist Meeting, San Diego, California, February 26–29 .

The Proposed Standard for Digital Cartographic Data. 1988. *The American Cartographer* 15, 1. American Congress on Surveying and Mapping.

Warnecke, L. 1992a. Building the national GI/GIS partnership. *Geo-Info Systems*, 2, 4: 16–23.

_____. 1992b. State Geographic Information Coordination Efforts and Groups, *Conference Proceedings - Maintaining Information NetWork, National Governor's Association*, January 18-23, Washington D.C., pp. 227-234.

Warnecke, L.; J. M. Johnson; K. Marshall; and R. S. Brown. 1992. *State Geographic Information activities compendium*. Lexington, KY: Council of State Governments.

Appendix 1

Issue Statement

An issue statement acts as a documentation and coordinating tool for technical advisory committees (TACs). There are two types of TACs: consensus groups that focus on data-specific items such as the USDA soils database (consensus groups); and subcommittees that focus on policy and procedures such as public access to automated information (subcommittees).

The issue statement has two primary functions. First, it acts as a documentation tool providing a corporate memory for the activity of the TAC, and secondly, as a management tool for the Council to coordinate the activities of both state and regional TACs.

The facilitator is a key factor to the success of a TAC. The membership, having full-time jobs within their own organizations, requires assistance in developing the issue statement, training in the methodology, and tools. Experience has indicated that the facilitator will be needed more in situations where there is no organization with a specified responsibility.

Template for Technical Advisory Committee Issue Statement

Date: Date that the issue statement was last modified.

Title: A brief descriptive title of the Technical Advisory Committee.

Chairman (Co-chairs):

The co-chairs, in conjunction with the facilitator, develop a strawman issue statement. This is used by the staff advisory committee to identify who in their organization (if any) should participate in the subcommittee.

Name of TAC Chair	Agency	Telephone

Activity:

A brief description of the activity or purpose of the TAC. The statement should be no longer than three sentences.

Agencies Involved:

A list of the membership of the TAC should be updated.

Name of TAC Member	Agency	Telephone

Problem Description:

Introduction

A one- to three-paragraph *general* description of the problem that is being addressed. No solution is to be included.

Problem Statement

A bulleted list of specific problems that need to be addressed. The TAC may choose to address only a subset of problems stated, but it is important that the most important problems are identified. This is vital to future activities where a different membership does not have the "corporate memory" of the original participants.

Potential Benefits:

A bulleted list of potential benefits to the agencies that may be achieved when the goal of has been met.

Ongoing Activities:

A description of other similar activities that are taking place. These descriptions should be briefly described in this subsection, and a one-page description should be included in an appendix.

Goals:

A one- to three-sentence description of what the TAC is trying to achieve. Goals are descriptions at the highest level. Although one goal is preferable, it is possible to have more than one. Goals should be achievable in a six-month to two-year time frame.

Objectives:

Objectives are a set of specific activities which, if achieved, would result in the completion of the goal. If there is more than one goal, each objective should be limited to one goal.

Actions:

Action items identify a set of conceptually and functionally related tasks that are developed to achieve objectives.

A general statement of what the action item is. There may be multiple actions per objective. Action items are numbered to match with specific objectives. The first number identifies the objective and the second number simply sequentially identifies actions. For example, Action 2.3 is the third action for objective two. If there are multiple goals, Action 2.3.2 would be the second action item for the third objective of the second goal.

Who: Each action item has an action work group. The work group has a work group leader(s) who is responsible for the activities and compiling the documents that are produced by the work group. The work group leader is the primary point of contact for the co-chairs.

What: The *what* is a set of definitive measurable tasks with specific end dates. Tasks are identified as either critical (must be completed before other tasks can begin), dependent (cannot begin until another tasks is completed) or independent (is neither critical nor dependent). Dates need to be given as to when each task is scheduled for completion

Tasks: One- to three-sentence description of the tasks. There will be multiple tasks for an action item. If there is only one, the action item should be justified as to why it is not included as a task in a more generic action.

Type. Item is identified as critical, dependent or independent.

When. Identifies when a task is to begin and when it is to be completed.

Supplementary materials. This section locates all reports and supplementary materials in the appendices.

When: Describes the expected beginning and ending dates for the action.

How Much: Estimates the costs of the action.

Appendix 2

Quality and Accuracy Report

The purpose of the template for the quality and accuracy report is to provide as thorough a documentation of the data as possible, allowing a potential user to read the report and determine the utility of the data for his/her needs. The intent is to follow the "truth-in-labeling" practices proposed by the work group that developed the Spatial Data Transfer Standards proposed by the U.S. Geological Survey. The following quotes come from "Testing the Interim Proposed Standard for Digital Cartographic Data Standards," Report of the testing phase, Cycle 4, National Committee for Digital Cartographic Data Standards Working Group II on Data Set Quality, p. 53, by N. Chrisman.

> We find "quality" to be a wide-ranging concern which can cover any issue affecting the use of cartographic data. The potential uses of digital cartographic data are so diverse that a fixed set of numerical thresholds could not adjust to the potential uses. In more circumscribed application areas (for example, a multipurpose cadastre or a forest inventory), a set of thresholds might be fruitful. Because these standards must serve the whole profession, we foresee a truth-in-labeling standard instead. The idea is to communicate actual numerical properties of the data in a way that potential users can make their own informed decisions of fitness.
>
> The truth-in-labeling concept may seem less rigorous in that it blesses the status quo. Any imprecise, inaccurate data base could meet the standard in the formal sense by proclaiming those imprecisions and inaccuracies. These standards place a substantial responsibility on the user to evaluate the quality report to ensure fitness for the particular application.

It is with this thought in mind that the templates have been created. This report format is better suited for user evaluation of the data because it formalizes the structure but still allows the basis to be textual in content.

A quality and accuracy report has also been developed for Raster Data. The development of a supplementary document, users manual, and a structured documentation tool (automation of the templates) is currently under development.

Template for Quality and Accuracy Report—Vector Data

A. Lineage

1. Description of source material(s)
   a. Name
   b. Scale (specify ratio)
   c. Datum
   d. Map projection checklist
   e. Media of source
   f. Condition of media
   g. Creator organization/individual
      Name — — — — — — — — — — — — — — — — — —
      Address — — — — — — — — — — — — — — — — — —
      Phone (— —) — — — — — — —
   h. Date of source material
      1. Time interval covered
      2. Update schedule
2. Derivation methods for data
   a. Methods of derivation
      1. preautomation compilation
      2. equipment
         a. model
         b. resolution
         c. tolerance of digitizer (if digitized)
         d. software
   b. Date of automation
      1. Initial date
      2. Update schedule
   c. Explanation of procedures used to digitize/scan/transform, etc., the data
      1. Description/decision rules
      2. Control points
      3. Name of transformation methodology
         a. software system and version used
         b. description of algorithm
         c. mathematics
         d. set of sample computations
   d. Software system and version used

B. Positional Accuracy

1. Line work completeness
   a. Date
   b. Value
   c. Method
2. Line work positional accuracy
   a. Date
   b. Value
   c. Method
3. Test reports (use one or more of the following: Deductive estimate, Internal evidence (geodesy), Comparison to source (overlay), Independent source of higher accuracy)
   a. Report
      1. type

        2. date of tests
        3. report

C. Attribute Accuracy

1. Continuous (define what document can be referenced here)
    a. Absolute measure of error reference in units of the measurement system
        1. Date
        2. Value
        3. Method of derivation of that number
    b. Test reports (use one or more of the following: Deductive estimate, Internal evidence (geodesy), Comparison to source (overlay), Independent source of higher accuracy
        1. Report
            a. type
            b. date of tests
            c. report
2. Accuracy of source materials
3. Categorical
    a. Test reports (if appropriate use one of the following)
        1. Deductive estimate
            a. date
                1. date of tests
                2. date of materials used
                3. rates of change
            b. report
                1. basis for the deduction
                2. deduction
        2. Independent samples
            a. date
                1. date of tests
                2. date of materials used
                3. rates of change
            b. map scale
            c. dates of materials used
            d. report (misclassification matrix)
                1. sampling procedures
                2. sampling locations
                3. results
        3. Tests based on polygon overlay
            a. date
                1. date of tests
                2. date of materials used
                3. rate of change
            b. map scale
            c. date of materials used
            d. report
                1. misclassification matrix
                2. explain relationship between maps

D. Logical Consistency

1. Tests of valid values
    a. tests performed
    b. date

c. results
2. Cartographic tests
   Does the provider feel the map is cartographically clean? ___ Yes ___ No
   a. Answer the following
      1. Do lines intersect only where intended?
      2. Were duplicate lines eliminated?
      3. Are all areas completely described?
      4. Have overshoots and undershoots been eliminated?
      5. Have all slivers been eliminated?
      6. Does documentation in the lineage section (d-5) describe why the data is cartographically clean?
   b. Tests
      1. date
      2. tests
      3. reference to documentation
      4. results
3. Topological tests
   a. test performed
   b. date
   c. software
      1. name
      2. version
   d. results
      1. Test for polygon coverage
         a. How many polygons are represented on the digital map product?
         b. Has a polygon closure been verified?
         c. Are Polygon-Ids assigned to each polygon on the digital map?
            1. Do polygons have more than one Polygon-Id?
            2. Are the Polygon-Ids unique?
      2. Test for line coverage
         a. How many lines are represented on the digital map product?
         b. Do the line segments have unique line segment values?
         c. Is the digital map topologically clean?
      3. Test for point coverage
         a. How many points are represented on the digital map product?
         b. Are the Point-Ids unique?

E. Completeness of Source Materials
1. Selection criteria
2. Definitions used
3. Other relevant mapping rules
4. Deviation from standard definitions and interpretations
5. Description of the relationship between the objects
6. Tests for taxonomic completeness
   a. procedures
   b. results

Appendix 3

Data Dictionary

The proposed data dictionary is a data documentation tool, but it also fulfills the requirements of the Spatial Data Transfer Standards and is essential for the development of data standards. It is not designed to be a systems data dictionary (aliases are not included), but the information should be fundamental to all data dictionaries.

Template for Data Dictionary

A. Entity Template
   1. Label
   2. Entity Authority
   3. Definition
   4. Point/Line/Polygon
   5. Quantity of Data
B. Attribute Template
   1. Label
   2. Attribute Authority
   3. Definition
      a. Description
      b. Measurement/Determination
   4. Domain value
      a. Domain Format
         1. Character Type
         2. Allowable Values (domain enumeration)
            a. Length
            b. Number of Significant Digits
            c. Units of Measure
      b. Categorical
         1. Value
         2. Meaning
      c. Continuous
         1. Range of Values
            a. Minimum
               1. Value
               2. Inclusive/Exclusive
            b. Maximum
               1. Value
               2. Inclusive/Exclusive
         2. Typical Value
   5. Other Editing Information

JOHN EVANS AND JOSEPH FERREIRA, JR.

27 *Sharing Spatial Information in an Imperfect World: Interactions Between Technical and Organizational Issues*

In today's unsettled, rapidly changing technological and organizational context, spatial data sharing is neither a purely technical problem nor a purely behavioral one. Despite innovations on both the technical and organizational fronts, both of these areas will remain "messy" (imperfect and dynamic) for some time, and we must learn to address such a context. In addition, too sharp a dichotomy between "technical" and "organizational" problems risks missing many aspects of the problem that are unique to spatial data. Rather than isolate technical from organizational issues, research efforts should focus on their overlap. We encourage (1) technical innovations intended for multiple actors with imperfect coordination; (2) organizational research tightly linked to current and upcoming technology; and (3) an emphasis on the pieces of the problem that are unique to spatial data. We offer three examples of such research and propose a cooperative research agenda based on experiments in spatial data sharing across sectors and levels of government.

THE OVERLAP OF ORGANIZATIONAL AND TECHNICAL RESEARCH

Incentives and impediments to spatial data sharing among organizations are often categorized as either "technical" or "organizational." For instance, an agency might not make use of an outside data source for "technical" reasons: it may lack suitable data-conversion software, compatible hardware, or a means to search through a sea of unknown data. At the same time, spatial information sharing is often obstructed by "organizational" issues: "turf battles," the need for reorganization, or institutional inertia. In response, some researchers have focused on technical topics such as data structures, formats, query languages, or client-server architectures; others, meanwhile, have concentrated on organizational issues: bottom-up or top-down development strategies, the role of coordinating bodies, or cost-recovery schemes.

In today's unsettled, rapidly changing technological and organizational context, spatial data sharing is neither a purely technical problem nor a purely behavioral one. For example, what users often describe as "turf battles" or "bureaucratic inertia" may sometimes be the outcome of data producers and users balancing (implicitly or explicitly) the uncertain, intangible benefits of shared spatial data against the very tangible time, effort, and expertise needed to make data sharing work effectively. The reluctance of some to use outside spatial data, and the refusal of others to give out spatial data, can be described not only as power struggles to be solved apart from technical innovations, but also as behavioral responses to difficult data-coordination problems that are intimately tied to the ever-changing capabilities of current technology and to the data's complex, multidimensional nature.

ADAPTING TECHNOLOGY TO "MESSY" ORGANIZATIONS

The Geographic Information Systems (GIS) community includes organizations with a variety of purposes, a wide range of information requirements (security, timeliness, accuracy), and unevenly distributed resources (time, expertise, and funds). Coordination among these organizations, and purposeful change within them, is often difficult. These factors have implications for research on spatial data sharing in such fields as geography, computer science, or database theory. It may be years before a majority of these organizations have the structure, priorities, and expertise necessary to adopt and take full advantage of such data-sharing innovations as spatial data standards, improved data structures and query languages, or "open" hardware and software. Research on technical aspects of data sharing ought to address the "messy" transition period when not all the relevant organizations are fully equipped or conforming to new standards and theories for spatial data sharing.

A further aspect of many organizations is unforeseen change. Since personnel, funds, and organizational relationships may come and go, it is crucial to choose information strategies flexible enough to withstand these changes. Technical strategies adopted early on may strongly constrain both organizational and technological futures. For this reason, loosely coupled, cooperating modules of software and hardware are often preferable to fully integrated systems. Research on such modules, and on modes of spatial data interchange among them, holds promise for dynamic organizational environments.

ADAPTING ORGANIZATIONS TO "MESSY" TECHNOLOGY

Technology for data interchange is far from mature. It, too, is in a "messy" (imperfect and dynamic) transition period. Just as technical choices should allow for organizational change, organizational strategies for data sharing must accommodate new modes of data retrieval and interchange, particularly as networked data comes of age. "Traditional" network services such as remote log-in, electronic mail, and file transfer (downloading) are becoming commonplace, and the resulting pressure grows daily to "turn on" electronic paths to data. But networked file systems (NFS), client-server architectures, and network authentication are relatively new technologies that are just beginning to gain widespread acceptance; and federated databases, high-speed networks, "object-oriented" operating systems, and various "intelligent agent" technologies are now in the prototyping stages. The newer technologies will change not only the cost and benefits of networked data sharing but will also induce structural changes in the system architecture, software "layering," and organizational arrangements that are involved.

For example, "dial up" connections to bulletin boards allow anyone with a personal computer and access to a phone to download data from a central repository. Consider the so-called TIGER street centerline files developed by the U.S. Bureau of the Census and the U.S. Geological Survey for the 1990 census. If there were a central bulletin board, a "dial-up" user could download such road network files (and associated data) for any portion of the country. But the TIGER data are encoded in a dozen separate file types for each of the country's 3,000+ counties, and each version of them occupies more than 20 gigabytes of storage space. The "dial-up" user would have to browse through menus, data dictionaries, and detailed instructions in order to determine which parts are of interest and how to download them. At typical phone network speeds (9600 baud), downloading a small county's data could take all night. Hence, it has been more practical for the government to distribute raw TIGER files via magnetic tape and CD-ROMs and then to expect a lively third-party industry in enhancing, segmenting, customizing, and repackaging these data for myriad other stand-alone uses.

Higher speed networks now make the downloading practical, and the use of NFS over high-speed networks enables one to make use of a centrally maintained "official" data set whenever it is needed without having to download it first. At MIT, for example, we make extensive use of Internet connections (with a bandwidth of a least 10 megabits per second) between Cambridge, Massachusetts, and Washington, D.C., in order to utilize data sets at the National Capital Planning Commission or the U.S. Geological Survey—without having to transfer the files to MIT as a separate initial step. Moreover, improved software engineering and client/server architectures over high-speed nets allow the nature of the data "sharing" to change. Instead of simply accessing raw data sets from afar, one can imagine storing them within a relational database manager and using structured query language (SQL) queries to extract relevant portions; or running applications on the remote machine that analyze, model, filter, and otherwise transform the raw data so that only the digested results are sent back. To be sure, mainframes have been providing variations on such services for years—but their "timesharing" model has been oriented to "dumb" terminals at the receiving end and to slow communication links that are practical for the exchange of only small amounts of text-oriented information.

Networked graphics workstations and other emerging technologies let us redefine the stage at which the information exchange takes place—and the spatial data infrastructure that can support it. Suppose we wanted to "explore" the downtown of a far-off city before we visited it. We could imagine doing GIS-like things on our general purpose desktop computer without collecting lots of information in advance, without being a GIS technology wizard, and without buying into an expensive, customized package that did nothing but travelogue multimedia shows. A local geographic browsing tool might initiate a dialog with remote "servers" that stored various images, geographic data, and processing tools. The local browser might construct maps, dialog boxes, and various "direct manipulation" interfaces to help us interpret the available data, zoom in on the right place, select appropriate views, construct relevant information, book our hotel room or dinner reservation, and examine mass transit routes and schedules. As the richness of the example increases, it becomes increasingly apparent that it can be practical (as an inexpensive and widely accessible resource) only if the underlying data are distributed and many of the mapping and data integration elements are constructed "on the fly." The accounting, licensing, and interorganizational arrangements needed to support such a spatial data infrastructure are far different from today's electronic data center, and they will have to evolve along with the technology if we are ever to get there.

Client-server architectures promise to alleviate much of the processing overhead associated with such information exchanges, while network authentication provides tight control over who has access to what data and applications. These and other emerging technologies will continue to mature and develop for some time, and they can radically alter the choice of strategies for data sharing and development, as well as institutional modes of work. Thus, even the most non-technical analysis and the most preliminary implementation plan must include flexibility to accommodate significant technological changes.

GIS technology itself will be "messy" for some time. First, popular packages are just beginning to incorporate general-purpose graphical user interfaces, and they lack many desirable techniques for spatial data analysis, manipulation, modeling, and data interchange. Second, the field is continually changing, so that what may be nearly impossible to accomplish today may become straightforward with a new release of the software. For instance, in recent years, a broad set of data translators has become a de facto requirement for any GIS. Already, direct software intercommunication ("live links") between GIS and off-the-shelf relational database management systems is becoming increasingly commonplace. This trend will let GIS software developers focus their efforts on spatial data processing and spatial analysis tools (rather than on general-purpose database management tools). It will also let an organization choose the database management systems best suited to its needs, and it will allow any part of an organization's data resources to be related more readily to geographic features in a GIS. As a result, organizations may find that live links between spatial and "traditional" databases lead to important ties between formally unconnected departments. Thus, any organizational plan for spatial data sharing must account for the "messiness" of GIS technology as well.

In an environment of continually shifting technical options for data interchange, the debate should not be over "top-down" versus "bottom-up" organizational implementation strategies but over how to implement and manage loosely coupled, overlapping, cooperating groups of people working on linked sets of information, given the capabilities of current and emerging technology for data interchange and GIS. For instance, specifying the nature and formats of data interchange among departments and agencies may be enough to ensure a reliable, high-quality flow of data among them. In some cases, a general mechanism—rather than a specific protocol—for feedback and updating may be more appropriate. The choice of participants in data sharing, and the procedures they follow, may often be key. For instance, the reliability, accessibility, and feasibility of data sharing can depend on whether data are shared among multiple departments that depend upon the data to feed a divergent set of multi-user applications, or whether the data are only informally shared among one department's technically able staff. Equally important is whether the interchange is highly structured, via formal requests, or ad hoc, on a "help yourself" basis, and whether the data are static histories of past events or dynamically changing records of ongoing processes.

SPATIAL DATA AND ORGANIZATIONAL BEHAVIOR

Research on organizations and institutional behavior contributes much to our understanding of how to implement automation, manage technology, and introduce innovations. An extensive literature exists, including numerous studies of organizational and information management issues related to computing in government (see, for example, Kraemer et al. 1989). However, most of this literature focuses on text-oriented data processing and automation issues. While it is relevant to many aspects of spatial data sharing, it tends not to focus on the specific spatial nature of the information being shared and the resulting influence on organizational incentives and impediments to data sharing.

First, spatial information presents a number of incentives to sharing that would not occur for simpler, alphanumeric data. For instance, an overlay relates information from multiple sources based on feature location, geometry, and topology. Compared to traditional table lookups on Social Security codes or license plate numbers, this is a far richer set of relationships among data. Through GIS overlays, for example, the utility of one piece of spatial data can increase dramatically when juxtaposed spatially with other pieces. Indeed, some spatial data sets may be useful only when combined spatially with others. Furthermore, interacting with a graphical representation of spatially combined data may add significant value to the shared data. By identifying spatial relations among data originating from multiple sources, GIS can result in a degree of "synergism" seldom found with traditional non-spatial data. A behavioral model devised in a world of text and numeric databases might not account for such incentives to spatial data sharing.

The spatial nature of GIS data may provide incentives to sharing; its complexity may also lead to several impediments. First, the GIS operations required, for example, to do a spatial overlay are more complex and (at present) less automatable than table lookups. Second, spatial data is especially prone to misunderstanding or misuse by outside users. Third, maintaining consistent, up-to-date geometry, topology, and attributes on spatial data from multiple sources is difficult since each layer can change independently of others. Many applications built upon shared data must reconstruct spatial features, segmentation, and spatial relationships each time changes occur in the underlying data. Fourth, most current parcel, road, and topographic base maps that are candidates for data sharing are sufficiently poor that any data development plan must include provisions for their enhancement or replacement. Often, agencies either accept the gradual deterioration of their spatial data due to inadequate update mechanisms, or they wait too long to begin an automation effort in the false hope that spatial data formats and processing tools will soon stabilize. For example, many towns and municipalities, without any GIS or automated mapping program, distribute hard copies of their "official" mylar base maps to the various municipal departments—engineering, assessing, planning, police, and so on. It is common practice for these departments to update information on these copies by hand without any reliable system for ensuring that all such edits and changes find their way back to the mylar masters.

RESEARCH NEEDS FOR A "MESSY" WORLD

The GIS community is just getting to a point where more than one agency has spatial data for a given area, and there exist alternatives to building what one needs (and protecting what one has). We are entering an "information rich" era in which high volumes of data will flow through ubiquitous communication networks, with interactive information filters (Malone and Lai 1992). Some have used an "information marketplace" metaphor to describe the pivotal role to be played by information flows of all kinds within a few years (Dertouzos 1991). The impact of these changes in the basic information infrastructure is being felt already, but it will be several years before (1) industry-wide rules and standards are established regarding data access and formats; (2) GIS technology matures so that interactive decision support tools take the place of today's data-centered, production oriented GIS tools (Clarke 1990); and (3) institutions adjust to radically different modes of work (Zuboff 1991). We propose that spatial data sharing is in essence a problem of inventing and building a "spatial information infrastructure" of people, organizations, and technology on top of today's GIS tools, data, and operations.

We suggest that in building this "spatial information infrastructure," key research areas will involve problems where (1) both technical and human factors are central, (2) technical approaches can greatly alleviate the problems behind behavioral symptoms, and thus (3) organizational changes, in concert with current and emerging technology, can have the greatest impact. Such research would seek not only to define data standards or interfaces or merely to categorize organizational behavior; rather, it would seek joint technical and organizational strategies for both data producers and data users to cope and grow effectively in a dynamic environment. Such research would also seek ways to use newly emerging technologies specifically to address difficulties of maintaining and sharing spatial data from many sources. The following section presents three examples of such research topics.

Sharing Spatial Data in a "Messy" World: Three Sample Research Topics

To illustrate research needs in spatial data sharing, this section describes three examples of research topics where both technical and organizational factors play pivotal roles: GIS in transit operations and management, "pipelines" for spatial data management, and federated spatial databases.

GIS in transit operations and management. Transit applications use a rich assortment of spatial data including road networks, transit routes and stops, ridechecks, pointchecks, passenger surveys, and demographic (census) data. Even when these data are used by many applications in a transit agency, our recent study (Azar and Ferreira 1991) indicates that (1) different departments often diverge in their system development strategies and their representations of networks and routes, and (2) each department tends to implement its GIS technology with the implicit assumption that their base maps and road networks can easily match those needed by others (or that others will change). This incoherence limits a transit agency's capacity to make effective use of GIS technology and network analysis tools beyond simple automation. Alternative development paths do exist that could enable more sharing of data and tools; however, they require far more coordination of development efforts and surprisingly complex strategies for database design and management (Ferreira and Azar 1992). For instance, building sustainable digital representations of transit routes and stops involves many subtle database design and data-structuring issues if one wants the routes and stops to maintain cross-reference links to road segments and demographic information in a regionally shared road network file. One must pay attention to segmentation and/or conflation issues and "one-to-many" problems that go far beyond digitizing a map of routes and attaching route labels.

Such complications raise questions about the practicality of "top-down" approaches for tying all mapping efforts to a metropolitan-wide "multipurpose cadastre" that goes down to the parcel level and has tightly controlled data elements and accuracy standards. By the time the base map is finished, many local and regional agencies will have enormous amounts of relevant data buried in a wide assortment of mostly incompatible applications. Instead, a regional government authority or state committee might accept the notion of a few carefully chosen base maps evolving at different scales, and over different time frames, for overlapping purposes. For instance, transportation-related agencies might maintain a road network file, derived from Digital Line Graph (DLG) or TIGER files, whereas each city or county might develop and maintain a parcel map that would, for some time to come, match neither the TIGER/DLG file (nor reality) with the precision that is ultimately desired. A "coordination committee" might define the data-building responsibilities of each department or agency, the "supported" tools for data interchange, updating, and so on, and a minimal set of data relationships among them. It would also identify and educate its members about the benefits and costs of increased data coordination and would help them design (or redesign) applications (and associated maintenance strategies) that could readily adapt to the new and improved base maps that will come later.

"Pipelines" for spatial data management. Incorporating outside information into a set of interconnected data items is a complex task. At present, this requires a highly skilled "data tamer" to convert and manipulate data, verify its geometric and attribute quality, maintain its consistency with related data, and update it over time. Maintaining consistency over time is particularly difficult when dynamic data are used to produce derivative works, since changes to the original, "upstream" data must be propagated across many steps of data processing and transformation. Many of the steps involved are difficult to automate and require judgment and complex user interaction. We have found that these judgments and data-conversion steps are analogous to a pipeline with a unidirectional "flow" of data layers from "raw" to "done" to "demo-ready." For instance, along the pipeline, data layers have increasing intelligence built in and support detailed queries more and more effectively. Given such a pipeline, incorporating a new data set consists of determining where it fits along the pipeline and how to merge it with data at comparable points along the pipeline. The following diagram illustrates this pipeline concept. Note that the flow of information is from left to right and that replacement of an item with a later/updated version requires attention to "downstream" effects (Figure 27.1). Spatial data-processing technologies—and organizations—tend to overlook the complexity of the information management needed to maintain the whole system, especially if it involves distributed or heterogeneous elements.

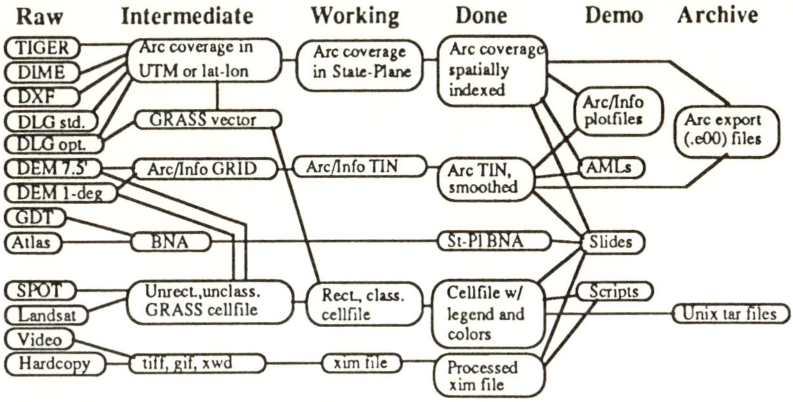

FIGURE 27.1
The "pipeline" metaphor for spatial data management

We have begun to experiment with such data pipelines on top of spatial data files (Ferreira 1992) by means of (1) a "data-database," describing each data layer's format, accuracy, topology, attributes, and so on; (2) a suite of interlinked software modules that uses the data-database to decide which conversion steps are needed, which ones are straightforward, and which ones will require user interaction; and (3) an interactive, graphical user interface, so as to avoid a lengthy learning period while ensuring a properly interactive, educational process. Indeed, such a system must not "shield" the user from data-processing details by "transparently" embedding expert judgments. Rather, it must illuminate the handling of interrelated, dynamic, heterogeneous spatial data, interactively guide a user through a series of steps, and pinpoint data dependencies, sources or degrees of error, and the like. This is analogous to the Khoros/Cantata graphical interface to image-processing software (Rasure and Williams 1991), in which the user constructs and manages interconnected "filters" that modify data flowing along a "pipe." Kirby and Pazner (1990) also describe ways in which such types of pipeline metaphors might enhance user-machine interaction in geographic data analysis. The pipeline metaphor for data management may be particularly appropriate in environments where both the technology and the organization consist of loosely coupled, cooperating units.

Federated spatial databases. This research area has arisen and matured in the world of management information systems (MIS) to cope with information resources spanning more than one organization. The federated, or composite, information systems approach is summarized in Sheth and

Larson (1990) and is applied to both technical and institutional aspects of spatial data sharing in Webster (1988) and Nyerges (1989a). The federated approach involves heterogeneity among data resources, autonomy in data maintenance, non-intrusive data access, and user-transparent query capability through global data representations ("schemas"). These concerns address many of the behavioral symptoms encountered in spatial data sharing since they allow data distribution and combination to occur with minimal changes in the way each organization represents, structures, and stores its own data. In addition, differences in data definitions (semantics) among organizations are the focus of efforts in "semantic reconciliation," summarized by Siegel and Madnick (1991, 1992) and applied to spatial data by Nyerges (1989b) and Worboys and Deen (1991). Based on this research, we are building a browser interface to heterogeneous spatial data that may be stored locally or distributed over wide-area networks. Based on the data-database described previously, this browser highlights the different data formats on menus and automatically accounts for differences in, for example, coordinate projections, so that all of the data can be viewed simultaneously (Evans et al. 1992). Such a browser could remove many of the barriers among organizational data resources and would enable "composite geographic information systems" in which an agency would be responsible for maintaining its own spatial data but could transparently access and cross-reference that of many others. This research, in turn, may lead to the design and testing of intelligent software agents that would query for spatial information across a heterogeneous wide-area network and perform routine tasks, or make "suggestions," based on their findings.

HOLISTIC APPROACHES TO SPATIAL DATA SHARING: A PRELIMINARY RESEARCH AGENDA

The three examples above illustrate research topics relevant to spatial data sharing that are technical in nature yet closely tied to organizational issues and institutional context, and outside the usual agenda of technical issues. A number of specific research projects could build upon these ideas. Further work is needed to develop research questions and research design in this area; however, the following are a few possible avenues to pursue.

- Study and interpret the incentives and impediments to spatial data-sharing efforts, with an emphasis on (1) the mix and overlap between technical and organizational issues, (2) the relation between symptoms and underlying issues, and (3) the aspects of the problem that are specific to spatial data.

- Construct data-sharing scenarios that use emerging technologies in contexts and in ways that are practical during the next few years. For instance, a university, a state agency, and a software vendor might jointly experiment with networked file systems (NFS) or authenticated networks to distribute geographic information to local governments. Learning how to map organizational types and skills to tasks and technology would be a key goal of this research. Consortia among universities, government agencies, and vendors of software or hardware could play an important role in research of this kind if we can construct the right mix of substantive problems, funding, technical and institutional skills, and professional motivation.
- Determine the additional requirements of building spatial data pipelines and federated spatial databases so as to identify and address the unique issues associated with geographic data (as opposed to numeric, textual, or unstructured image data). In this effort, one key challenge will be to define spatial data "semantics" and a set of rules for comparing and reconciling semantics between separate data sets. The resulting composite information systems will depend for their proper use on an interactive, decision-oriented, graphical user interface; this will require many rounds of prototyping and evaluation, in carefully chosen settings and with informative peer review.
- Conduct experiments in data sharing, both across hierarchical levels (city, state, country) and across sectors (tax assessment, environmental protection, transportation planning) that emphasize alternative pathways for information infrastructure development. This research would seek to (1) learn how to build loosely coupled, cooperating organizations linked by information interchange, and (2) evaluate and measure what "works" in terms of fostering sustainable, productive pathways for organizational and technological learning.

REFERENCES

Azar, Kamal, and Joseph Ferreira. 1991. Standardizing transit applications of GIS technology: a corridor study example. URISA 1991 *Proceedings*, vol. 1, 110-121.

Clarke, Martin. 1990. Geographical Information Systems and model-based analysis: towards effective decision support systems. In H. J. Scholten and J. C. H. Stillwell (eds.), *Geographical Information Systems for urban and regional planning*. Netherlands: Kluwer Academic Publishers, 165-175.

Dertouzos, Michael L. 1991. Building the information marketplace. *Technology Review* (January), 29-40.

Evans, John, Joseph Ferreira, and Philip Thompson. 1992. A visual interface to heterogeneous spatial databases based on spatial metadata. Fifth International Conference on Spatial Data Handling, *Proceedings*. August.

Ferreira, Joseph. 1992. *Computer-based land use planning projects for the National Capital Planning Commission*. Work in progress by the Computer Resource Laboratory, MIT School of Architecture and Planning, National Capital Planning Commission Contract NCPC 91-02.

Ferreira, Joseph, and Kamal Azar. 1992. Building and managing sharable databases for transit agency GIS applications. Seventy-first Conference of the Transportation Research Board, Session 194, *Proceedings*. January.

Kirby, K. Chris, and Micha Pazner. 1990. Graphic map algebra. Fourth International Symposium on Spatial Handling, Zurich, Switzerland, *Proceedings*.

Kraemer, K. L., J. L. King, D. E. Dunkle, and J. P. Lane. 1989. *Managing information systems: change and control in organizational computing*. San Francisco: Jossey-Bass Publishers.

Malone, Thomas W., and Kum-Yew Lai. 1992. Toward intelligent tools for information sharing and collaboration. In Robert P. Bostrom et al. (eds.), *Computer augmented teamwork: a guided tour*. New York: Van Nostrand Reinhold.

Nyerges, Timothy L. 1989a. Information integration for multipurpose land information systems. *URISA Journal* 1, 1: 27-38.

_____. 1989b. Schema integration analysis for the development of GIS databases. *International Journal of Geographical Information Systems* 3, 2: 153–184.

Sheth, A. P., and J. A. Larson. 1990. Federated database systems for managing distributed, heterogeneous, and autonomous databases. *ACM Computing Surveys* 22, 3.

Siegel, M., and S. Madnick. 1991. A metadata approach to semantic reconciliation. Working Paper #3252-91-MSA, Sloan School of Management, Massachusetts Institute of Technology.

_____. 1992. Context integration: sharing the meaning of data. *ACM SIGMOD Record* 20, 2.

Webster, C. 1988. Disaggregated GIS architecture: lessons from recent developments in multi-site database management systems. *International Journal of Geographical Information Systems* 2, 1: 67–79.

Worboys, M. F., and S. M. Deen. 1991. Semantic heterogeneity in distributed geographic databases. *ACM SIGMOD Record* 20, 4.

Zuboff, S. 1991. Informate the enterprise: an agenda for the twenty-first century. *National Forum* (Summer).

GERARD RUSHTON AND STEVEN FRANK

28 *Sharing Spatial Data Among Social Scientists*

Social scientists are now able to access social and economic data at finer levels of spatial resolution than before and to flexibly aggregate such fine resolution, or small-area, data. Using Topologically Integrated Geographic and Encoding Reference (TIGER) Line files – the geographic base file developed by the U.S. Census – many applied social scientists are address-matching data from administrative record systems and providing spatial decision support for many human services management problems. Spatial data sharing is further facilitated with the development of state data centers by the U.S. Census Bureau and with the activities of the Inter-University Consortium for Political and Social Research.

SIGNIFICANCE OF IMPROVED SPATIAL DATA TO THE SOCIAL SCIENCES

There is virtually no literature that explicitly addresses the relationship between spatial data, Geographic Information Systems (GIS), and the advancement of the social sciences although, frequently, researchers conclude their research studies expressing the limitations of their findings caused by poor spatial resolution of their data. Some social scientists recognize that important scientific questions could be better addressed if the availability and characteristics of digital spatial data were improved, if new methods of analysis were developed to process such data, and if the typical geographical scale for which theory is constructed and tested were to become a question to be addressed rather than an exogenously derived constraint on empirical work.

TRADITIONAL, CENSUS-BASED SPATIAL DATA SHARING

Until recently, when social scientists talked of sharing spatial data, they were generally referring to social and economic data spatially aggregated for different geographic entities. In most countries, the geographic units commonly used were those officially recognized in their national censuses. Spatial data was seen as a set of tables with geographic units as rows and their attributes as columns. The development of the microfiche in the 1960s permitted the widespread dissemination of large volumes of such data. Search and retrieval depended on nominal geographic identifiers and, though suitable for manual search for specific information on a few geographic areas, this system was inadequate for most computer-based analyses. For this purpose in the United States, the State Data Center program of the U.S. Census distributed computer tapes and, more recently, diskettes, CD-ROMs and, in some cases, on-line data (U.S. Bureau of the Census 1992b).

Almost all national censuses make social and economic information available at different, hierarchically organized, geographic units. For urban areas in the United States this is the block, block group (BG), Census tract, or block numbering area (BNA), county, metropolitan area (MA), and urbanized area (UA). For rural areas in the United States, the geographic units are the township (Twp), place, and county (Bureau of the Census 1992a). In the United States, county-level data are also widely distributed. Some widely used collections of data include *County Business Patterns* and a county-level file of health data. The U.S. Census Bureau also has established an on-line database (CENDATA).

SMALL-AREA SPATIAL DATA

Social scientists in the past decade increasingly have asked for "small-area" data. While no accepted definition of "small area" exists, the phrase is used in contexts where the scientist is arguing that making spatial data available for smaller geographic units (finer geographic resolution) than currently available will permit new questions to be addressed or will give better answers to old questions. Many social scientists now believe that greater spatial resolution of data advances the social sciences and will be helpful in the planning and administration of social service systems. Traditionally, the federal government was not persuaded that small-area data was important. Garrison and Hummon (1973) quoted from a 1971 report of the Presidential Commission on Federal Statistics (Federal Statistics 1971, vol. I: 126-7):

> Many people in state and local governments have questions that cannot be answered with existing data. As with the federal policy makers, however, their questions are often proved to be inchoate, they generally lack analytic models, and there is substantial evidence that the problems in question have not been thought through. A role for the federal government in providing statistics can hardly be defined when technical issues of what to measure have not been resolved.

Since then, as described below, the TIGER data system has become the tool for generating small-area spatial data, although it was developed as a tool for implementing the 1990 Census and other censuses of the Census Bureau rather than as a tool for generating small-area statistics.

FLEXIBLE GEOGRAPHIC AGGREGATIONS FROM SMALL-AREA SPATIAL DATA

A common problem traditionally faced by social scientists is that the areas for which they want data are frequently different from the geographic units for which the data is available. Compounding this problem is the well-known finding that many conclusions of social science research are restricted to the geographic scale of the spatial data used to confirm them. Known as "the modifiable areal unit problem" (Openshaw 1983), in its various forms "it plagues GIS analysis" (Heywood 1990, 850). Heywood notes that "in the United Kingdom, there are at least sixty different spatial units for which social data are collected." Until recently, researchers could not examine data for geographic regions that cut across the smallest Census geographic units. For example, until the most recent (1990) Census in the United States, it was not possible to find social and economic information for school districts because in most parts of the country school district boundaries do not conform with any Census geography. The boundary of a school district, for example, may bisect townships in rural areas or cut through Census tracts and block groups in urban areas. The Census was not able to tabulate data for such geographic units because, with few exceptions, its tabulation process involved aggregating data from tabulations available for smaller areas (U.S. Bureau of the Census 1992a, 1992b).

The problem of aggregating spatial data for regions not predefined by the Census Bureau has been solved quite recently with the development of flexible geographic aggregation methods made possible by the Census Bureau's TIGER system. This system is the automated geographic database used by the Census Bureau for producing 1990 Census maps. The TIGER Line files contain, for each segment of road, information on all geographic area codes for the right and left sides of the segment as well as latitude and longitude coordinates of the endpoints and shape points of the segment. Associated with each segment in urban areas are the address ranges and ZIP codes. After attaching voting district boundaries to other Census-defined boundaries, the U.S. Census Bureau prepared a Redistricting Data File as required by Public Law 94-171. This ability to code new geographic regions in the TIGER Line file that are independent of official Census-delimited geographic regions, combined with more intelligent search and retrieval systems embedded in software attached to the data files, increases the usefulness of census information and permits social scientists to investigate questions about the new geographic units so formed. These regions are functional regions with respect to a particular activity. In theory, if not yet in practice, it is possible for the Census to aggregate its data for any regions that are defined in the TIGER Line files. It is possible in the future that users of the Census will define boundaries that interest them in the TIGER file and that the Census will make custom aggregations of their data to these new regional entities.

The issue of privacy of individual records, the confidentiality of which the Census guarantees, requires that custom aggregations be performed within the Census itself. It would be possible, for example, to define in terms of a collection of streets, the area of southwest Los Angeles affected by the disturbances of 1992 and to relate the social economic data of that area to the 1993 building surveys of rebuilt, repaired, and demolished properties (*New York Times* 1993). On a more national scale, if the country were to be divided into new managed health care regions—as health care policy makers in the first year of the Clinton administration frequently hinted was likely to become official policy—it would be possible to code these boundaries in the TIGER Line file and thus generate for these regions information from the Census on any of the social and economic variables desired. This geographic encoding technique permits the retroactive generation of aggregated Census data for new regions not even defined at the time of the Census itself. Typical of new software products linked to such new Census databases is the "School District Data Book" (SDDB) developed for the National Center for Education Statistics of the U.S. Department of Education by The MESA Group (at the time of this writing scheduled for release on CD-ROM in 1993). The SDDB software has basic features of data browsing, data look-up/query, printing, and file output as well as analytical features of spreadsheet operations and thematic mapping. It also contains data dictionary access and database manipulative command structure (NCES 1992). Other TIGER boundary files are described in FGDC 1992c, section 2, p.25.

A second method for making custom spatial aggregations of census data involves forming regions as aggregations of block data, the block being the smallest geographic unit for which the Census Bureau tabulates data. It is generally bounded by streets, political boundaries, and other visible features. Several GIS products display block-level Census data and permit new databases to be developed with custom aggregations of blocks. For many social science applications, this level of geographic specificity is adequate for many questions. A survey of articles in social science journals in which Census data was used, however, turned up no example in which a researcher had aggregated block-level data to geographic areas of interest to them, although many noted the lack of congruence between the level of Census geography used in their analyses and the geographic scale of the phenomena they were investigating (see, for example, Hughes 1989).

Many social scientists would agree that the customization of regions would improve the results of their analyses. Most of them are still simply resigned, however, to a belief that this will never happen. Researchers accept arbitrary regions as if they were logical spatial units for analysis. Lipietz's comments (1980, 8) are typical of this resignation:

> We have relied on analysis of the census by region, by branch of economic activity, by skill, and by socio-professional category. I will not enlarge here on the difficulties resulting from the crudeness of the statistical divisions. We have confined the study to a sample of eleven regions which we regarded, a priori, as representative of the different regional types. For these regions we have worked out. . . .

Thus, many social science researchers throughout the world still face spatial data aggregated in large regions defined by criteria unrelated to the purpose of their research. In other cases, they are either not familiar with methods of aggregating block-level Census data to make custom-defined regions or do not have access to GIS software to perform these tasks. For users unable or unwilling to aggregate block-level data, the U.S. Census has a User Defined Areas Program (UDAP) in which the user may define any groups of Census blocks (at the user's discretion) to receive a standard set of data sets from the 1990 *Census of Population and Housing* as an additional cost option (FGDC 1992c, Sec. 2, p.27).

THE APPLIED SOCIAL SCIENTIST

Improvements in spatial data benefit the social scientist who is engaged in practical problem-solving tasks. For example, there is widespread recognition that improved spatial data can be used to increase the effectiveness and efficiency of many service delivery systems, both public and private, that depend on geographically distributed resources organized to meet the needs of geographically dispersed populations. Applied social science research in urban areas relied, traditionally, on Census data for information about the small-area social and economic characteristics of people. They are poorly served by this data source for the following reasons:

- the decennial nature of the Census
- the inflexible spatial aggregation units for particular data
- the many social and economic transactions of interest in the city that are not recorded in the Census
- the rapidly changing nature of some relationships that require real-time—or close to real-time—monitoring
- increasing doubts about the reliability of Census data

Hughes (1989, 190), in a critique of Census-based studies of urban social problems, noted that:

> Most of what we know about the inner city is based on census-tract or similar neighborhood data. These data provide a picture of the co-incidence of characteristics such as race, poverty, and so on in a single neighborhood. What tends to happen, however, even in the most sophisticated analyses, is a subtle form of the ecological fallacy; the attributes of a shared space are believed to imply shared attributes among individuals occupying that space.

Recent years have seen the development of a wide range of spatial decision-support systems for facility location, route planning, spatial administration, emergency service management, and many other purposes. All depend on the availability of spatially referenced data. The common geo-code for such administrative data systems is the street address. Many software programs are able to import TIGER Line files and match addresses to the street network. From this address-matching process distances can be computed through the street network between, for example, patients and hospitals, the scene of crimes and police patrol routes, the homes of children and alternative schools they could be assigned to attend, the homes of the poor, and the path of a hurricane. These geographically referenced questions are different from the "where is . . . " kind of question; they require a more complex search for a location on a street network, a search for a class of events on the same street network, and finally a spatial search of the relationship between the two. Until TIGER Line file databases became available in 1990 and widely shared through their availability in the public domain, such questions could not be answered easily. Most countries do not yet have such street centerline data files and, as with other base geographic data, many do not make them available in the public domain (Rhind 1992). Heywood (1990) makes a strong argument for access to spatial information as a basis for the development of societies in which one social group does not dominate another because of its more privileged access to basic spatial information.

Four recent technical developments permit the monitoring of small-area social-economic data in U.S. cities. The first three are (1) the development of digital maps of cities containing address information as well as associated codes linked to Census geography; (2) the development of household "geo-files" linking home addresses to digital map products; and (3) the development of special-purpose, administrative record systems describing attributes of people, service delivered, and information on the service provider. These information systems contain geographic identifiers for both recipient and provider of the service (Table 28.1).

TABLE 28.1

Example Service Providers and Recipients

Service Recipient	*Service Provider*
School children	Schools
Hospital services	Hospitals
Emergency episode	EMS emergency dispatch
Crime episode reports	Police response
Fire report	Fire equipment deployed
Utility outage report	Utility response
Job trainee	Job training center
Job seeker	Job information center
Transit trip	Public transportation system
Van pooler	Van pooling system
Child day care	Day care centers
Employed worker	Employer

The fourth technical development is the implementation of advanced software systems that can be used to spatially integrate the other three data systems. Examples of such applied work can be found in Bylinsky 1989; Demark 1991; and Garcia 1990. Examples of other applications of such street centerline databases to the resolution of urban problems are described in chapter 5, National Research Council (1993). A review of urban GIS users where data was shared found that these applications provided (Executive Forum 1990, v):

> tangible economic benefits in the form of cost savings, improved productivity, and elimination of duplicate functions. Intangible economic benefits include:
>
> - an improved ability for daily interdepartmental coordination of government services (e.g., public transit routing and public works maintenance).
> - an improved ability to respond to unexpected events which require coordination of multiple agencies (e.g., planning and delivery of government disaster recovery services).
> - improvements in the quantity and quality of information available for public policy analysis (e.g., better able to measure need for public services and effect their delivery because there is more data compatibility between departments).
> - support for development of an integrated enterprise information system by providing tools and procedures for data integration based upon common definitions of location.

Without detailing here the evidence and arguments (see Executive Forum 1990), it seems likely that very soon most North American cities, as well as many other cities around the world, will have spatial data at least at the level of the street and block face, if not also at the land parcel (cadastral) level, and operational GIS/LIS/AM–FM (Geographic Information Systems/Land Information Systems/automated mapping-facilities management) software systems designed to perform much of the routine administration of many of the functional departments of a modern city (de Neufville and Croissant 1990; National Research Council 1990, 1993). The Urban and Regional Information Systems Association (URISA) has been active in promoting spatial data-sharing activities within local governments and among local, state, and national governments. Their *Urban GIS/Data Sharing Guidebook* (Executive Forum 1990) addresses such issues as costs and benefits of spatial data sharing, identification of small- and large-scale spatial data, administrative structures for data sharing, and information sources on data sharing.

AN EXAMPLE: SPATIAL DATA-SHARING REQUIREMENTS FOR HEALTH POLICY RESEARCH

In health policy research, the geographic distribution of resources and health services needs is widely recognized as significant. Yet, the ability of social scientists to investigate important questions is limited by the poorly selected and non-standard geographic referencing systems of the custodians of the different data sets that need to be brought together. Systems have recently been developed for making data queries and choropleth maps for cancer incidence data, such as that made by the National Cancer Institute (1992). This "State Cancer Control Map and Data System" developed in the Map Program of the Division of Cancer Prevention and Control, Early Detection Branch, is an interactive map program that allows the user to construct state maps illustrating the geographic distribution (on a county basis) of cancer mortality based upon the user's choice of such parameters as type of cancer or age, sex, and race of the victim. Other examples include the Cadman and Associates (1992) Pandora system for viewing and analyzing the spatial flows of patients to hospitals. Hospital service areas and relative market shares of competing hospitals, when related to demographic base data and physician practice data, are valuable for identifying geographical accessibility problems as well as opportunities for service development at specific sites.

In addition to viewing data spatially, analysis systems have been developed to search health incidence data spatially to identify areas with greater than expected values on the basis of known demographic factors that affect incidence rates (Openshaw et al. 1987). However, systems need to be developed for investigating questions where different resources, with possibly different geographic referencing systems, need to be integrated. A recent study, for example, investigated the geographic distribution of low birth weight babies in Iowa by computing the relative distances of the mother's residence to the closest doctor, to the closest hospital with obstetric services, and to the closest hospital with more than fifty births per year (Armstrong et al. 1991). This study required access to five different spatial data sets from four different data custodians. These were: (1) locations of centroids of all five-digit zip areas; (2) zip area demographic data; (3) physician data; (4) hospital data; and (5) birth data. Although some had better geographic referencing data than others, the study had to use the one common geo-referencing system available in all five sets: the five-digit zip code. In many areas, however, the size of this area is so large relative to the geographic scale of movement thought to be relevant to the health services choices of pregnant women, that results may not show significance of geographical organization of resources when, in fact, the distribution is relevant to the explanation of outcome differences. This "geographic aggregation problem" is now known to be relevant to much of the research on the geographical organization of service systems (Rushton 1989; Current and Schilling 1990). The National Vital Registration Data, developed by the

Centers for Disease Control, describe natality and mortality data for selected geographic areas. Unlike the U.S. Bureau of the Census, its records are not geographically coded to take advantage of the TIGER Line files. Consequently, data are available for selected geographic areas only—using FIPS (Federal Information Processing System) codes—and, unlike census data, cannot be aggregated to user-defined geographic areas (FGDC 1992c). This limits their usefulness in investigating many health policy research problems. If, for example, one wishes to investigate changes in infant mortality or birth rates in southwest Los Angeles, it is impossible to find this information from current data sources.

ACCESS TO SOCIAL AND ECONOMIC DATA

For thirty years, the Inter-university Consortium for Political and Social Research (ICPSR) has served the data-sharing needs of social scientists by providing:

- a central repository and dissemination service for machine-readable social science data;
- training facilities in basic and advanced techniques of quantitative social analysis; and
- resources that facilitate the use of advanced computer technology by social scientists (ICPSR 1992, vii).

ICPSR's Data Archive receives, processes, and distributes machine-readable data on social phenomena occurring in more than 130 countries. In addition to a large holding of census data, including many historical censuses, the archive contains a very large collection of survey research data. It also contains two separate archives: the National Archive of Criminal Justice Data and the National Archive of Computerized Data on Aging. Descriptions of data in the archive usually define the spatial units of measurement in any data set. Miller (1988) used data from the archive to construct his "Great American History Machine"—a pioneering computer-based educational program that permits students to make their own choropleth maps at a county-level scale of U.S. census variables from 1840 to 1980. ICPSR's data is made available to the 360 member colleges and universities, generally at no cost, and is available to other institutions with a charge. Data is available on magnetic tape, diskettes, and CD-ROMs. More recently, a limited number of files were made available on Internet by network file transfer.

Social scientists encounter problems in sharing data not generally faced by physical scientists. These include issues of privacy and access rights. National censuses always guarantee that data given by individuals will never be released to the public in a form from which information about the individual could be recovered. The U.S. Census organizes micro-data files. These files describe unidentifiable individual housing-unit records with sample data for large geographic areas (100,000+ population). Commercial organizations in the United States have no such restriction on the release of information on the public. Discussions of privacy issues are found in Archer and Croswell 1989; Blakemore 1991; Epstein 1991; Hakim 1990; Onsrud 1992a, 1992b; and Rhind 1992.

METADATA

Many spatial data sets have been cataloged in special library holdings maintained by various public and private agencies. Many other data sets sit uncataloged at various anonymous FTP (file transfer protocol) sites on the Internet. But knowledge of the availability of these catalogs and data sets can be gained only through informal methods, such as word of mouth or through occasional postings to professional journals or electronic news services such as GIS-L.

Cataloging spatial data is in itself difficult. One problem facing the cataloging of spatial data is that "systematic information about data . . . goes beyond the descriptive fields used by the library world" (Burnhill 1991, 16).

As spatial data sharing increases, there will be an increased need for accurate descriptions of content to accompany the exchange so that users can determine for themselves the fitness of the data for their future use (Koeppel 1992; Onsrud and Rushton 1992; Salge et al. 1992; Williams 1991). Metadata, or data about data, is organized to answer the questions of what, where, how current, how good, and how complete. Draft standards for spatial metadata content recently have been distributed for comment (FGDC 1992a; 1992b).

CONCLUSIONS

Social scientists in the United States have benefited from developments in the organization and dissemination of U.S. Census products. Recognition by the Census that the same geographic base files that are used to produce the many different spatial aggregations tabulated by the Census can be used to make new, flexible, spatial aggregations was a major development in providing more relevant spatial data to social scientists. The same (TIGER) geographic base files are extremely valuable to social scientists whose data are commonly geo-referenced by street addresses and zip codes. The address-matching capabilities of GIS software applied to data from administrative records permit individual researchers to investigate many interesting questions about the spatial organization of human activities. The time is right for institutions to develop effective mechanisms for sharing spatial data from disparate sources with social scientists to enhance knowledge of the spatial relationships in human data.

REFERENCES

Archer, H., and P. L. Croswell. 1989. Public access to geographic information systems: an emerging legal issue. *Photogrammetric Engineering and Remote Sensing* 55: 1575–81.

Armstrong, M. P., G. Rushton, and P. Lolonis. 1991. Relationships between the birth weights of Iowa children and geographical accessibility to obstetrical care. Unpublished report prepared for the Iowa Department of Public Health, Des Moines, Iowa.

Blakemore, M. 1991. Access and security issues in the provision of geographic information. In E. Medyckyj-Scott et al., eds., *Metadata in the geosciences.* Loughborough, England: Group D Publications, Ltd.

Burnhill, P. 1991. *Metadata and cataloguing standards: one eye on the spatial Metadata in the Geosciences.* Ed. Ian Newman David Medyckyj-Scott Clive Ruggles and David Walker. Loughborough, England: Group D Publications, Ltd., 13-38.

Bylinsky, G. 1989. Managing with electronic maps. *Fortune,* April 24.

Cadman and Associates. 1992. Pandora: a system for analyzing hospital service areas. Lyme, New Hampshire.

Current, J. R., and D. A. Schilling. 1987. Elimination of source A and B errors in p-median location problems. *Geographical Analysis* 19: 95-110.

de Neufville, R. and J. Croissant. 1990. A policy for technology leadership. *URISA Journal* 2, 2: 7–15.

Demark, P. Van. 1991. School bus routing package survey. *URISA News* 114 (June): 7.

Epstein, E. F. 1991. Legal aspects of GIS. In D. J. Maguire, M. F. Goodchild, and D. W. Rhind, eds., *Geographical Information Systems.* Essex, England: Longman Scientific and Technical: 489–502.

Executive Forum on Local Government Geographic Information System (GIS)/ Data Sharing. 1990. *Urban GIS/data sharing guidebook.* National Computer Graphics Association and Urban and Regional Information Systems Association.

FGDC (Federal Geographic Data Committee). 1992a. Information exchange forum on spatial metadata, Washington D.C., U.S. Geological Survey, June 16–18.

______. 1992b. Draft: content standards for spatial metadata. Washington D.C., U.S. Geological Survey, November 3.

______. 1992c. *Manual of federal geographic data products*. Washington D.C.

Federal Statistics. 1971. *Report of the President's Commission*. two vols. Washington, D.C.: U.S. Government Printing Office.

Garcia, J. 1990. California school district gets a lesson in GIS. *Government Technology* (October): 34.

Garrison, W. L., and N. Hummon. 1973. Small area data systems for social science and social decisions. Unpublished paper.

Hakim, C. 1990. Free access to public data. *ESRC Data Archive Bulletin* 44: 1.

Heywood, I. 1990. Geographic Information Systems in the social sciences. *Environment and Planning A* 22: 849–852.

Hughes, M. A. 1989. Misspeaking truth to power: a geographical perspective on the "underclass" fallacy. *Economic Geography* 65: 187–207.

ICPSR (Inter-university Consortium for Political and Social Research). 1992. *Guide to resources and services 1992–1993*. Ann Arbor, MI: Inter-university Consortium for Political and Social Research (ICPSR).

Koeppel, I. J. 1992. Metadata for effective AM/FM/GIS utilization. AM/FM International Conference *Proceedings* 15: 186–196.

Lipietz, A. 1980. Inter-regional polarisation and the tertiarisation of society. *Papers, Regional Science Association* 44: 3–17.

Miller, D. W. 1988. The great American history machine. *Academic Computing* (October): 28.

National Cancer Institute. 1992. State Cancer Control Map and Data System, version 8.4. Division of Cancer Etiology, Clinical Epidemiology Branch, Atlanta, Georgia. p. 12.

National Center for Education Statistics. 1992. *School district data book: overview and invitation to comment*. Washington, D.C.: U.S. Department of Education.

National Research Council. 1990. *Spatial data needs: the future of the National Mapping Program*. Washington, D.C.: National Academy Press.

______. 1993. *Toward a coordinated spatial data infrastructure for the nation*. Washington, D.C.: National Academy Press.

New York Times. 1993. Rebuilding lags in Los Angeles a year after riots. May 10: 1.

Onsrud, H. J. 1992a. In support of open access for publicly held geographic information. *GIS Law* 1, 1: 3–6.

______. 1992b. In support of cost recovery for publicly held geographic information. *GIS Law* 1, 2: 1–7.

Onsrud, H. J., and G. Rushton. 1992. *Institutions sharing geographic information*. Technical Report 92-5. Santa Barbara, CA.: National Center for Geographic Information and Analysis.

Openshaw, S. 1983. The modifiable areal unit problem. *CATMOG 38*. Norwich, England: Geo Books.

Openshaw, S., M. Charlton, C. Wymer, and A. Craft. 1987. A Mark 1 geographical analysis machine for the automated analysis of point data sets. *International Journal of Geographical Information Systems* 1: 335–358.

Rhind, D. 1992. Data access, charging, and copyright, and their implications for GIS. *International Journal of Geographical Information Systems* 6: 13–30.

Rushton, G. 1989. Applications of location models. *Annals of Operations Research* 18: 25–42.

Salge, F., N. Smith, and P. Ahonen. 1992. Towards harmonized geographical data for Europe: MEGRIN and the needs for research. Fifth International Symposium on Spatial Data Handling, *Proceedings* 1: 294-302. ed. E. Corwin, D. Cowen, and P. Bresnahan. Charleston: IGU Commission of GIS.

U.S. Bureau of the Census. 1992a. *Maps and more: your guide to Census Bureau geography*. Washington, D.C.: Data User Services Division.

_____. 1992b. *Census, CD-ROM, and you: new horizons for microcomputer users of Census Bureau data*. Washington, D.C.: Data User Services Division.

Williams, R. J. 1991. The definition and description of data quality through truth in labeling. Symposium on Spatial Database Accuracy, University of Melbourne, Melbourne, Australia, *Proceedings* 1: 240-251.

Michael F. Goodchild

29 *Sharing Spatial Data Among Physical Scientists*

This chapter begins by asking why physical scientists need spatial data at all and finds a recent explosion of need largely attributable to renewed interest in modeling the global physical environment. It asks what kinds of spatial data are most valuable to physical science and identifies some of the characteristics that make data valuable and suitable for sharing. It then moves to a discussion of the context provided by the culture of physical science, with its customs and myths, and makes comparisons with other cultures in which spatial data plays an important role. Finally, it looks at changes underway in the culture of science that are affecting data sharing, particularly those attributable to the introduction of new technologies of communication.

INTRODUCTION

The term "physical science" encompasses an extraordinarily broad collection of fields, from the more traditional sciences of physics, chemistry, and biology to more recent arrivals like biochemistry, geophysics, and oceanography. Its stereotypes range from Hollywood's mad scientist to the white-coated laboratory worker and the bearded field geologist. Its methods range from sophisticated field instrumentation through statistical analysis to numerically intensive supercomputing and scientific visualization. With such diversity, it is at first sight difficult to find patterns in data sharing or suitable frameworks for an overview essay. Nevertheless, there are clear commonalities in the roles played by spatial data in the physical sciences, in the culture of science that forms the context for sharing, and in the new technologies becoming available to support more effective sharing.

WHY DO PHYSICAL SCIENTISTS NEED SPATIAL DATA?

Raw data is the feedstock of physical science, which turns observations about the natural world into laws, theories, and generalizations about the physical behavior of systems. One of the most fundamental forms of generalization concerns space: the laws of physics and chemistry, and indeed all physical sciences, are laws at least in part because they apply independently of location. The laws of thermodynamics apply everywhere, and they would not merit the title of laws if they did not. So fundamental is this principle, in fact, that we rarely require demonstration. Newton's laws of motion were not tested in different geographical locations before they became widely accepted, although they would have been quickly rejected if they had turned out not to apply in all locations. Thus, although physical science is built on observation, the observation of place is rarely important. Einstein needed no spatial data to develop the special theory of relativity, and Darwin had very little need of spatial data in developing the theory of natural selection, although maps provided a framework for many of his observations.

Despite its irrelevance to sciences like physics and chemistry, spatial data has played a somewhat greater role in the development of the Earth and environmental sciences. Geographical variation provides a range of conditions or settings for physical processes, and it is possible therefore to deduce laws about physical processes by observing their actions in geographic space. Geomorphologists, geologists, and geophysicists rely heavily on field observation to build knowledge and understanding, as do ecologists. There are even rare instances where laws have been deduced from analysis of maps, such as the Horton law of stream numbers (Horton 1945; Shreve 1966; Abrahams 1984), and maps provide much of the raw data for quantitative channel morphometry and fractal modeling of terrain (Klinkenberg and Goodchild 1992).

Such examples are rare, however, and there is general acceptance that geographic form provides only limited insight into the processes that form and shape the landscape (Pitty 1985). The prospect of deducing process from form is limited by the unknown effects of initial conditions and by the tendency of many dynamic physical systems to equifinality—the possibility that more than one process operating on the same set of initial boundary conditions could lead to the same eventual form. Over the last half century, there has been a strong tendency in the environmental sciences toward process studies in the laboratory or under controlled field conditions and away from deduction based on observations of physical form in the natural environment. Under this process-oriented paradigm, spatial data is used to establish sampling frames, or to find field sites for detailed observations and experiments, but rarely as direct input to models or analysis.

The ultimate value of an understanding of environmental processes lies in prediction, or the ability to forecast the outcome of changing conditions. Although a purist would argue that we have a fundamental understanding of all environmental processes at the level of physics and chemistry, our ability to aggregate processes to the level required for environmental prediction is very limited. A hydrologist wanting to predict the operation of a drainage basin under a range of precipitation scenarios could, in principle, break the basin into parts so small that each part's operation is fully described by the laws of physics, but would then lack the necessary data on boundary conditions to reaggregate the basin. This is also true if the basin is subdivided to the level required for prediction using known laws of hydrology, as those laws have substantial uncertainty, and again the spatial data necessary for reaggregation may be missing. Thus, much drainage basin hydrology relies on empirical laws that provide rough prediction but have no basis in either physics or hydrology, and apply only to a limited geographic area (Maidment 1993).

The value of spatial data for physical science is not derived from its role as a feedstock for fundamental understanding, which is very limited; instead, spatial data become essential in converting the results of fundamental physical science into practical use. Spatial data is essential to predict outcomes when physical processes operate over diverse geographical landscapes. However good our understanding of drainage basin hydrology, we cannot predict the behavior of a single real basin without spatial data on its geometry, topography, soil characteristics, and so on.

Nowhere is this role of spatial data more apparent than in global change research (Steyaert 1993). If the Earth were a uniform sphere, then the effects of a doubling of atmospheric CO_2 could be predicted from its basic physics. But the irregular nature of the land surface, ocean depth, and a host of other forms of geographic variation, coupled with the nonlinearities and instabilities in the underlying physical processes, all contribute to the conclusion that accurate spatial data is essential to accurate prediction of global climatic change. Global climate models require uniform coverage of data over the sphere, with uniform spatial resolution.

Science has traditionally valued understanding more than prediction, arguing that general laws of the behavior of physical systems are more valuable than the ability to predict specific outcomes. But in global change research, "understanding" of the operation of the Earth system does not simply mean a knowledge of the general laws that govern the system—its meaning has moved much much closely to prediction. Eddy (1993), quoting

Broecker, refers to this as "finding the plans" of the Earth system, in a charming analogy to the problem faced by an engineer who is required to deal with an emergency at a refinery without knowing how it works.

Traditionally, scientists have collected their own data, from the laboratory or the field. No self-respecting geomorphologist or ecologist would attempt to generalize or develop understanding from data that he or she had not had some personal involvement in collecting. Where data have been shared, it has been through personal networks and within informal groups. Science is territorial, and there is little incentive to share data from one group's field territory with another group in another territory. In some disciplines it is even common for the academic hierarchy to be reflected geographically, each graduate student being assigned one section of the professor's territory.

The emergence of global change as a significant and well-funded research area over the past decade (Committee on Earth Sciences 1989) has created a new set of priorities for data. If we are to understand the Earth system, we require spatial data and cannot rely on traditional field science to provide the necessary uniformity of coverage, spatial resolution, and quality control. Remote sensing provides at least part of the answer, but science has no tradition of collective efforts to organize data collection on this scale, or the necessary financial resources. Happily, the solution has been provided in this instance by government, which has taken on a limited role as data collector for science. Yet this role now being played by NASA (the National Aeronautics and Space Administration), as a collector of scientific data for the United States and international scientific communities through programs such as the Earth Observing System (EOS), has very little historic precedent. Precursors such as Landsat, or SPOT, or the topographic mapping program of the U.S. Geological Survey, have provided abundant spatial data, but for many different purposes—many of which have no obvious or essential connection with science. Moreover, they have done so in some cases at a price that reflects the cost of data collection, which is prohibitive for many scientific projects.

WHAT KINDS OF DATA ARE NEEDED?

As noted earlier, the traditional role of spatial data in physical science has been to allow scientists to locate experiments or observations on the surface of the Earth and to determine proximity or coincidence between observations. A geomorphologist uses a topographic map to record field interpretations, such as identifications of glacial features; a geologist might use the same map to record field observations of surficial geology. Modern spatial data technologies, such as remote sensing or GPS (Global

Positioning System), perform essentially the same function. The spatial data performs the function of a *frame*, allowing the scientist to position an observed feature and its attributes with respect to easily identified features shown on the topographic map, or with respect to the Earth's geographic referencing system.

Within the digital context of Geographic Information Systems (GIS), scientists now routinely use similar frame data sets to provide a geographical frame of reference, or "backcloth," to outputs from simulation models, or mapping projects. Within the United States, such frame data sets are readily available in many different scales and formats, particularly:

- DLG (Digital Line Graph) files of topographic features at scales of 1:24,000 (partial coverage), 1:100,000 (full coverage of conterminous United States), and 1:2,000,000; and
- Raster imagery derived from Landsat (30 to 60m spatial resolution) and AVHRR (1km resolution) satellites.

In addition, digital orthophoto quadrangles (DOQs) are becoming available under a program jointly sponsored by the U.S. Geological Survey and the Soil Conservation Service (U.S. Department of Agriculture). These provide corrected (vertical) digital images of the land surface with a spatial resolution of approximately 1m and a positional (registration) accuracy of approximately 6m. With the notable exception of recent LANDSAT imagery, all of these are available at the cost of reproduction. A useful guide to federal geographic data products has been published recently (Federal Geographic Data Committee 1993).

At the global level, there is similar interest in the use of frame data sets. In addition to a multitude of data sets at smaller scales, the Digital Chart of the World (DCW) (available from the Eros Data Center, U.S. Geological Survey) now provides a frame with uniform coverage of the Earth on CD-ROM at a scale of 1:1,000,000, showing coastlines, hydrographic features, roads, railroads, towns, and crude topography. DCW can support simple analyses, for example, by allowing predicted mean temperature from a global climate model to be calculated by nation, state, or sampled at the locations of towns.

Despite their usefulness, the function of frame data sets is more cosmetic than fundamental to science. To be effective, it is desirable to have uniform coverage, but considerable distortion or incompleteness might be tolerable before the value of the frame data is lost. For example, it would not matter greatly if in displaying the predictions of a global climate model over a backcloth of international boundaries, the internal boundaries of Europe were missing or misplaced.

Moreover, frame data sets are useful for a variety of purposes besides those of science, and the private sector may be just as effective a supplier as government. There is increasing pressure on governments to recover more of the costs of spatial data production from users, and this may have an adverse effect on scientific users, who have a comparatively low ability to pay. In addition, the U.S. policy of providing data to the international global change research community at the cost of reproduction may be increasingly controversial (Rhind 1992a, 1992b).

By contrast, low cost, completeness, and accuracy are essential in the types of spatial data used as direct input to scientific models, here referred to as *measurement* data. Whether the data are used to "roll out" the predictions of a climate model over the variable topography of the Earth surface, or to "scale up" observations on ecological conditions at sample sites to global coverage, the value of the exercise is directly related to the quality of the spatial data used (Burrough 1992). Although remote sensing appears to offer uniform coverage, optical remote sensing is differentially affected by cloud. Direct field observation is affected by problems of physical access, and particularly cost.

At best, spatial data can only approximate real geographical variation, at a finite level of resolution. Besides resolution in the spatial domain, the accuracy of data is affected by sampling resolution in the temporal domain, particularly in remote sensing, since the spectral response of a given scene varies by time of day as well as by season. Accuracy is also affected by resolution in the thematic domain, for example through the precision of the taxonomy used to classify soils, or land cover, or through the spectral resolution of a satellite-borne sensor. All of these factors affect the appropriateness of spatial data for input to a particular model application.

Ideally, data should be captured at the finest possible spatial, temporal, and thematic resolution. In practice, resolution is constrained by factors of cost and the enormous volumes of data involved: it is estimated that the EOS satellites will produce terabytes (1 terabyte = 10^{12} bytes) of data per day. High spatial resolution is essential for mapping and the creation of frame data sets, so that clearly defined objects such as roads and rivers can be recognized. The value of frame data is not seriously compromised by coarse temporal resolution, since most features used for spatial reference on the Earth's surface are essentially static, and we are familiar with the problems of keeping maps up to date. Similarly, high thematic resolution is not essential for simple mapping, where panchromatic imagery may be perfectly adequate for most purposes.

On the other hand, thematic (or spectral) and temporal resolution have become increasingly important for measurement data, particularly in support of global environmental modeling. Models of atmospheric circulation or evapotranspiration require frequent observations in narrow parts of the electromagnetic spectrum, and modelers are often willing to lower expectations about spatial resolution in order to obtain higher resolutions in the temporal and spectral domains. Thus, it seems that EOS will not provide a significant improvement in spatial resolution relative to current Landsat and SPOT imagery, despite the enormous volumes of data anticipated, because spectral and temporal resolution have been given higher priority in the interests of environmental modeling. The resolution requirements of frame and measurement data are distinctly different, and it is increasingly difficult for one program to satisfy both.

These spatial measurement data perform a range of functions in models. In some, they provide a source of direct measurements of model variables; for example, satellites are used to measure energy fluxes in studies of the Earth's energy cycle. In other cases they provide measurements used to verify models; for example, models of ocean circulation and temperature distributions might be verified using satellite-derived measurements of ocean temperature. In other cases, the role of spatial data is to stratify the surface of the Earth, to provide values of model parameters appropriate to each stratum. For example, the surface might be stratified by land use, and each class of land use replaced by an appropriate value of an albedo parameter in a radiation model (for a range of examples of the application of spatial data in environmental modeling, see Goodchild, Parks, and Steyaert 1993).

In many of these examples, the spatial data provide a source of objective measurements for the model. In other cases, however, where the spatial data provide a source of stratification, there is likely to be a substantial element of subjectivity (Loveland and Ohlen 1993). No system of land-use classification is perfect, and thus strata will never be perfectly homogeneous. Such classifications attempt to reduce complex, multidimensional spatial variation in a generic fashion that does not necessarily anticipate the needs or sensitivity of the eventual model. For example, U.S. soil maps provide an approximation to the complex, continuous spatial variation of soils but are not necessarily optimal as a stratification for land-atmosphere transfer models. Yet, climate modelers must rely on standard soil maps (e.g., Xue and Sellers 1993) for stratification because any special-purpose alternative would be impossibly expensive to create. Again, a government agency (the Soil Conservation Service in this case) finds itself in the position of supplying raw data to the scientific modeling community because no alternative source of complete, low-cost data exists. Certainly, the need for complete data of assured quality could not be met by informal sharing among scientists.

DATA SHARING AND THE CULTURE OF SCIENCE

Although it celebrates objectivity and the power of scientific reasoning, the scientific community is nevertheless human and has its own codes of behavior like all human societies. Traditionally, the culture of physical science has been dominantly male, individualistic, fiercely competitive, and territorial (for a view of the influence of gender in science, see Rosser 1990). It has been relatively unsupportive of institutions and collective action, and the political lobby for science in the United States remains far weaker than those of other communities of similar size, such as the National Rifle Association.

In spatial data policy, the scientific community has been willing to argue strongly for government sponsorship of programs like the EOS, but has been impatient with arguments over costs and benefits and unwilling to pay for data at levels equal to the cost of collection. There is no tradition of economic valuation of data within the scientific community; in a caricature of the community's view, data should be collected at public expense and made available at the cost of reproduction, in view of the overwhelming evidence that the research it supports is in the public interest the survival of the planet depends on it.

In contrast to its lack of formal structures, the scientific community seems quite favorably disposed toward informal interaction. Traditionally this occurred within departments, or within universities, and at occasional conferences, but recently electronic mail has been adopted wholeheartedly by the community and has radically altered patterns of communication and influence.

Nowhere are the rites and customs of the scientific community more in evidence than in the area of publication, and intellectual property in general. Scientists write in disciplinary dialects that are often almost unintelligible to readers from other cultures. They observe rules about authorship and the order in which authors are listed. Authorship rules may even reflect ownership of data or geographical territory; if one scientist obtains data from another and reanalyzes it, the original owner of the data can expect to be listed as an author of any publication of the results. These rules are not enforced by any institutional structure and may even be unwritten. But informal communication among scientists and the role of peer review in career advancement are sufficient incentives to play by the rules.

Attitudes to intellectual property present something of a paradox. On the one hand, science is altruistic, dedicated to the advancement of human knowledge; scientific data, like scientific knowledge, should belong to all. Science owes much to the monastic tradition—of the celibate cut off from the common world in a life devoted to higher things. On the other hand, science is competitive, and advancement can depend on being first with a theory, or even first with a measurement (see Rosser 1990, for a discussion of the extreme consequences of competition). The intellectual property customs of the scientific community act to preserve a delicate balance between these two opposing principles, allowing the scientist to feel part of a collective enterprise but at the same time an owner of a certain amount of personal intellectual property.

In this culture, with its delicately balanced customs developed over decades, if not centuries, it is difficult to imagine how one could promote greater sharing of data. Institutional solutions are unlikely to succeed in an individualistic culture that is unsupportive or suspicious of institutions. Government-sponsored programs can provide the necessary spatial data for the developing field of environmental modeling and can frame data for many purposes, but such programs will be accepted only if they provide data at the cost of reproduction and thus in effect contribute to society's support of science.

But is there a problem? Is there clear evidence that the current system encourages or even condones redundancy and inefficiency in spatial data? Data are often costly to collect, but it is common for scientists to create consortia to share the costs of data collection. Research funding agencies like the National Science Foundation (NSF) have sponsored projects aimed at reducing data costs through the sharing of expertise and equipment. UNAVCO (the University Navstar Consortium), for example, now a part of UCAR (the University Corporation for Atmospheric Research in Boulder, Colorado), supports the geophysics community by providing expertise, training, and maintenance of high-precision GPS equipment for observation of crustal movement. ICPSR (the Inter-university Consortium for Political and Social Research) in Ann Arbor, Michigan, is supported by NSF as a common resource for social science data. Programs like these, targeted at particular sectors of the scientific community and linked to individual scientists through informal communication networks (with formal institutional affiliations where necessary), seem compatible with the culture of science and seem to work within their limited objectives.

But the culture of science is changing, and many of its old assumptions are being questioned. It is clear that the scientific culture of fifty or even twenty years from now will be almost unrecognizable. The next section looks at some of these changes and the impacts they are likely to have on spatial data sharing in the next decade.

THE CHANGING CULTURE OF SCIENCE

The message of President Clinton's 1993 inaugural address was quite clearly "change or be changed." Is the current male-dominated, individualistic culture of science essential to its progress, or are alternative cultures possible? Will the efforts to involve more women and minorities in science produce a cultural change? More significantly, perhaps, are changes inevitable because of other changes already occurring in the subject matter of science and in its underlying methodology, particularly in areas that rely on spatial data?

The science of Newton and Einstein was strictly positivist. The purpose of scientific research was the discovery of laws or generalizations about the natural world: where more than one such law was equally capable of explaining natural behavior, the simplest one was accepted, and a law was rejected whenever its predictions were inconsistent with observation (Popper 1968). Reproducibility was a fundamental requirement of research, and scientific writing respected the principle that experiments be described in sufficient detail to permit another scientist to replicate them. Finally, economics had no relevance to science – the quality of scientific knowledge could never be compromised by the costs of data collection.

This set of methodological principles is seriously compromised by much of the research in physical science making use of spatial data. When a global climate model is built, its validity derives from the builder's belief that the model replicates the processes that affect climate, not from comparisons between its predictions and reality. But it is impossible to capture all such processes in any finite model. Moreover, such processes operate in continuous space but must be modeled by discrete approximations. The data used to support the model are also approximate. Further, it is not possible to determine the degree of approximation present either in the modeled processes or in the data, or to predict the impact of those approximations on the results of the model. Thus the modeler has no way of determining the uncertainty associated with results, beyond subjective assessments based on poorly defined expectations. In short, we have no way of verifying such models. Where predictions of the same phenomena are available from more than one model and the predictions agree in some gross sense, we might feel that some sort of verification had been achieved, but that feeling would have no objective basis.

In this new world of environmental modeling, the quality of results is clearly linked to the costs of data. There are no right or wrong predictions--

the more resources society is willing to pour into data collection, the more reliable the results.

The models now being constructed and used to forecast future climates, and their impacts on ecological communities and human societies, are enormously complex computer codes with thousands or even millions of instructions. The constraints of the publication system simply do not allow every aspect of the model to be reported, and so replication of results is no longer possible. Moreover, models sometimes fall on the wrong side of the delicate balance of arguments over intellectual property—not all investigators are willing to make all parts of their models fully available (in source code form) to every other scientist. Advanced computer-based tools such as GIS or object-oriented modeling systems do not always reveal every aspect of their internal operations to the user, and it is common, for example, for a scientist not to know the precise details of a spatial interpolation algorithm that is being used to resample spatial data, particularly if the algorithm is embedded in a GIS developed by the private sector.

In short, the world of physical science based on spatial data is not the world of Newton or Einstein but is characterized by much more complex, confused, and ambivalent attitudes to the principles once held sacrosanct by all scientists. Adjectives that used to describe science, and still describe many scientists—"hard," "rigorous," "objective"—are increasingly replaced by words more consistent with cooperation, collaboration, and sharing. More frightening to many traditional scientists, perhaps, is the message that for complex environmental models we may have no alternative to rigorous verification but faith and trust.

In areas like environmental modeling, the culture of science is being affected by the need to deal with a complex and difficult subject matter that spans many disciplinary boundaries. At the same time, it is also affected by changes in the demography of science and the need to attract women and minorities into science, and by new technologies. The impact of word processing on science has been profound, and yet it has occurred entirely within the past ten years. We now have the technological means to flood the world with written material as never before, and the system of academic rewards continues to urge us to do so. We have the ability to distribute enormous volumes of papers, data sets, and messages over most of the globe at the speed of light. Only the human ability to absorb remains constant.

By early 1993, the U.S. Internet backbone (NSFNet) was operating at a bandwidth of 45 million bits per second (T3), and access to the international network had reached almost all of the world's scientific community, making it possible for the average scientist to have the same kind of informal conversation with a colleague half way around the world that they once would have had over morning coffee with their departmental colleagues—in fact, more easily, as there was no longer any need to wait for morning coffee. National policy was clearly committed to the further development of electronic communication through programs like the National Research and Education Network (NREN) (Kahin 1992), the National Information Infrastructure (NII), and the National Spatial Data Infrastructure (NSDI) (Tosta 1992). Parallels were being drawn between the investments made in the interstate highway system beginning in the 1950s and the new potential of an "electronic highway" system built at least in part at public expense.

In the new world of electronic communication that has emerged in the last five years, distance and costs are no longer impediments to communication. Departments of colocated researchers are no longer the primary source of intellectual stimulus, and the disciplinary barriers that they reinforced are no longer as relevant, except through their role in career advancement. Instead, we are seeing the emergence of "invisible colleges"—virtual groupings of researchers with similar interests—in the form of electronic list servers and discussion groups.

In the field of GIS and spatial data, the new electronic highways are providing a uniquely appropriate solution to the problem of data sharing. A list server like GIS-L, maintained at the Buffalo site of the National Center for Geographic Information and Analysis (NCGIA) and read daily by perhaps 2,000 people worldwide, is an immensely valuable way of finding data. A message posted in the morning on the East Coast of the United States, requesting information on digital spatial data for soils in Canada, might have a dozen responses by the same afternoon. The approach is informal, a major factor in its acceptability to the culture of science.

Also informal, but relying on automated search rather than the willingness of individuals to respond to electronic mail, are the network servers such as Gopher and Wide Area Information Server (WAIS) (Krol 1992; LaQuey 1992). These allow the user to initiate a search for data over the entire network and to rely only on each researcher's or institution's willingness to announce the data's availability by registering it with the system. More advanced systems allow the data to be documented and accessed remotely and transferred using protocols such as FTP (file transfer protocol). The most advanced systems might, in the future, provide support for the user who wants the entire search, access, and use to be transparent, so that a model might use data on soils from one host, and data on topography from a second, and be run using code obtained from a third host on a system located at a fourth location, all without any attention from the initiator of the request.

These more advanced options require standards, or formally agreed-upon rules that allow data to be integrated from different sources and processed by different models. But we have already argued that formal structures are difficult to establish in the culture of science, and not surprisingly there are few, if any, effective standards for spatial data. "Such efforts are almost always slow, laborious, political, petty, boring, ponderous, thankless, and of the utmost criticality" (Kleinrock 1992, 176). It is unlikely that the new federal standard for spatial data—FIPS 173—will have much immediate effect on science, as it is aimed much more at transfers of data between federal agencies for administrative rather than scientific purposes. More likely, standards in the scientific community will be of an informal, pragmatic nature, and such standards as emerge will be determined by the availability of GIS or modeling software.

In summary, several trends are likely to affect the culture of science in the next few years, and particularly the part of the scientific community that needs spatial data. We are moving away from the traditional paradigms of science into ones that are more visual, collaborative, and multidisciplinary, and less consistent with positivist objectives. The scientific community is being affected by new technologies for publication and communication, and access to data, and its demographics are changing also. It is clear that the scientific culture of ten years from now will provide a very different environment for sharing of spatial data but one that continues to emphasize informal arrangements over formal ones.

CONCLUSIONS

Despite the overriding spirit of altruism that underlies science, a scientist's data have always been part of his/her intellectual property, a principle that derives in part from the scientist's personal involvement in data collection. Spatial data have become important to physical science only recently, with increased interest in predicting the behavior of the physical environment, and science has come to rely on government to play the role of spatial data provider because of the high costs involved and the need for uniform coverage of assured quality.

Spatial data can be usefully divided into two major groups, frame and measurement, the former providing a reference framework for scientific activity, and the latter providing input to models. While both types are important to science, their characteristics and the ways they are used are different, creating a tension of priorities within agencies that supply spatial data to the scientific community.

The picture of the scientific community painted in this essay has emphasized its individuality, competitiveness, and informality. Spatial data are shared whenever there is an opportunity to do so that is perceived as being in the interests of both parties. It would be difficult to demonstrate that this encourages redundancy or inefficiency, or that it could be improved by intervention. On the other hand, new technologies are rapidly changing and expanding the informal networks of individual scientists, and providing an environment much more favorable to data sharing. New groups are entering science and changing its demography, and they may bring with them different cultural attitudes that question the current norms. Finally, the subject matter of environmental modeling is embedded within a different paradigm of science that may also undermine the assumptions of the scientific culture that currently makes use of spatial data.

REFERENCES

Abrahams, A. D. 1984. Channel networks: a geomorphological perspective. *Water Resources Research* 20: 161–188.

Burrough, P. A. 1992. Development of intelligent geographical information systems. *International Journal of Geographical Information Systems* 6, 1: 1–12.

Committee on Earth Sciences. 1989. *Our changing planet: a U.S. strategy for global change research.* Washington, D.C.: Federal Coordinating Council for Science, Engineering, and Technology, Office of Science and Technology Policy.

Eddy, J. A. 1993. Environmental research: what we must do. In M. F. Goodchild, B. O. Parks, and L. T. Steyaert, eds., *Environmental modeling with GIS.* New York: Oxford.

Federal Geographic Data Committee. 1993. *Manual of federal geographic data products.* Washington, D.C.: U.S. Department of the Interior, Federal Geographic Data Committee.

Goodchild, M. F., B. O. Parks, and L. T. Steyaert, eds. 1993. *Environmental modeling with GIS.* New York: Oxford.

Horton, R. E. 1945. Erosional development of streams and their drainage basins. *Bulletin of the Geological Society of America* 56: 275–370.

Kahin, B., ed. 1992. *Building information infrastructure.* New York: McGraw-Hill Primis.

Kleinrock, L. 1992. Technology issues in the design of the NREN. In B. Kahin, ed., *Building information infrastructure.* New York: McGraw-Hill Primis, pp. 174–198.

Klinkenberg, B., and M. F. Goodchild. 1992. The fractal properties of topography—a comparison of methods. *Earth Surface Processes and Landforms* 17, 3: 217–234.

Krol, E. 1992. *The whole Internet: user's guide and datalog.* Sebastopol, CA: O'Reilly and Associates.

LaQuey, T. 1992. *The Internet companion: a beginner's guide to global networking.* Reading, MA: Addison-Wesley.

Loveland, T. R., and D. O. Ohlen. 1993. Experimental AVHRR land data sets for environmental monitoring and modeling. In M. F. Goodchild, B. O. Parks, and L. T. Steyaert, eds., *Environmental modeling with GIS.* New York: Oxford.

Maidment, D. F. 1993. GIS and hydrologic modeling. In M. F. Goodchild, B. O. Parks, and L. T. Steyaert, eds., *Environmental modeling with GIS.* New York: Oxford.

Pitty, A. F. 1985. *Geomorphology: themes and trends.* Totowa, NJ: Barnes and Noble.

Popper, K. R. 1968. *The logic of scientific discovery.* New York: Harper and Row.

Rhind, D. W. 1992a. War and peace: GIS data as a commodity. *GIS Europe* 1, 8: 24–26.

______. 1992b. Data access, charging and copyright and their implications for geographical information systems. *International Journal of Geographical Information Systems* 6, 1: 13–30.

Rosser, S.V. 1990. *Female-friendly science.* New York: Pergamon.

Shreve, R. L. 1966. Statistical law of stream numbers. *Journal of Geology* 74: 17–37.

Steyaert, L. T. 1993. A perspective on the state of environmental simulation modeling. In M. F. Goodchild, B. O. Parks, and L. T. Steyaert, eds., *Environmental modeling with GIS.* New York: Oxford.

Tosta, N. 1992. The National Spatial Data Infrastructure. *Geo Info Systems* 2, 10: 30–35.

Xue, Y., and P. J. Sellers. 1993. A brief review of the simple biosphere model. In M. F. Goodchild, B. O. Parks, and L. T. Steyaert, eds., *Environmental modeling with GIS.* New York: Oxford.

Section V

Fundamental Questions and Future Research

Fundamental Questions and Future Research

As stated in the introduction to this book, one major goal of the volume has been to provide insights on opportunities, options, and potential pitfalls for those organizations and individuals seeking to gain from the sharing of geographic information. Another has been to discuss, debate, and formulate research activities which, if accomplished by the at-large academic community, would be valuable to individuals, businesses, and governments sharing or attempting to share geographic information. Throughout this book the various authors have made numerous recommendations on actions or activities (including but not limited to research activities) that would enhance data sharing among federal, state, regional, and local levels of government. They have also made recommendations directed at enhancing access by private businesses, scientists, educators, and the general public to the geographic databases maintained by various levels of government.

The following outline summarizes many of the research issues and activities recommended by the contributors to this volume and includes extracts from their contributions. For more detailed discussions of the issues listed and for additional areas of inquiry, please consult the earlier chapters and the report of the National Center for Geographic Information and Analysis (NCGIA) Specialist Meeting on Institutions Sharing Geographic Information (Onsrud and Rushton 1992).

I. ORGANIZATIONAL ASPECTS OF SHARING GEOGRAPHIC DATA

A. General Issues

From an organizational research perspective, the question of what factors, conditions, and processes promote or allow "sharing of geographic information" should always be viewed within the context of what promotes or allows effective "use of geographic information." With this as a basis, the following fundamental questions arise.

1. What are the relationship variables that affect sharing of geographic data among people within an organization, among units of an organization, or among organizations? What is the relative importance of each variable standing alone and in relation to the others?
2. Can content and process models be developed that will be useful and practical tools in implementing sharing relationships and policies? Will such models also be useful in developing strategies for overcoming ineffective or less than optimal levels of sharing?
3. Should such models extend from comprehensive consideration of a wide range of behavioral, organizational, technical, and political factors and processes viewed firsthand in Geographic Information Systems (GIS) data-sharing environments, or may such models extend from frameworks and critical subsets of factors already isolated by other organizational theory and technology transfer research efforts? If the latter, what organizational research experience bases, frameworks, and subsets of factors and processes are most promising? What time-dependent stage models are most promising?
4. How have networking capabilities and data in digital form affected the relevant variables in such models?
5. What research methodologies are appropriate for building useful models of organizational and interorganizational behavior? Are models of practical use already available, are they likely to be available in the near future through transfer of methodologies and generalizable principles already developed in other fields, or is investment in a long-term cumulative research tradition in GIS sharing environments required in order to develop models that will be of practical use in understanding complex interorganizational dynamics?
6. What portion of the causes of success in sharing or success in use of geographic information may be incorporated in a practical model or combination of models, and what portion is idiosyncratic to the specific environment in which the change in behavior is being attempted?

B. Specific Issues

1. ORGANIZATIONAL

   These issues may be thought of as referring to those macro-level constructs or issues that pervade an organization at all levels. To what extent does the family of factors in each of the following areas affect the ability or willingness of a government agency or a private organization to share GIS data and information:

   a. Bureaucratic practices and standard operating procedures
   b. Cooperation required or allowed across departments
   c. Type of organizational structure and its extent of flexibility or rigidity (e.g., nature and level of hierarchy, accountability, and autonomy)
   d. Management state of the organization (e.g, congruent or mixed, and at what level?)
   e. Corporate culture (i.e., internal but overarching beliefs, attitudes and policies present in the organization)
   f. Political environment (i.e. internal competition for scarce corporate resources)
   g. Technological sophistication of the organization
   h. What factors, processes, and structures identified as critical in each of these areas are alterable by management in order to effectuate change?

2. INTERORGANIZATIONAL

   a. What are the likely interorganizational dynamics and responses if attempts at sharing GIS data occur on an ad hoc or case-by-case basis?
   b. What are the likely dynamics and responses if sharing arrangements are attempted through coordination on a widespread basis? What is the likely outcome of the use of standard management techniques in gaining interorganizational cooperation on sharing, such as establishing a long-term "shared vision" for geographic data sharing and then "setting targets" at successive intermediate levels for sharing among federal, state, and local governments and the private sector?
   c. To what extent is investment in technological infrastructure and communications facilities a means for enabling interorganizational sharing, and under what conditions?
   d. What are the factors and dynamics affecting sharing for routine, operational, and transaction tasks versus sharing for joint project tasks?
   e. How does treatment by an organization of geographic information as a "tool" versus geographic information as a "product" affect the willingness and ability of an organization to share geographic information?

f. What are the roles and functions that coordinating bodies can effectively supply at institutional or interorganizational levels in allowing or promoting the sharing of geographic data?
g. What are the roles and functions that coordinating bodies can effectively supply in allowing or promoting the sharing of geographic data across electronic networks?
h. What is the most effective framework or organizational form for coordinating or facilitating geographic data sharing among organizations?
i. What role might group decision-making regimes play in arriving at mutually beneficial decisions regarding sharing?
j. What effect do organization survival factors have on the ability and willingness to share?

3. BEHAVIORAL

Behavioral issues may be thought of as referring to micro-level constructs or issues that arise at the individual person level in the organization. Many researchers have pointed to behavioral issues as principal drivers of action within organizations.

a. Are data-sharing "champions," "re-inventors," and "opinion leaders" necessary or advantageous to successful sharing? If so, in what ways, to what extent, and how may they be supported or, if nonexistent, created?
b. What is the effect of professional background, academic training, and technical expertise of individual managers or critical personnel on the willingness or ability to share?
c. What is the likelihood of unforeseen or unintentional behavioral changes as a result of successful or unsuccessful sharing attempts, and what are likely impacts of those changes (e.g., greater or lesser cooperation among sharers on other matters of importance to the organization)?

II. LEGAL, ECONOMIC, AND CULTURAL ISSUES IN SHARING GEOGRAPHIC DATA

A. General Issues

1. What legal and economic arrangements will optimize the sharing and use of geographic information among agencies and organizations if adopted by them?
2. How does uncertainty about legal issues relevant to the handling of electronic information (i.e., copyright, liability, privacy, open records issues, and the like) affect organizational, interorganizational, and personal behavior?
3. What cultural or social conditions, attitudes, and behavior enhance or distract from the effective sharing of geographic information?

4. What mix of legal, economic, and cultural conditions encourages widespread cooperation among organizations in developing, sustaining, and diffusing geographic data and information?

B. Specific Issues

1. LEGAL

   What balance of laws relative to each of the following areas and their interactions permits the most effective long-term sharing and use of geographic information under socially desirable conditions?

   a. *Freedom of information and public access to records*
      To what extent should individuals have access to the records and databases of government?
   b. *Privacy*
      To what extent should individuals be free from electronic intrusions by government, the commercial sector, and other private parties?
   c. *Work product protection*
      To what extent should copyright, contract, licensing, and similar legal principles be altered in order to respond to changing technological conditions? What changes would aid in optimizing widespread opportunity of access to GIS and other electronic databases?
   d. *Liability*
      What distribution of liability burden in regard to the use and misuse of geographic databases would optimize appropriate social uses of such databases?

   To what extent can comparative analysis of the actual ramifications of disparate policies being followed in various jurisdictions relative to each of the above issues divulge appropriate balances of law?

2. ECONOMIC

   a. What is the state of economic theory in regard to publicly held geographic data and information?
   b. In the context of public goods versus private goods arguments, which of the following will result in greater use and sharing of the geographic databases developed and held by government: Distribute digital data to any individual for free, at the marginal cost to respond to the data request, or at some amount greater than marginal cost in an attempt to generate a revenue source for government?
   c. What is the likely impact of various economic approaches on the long-term stability and viability of arrangements for the control and distribution of data

and information? Is there rational economic justification for local, state, and federal agencies to follow differing approaches in regard to the distribution of geographic data to the public? What effect is each approach likely to have in expanding or narrowing current socio-economic gaps in current and future populations?

3. CULTURAL
   a. What are the cultural biases inherent in GIS technologies and databases?
   b. Is the widespread use of certain forms of geographic data collection, analysis, and representation having a homogenizing effect on the way varying societies would otherwise handle geographic data?
   c. Are different societal attitudes toward the proper role of government in the handling of data significant in affecting the ability and willingness to share geographic data?
   d. Because geographic information is of greatest interest to those present in a geographic area, do cultural differences among local communities, among states, or among nations have minimal effect on the use and sharing of geographic databases?
   e. Are cultural differences in attitudes and behavior among private citizens, business organizations, local governments, regional and state agencies, and federal agencies that are all concerned with the same or overlapping geographic areas substantial impediments to the more effective use and sharing of geographic databases?
   f. What is the likely distribution of social benefits and costs of GIS databases in respect to urban versus rural communities, large versus small agencies, and poor versus wealthy governments or countries?
   g. Are cross-national comparative research studies an effective means of divulging those policies and strategies that promote increased equity in the distribution of the beneficial consequences of GIS?

III. METHODOLOGY AND SUBSTANTIVE CASE STUDIES

A. General Issues

1. What are the most appropriate research methods for divulging factors, processes, and conditions favorable to optimal use and sharing of geographic information, and under what circumstances?

2. What measures are most appropriate in assessing the extent of use or sharing of geographic data under alternative sharing arrangements? What specific parameters or surrogates are rational and useful in measuring or evaluating "effective use," "optimal use," and "use success"? What measures are most appropriate in assessing organizational and social impacts?
3. Due to lack of a substantial cumulative research tradition on behavioral, organizational, and institutional issues within the GIS field itself, should the initial primary heuristic for the research community in analyzing use and sharing issues be case studies? Or are bodies of theory generated from data use and sharing experiences in other fields already sufficiently detailed and appropriate for testing in GIS environments through large-scale techniques such as mail questionnaires or phone surveys?
4. What sequential series of studies might best be suggested for advancing the current knowledge base relative to factors, processes, and conditions favorable to optimal use and sharing of geographic information?
5. To what extent should case study methodology be used to both build and test theory? In the context of case study methodology, to what extent should descriptive studies, comparative studies, literature reviews, focus groups, in-depth field observations, and ethnographic techniques play a role in aiding "theory generation"?
6. Do clusters of case studies of sharing or use focused on particular institutional settings, type of spatial data shared, or types of problems being addressed by the decision maker better enable "theory testing" due to the greater ease of making cross comparisons?

IV. METADATA ISSUES IN SHARING GEOGRAPHIC DATA

Effective sharing of geographic data on a widespread basis requires that individuals and organizations must be able to determine whether data already acquired by others is appropriate for their proposed use. For the following questions, geographic or spatial "metadata" is defined as digital information that allows potential users of digital geographic data to understand that data's fitness for use.

A. General Issues

1. In order to provide maximum opportunity for sharing geographic information, is there a need to develop a systematic metadata regime? If so, what mix of standards, technical capabilities, technical infrastructure, and organizational frameworks will best support a metadata regime of maximum utility?
2. If there is a need to develop a metadata regime, is there also a need to strike a balance between exhaustive completeness for those who need detailed information on the data and cryptic

efficiency for those who may be experts in use of the data? Or may a regime be designed that satisfies both extremes of demand for metadata?

3. Should a metadata regime for cataloging differ from a metadata regime for data transfer? If so, what should be the functional differences? May such regimes be combined?
4. Where should metadata reside—with the data set or in some depository—and who should maintain it? When the need arises to strike a balance or make a decision in regard to designing a metadata regime, what criteria should be used in making such decisions (e.g,. considerations of efficiency, long-term stability, incentives for stakeholders, concerns with competing interests over control of information, maximization of library patron access as opposed to scientist access or agency access, and so on)? Who should be the final arbiter in making such decisions?
5. Is it possible to determine geographic metadata requirements such as contents, methods of distribution, and usefulness directly from users and potential users? Is there a need to rank different types of geographic data by their suitability for description through metadata, and thus their suitability for sharing under various scenarios?
6. What kinds of standards can be formulated to guide the compilation, reporting, and use of geographic metadata? What is the role of current and proposed standards, such as the Content Standards for Spatial Metadata being considered by the Federal Geographic Data Committee and the Spatial Data Transfer Standard (FIPS 173)? What knowledge may be drawn from the experiences of the remote sensing community in cataloging and developing access tools for spatial data?
7. Given the purported need, why is there currently so little metadata in the GIS field? Will the existence of standards serve as a sufficient incentive for the widespread creation and maintenance of metadata? What additional impediments are present or incentives are lacking that may be hampering metadata development?
8. To what degree will existing and proposed electronic networks be able to support a metadata regime and the transfer of geographic datasets? What practical constraints are likely to be imposed by network limitations?
9. Under what mix of metadata and infrastructure arrangements is the greatest long-term use and sharing of geographic data likely to occur?

V. INFRASTRUCTURES FOR SHARING GEOGRAPHIC DATA

A. General Issues

Effective sharing of geographic data on a widespread basis requires that individuals and organizations must be able to determine whether desired data has already been collected by others and whether that data is fit for their proposed use. The ability to access and acquire the data efficiently must also be provided. A spatial data infrastructure might be defined as the distribution system that enables effective access to digital geographic data, the coordinating and control structures that develop and maintain data sets, and the geographic data sets accessible through the system.

1. What are the reasonable alternative means of creating and institutionalizing a national spatial data infrastructure? What is the overarching goal or vision among the numerous competing goals that such alternatives should be judged against?
2. What form of nationwide spatial data infrastructure is likely to be most conducive to improved sharing, access, and use? To what extent is it desirable to embed it in the nation's overall current and future information infrastructure?
3. How is geographic data different from other types of data in respect to its handling over networked systems? To what extent should it be differentiated from other electronic information or integrated with other electronic information?
4. In order to optimize widespread use and sharing of geographic data, should any future spatial data infrastructure be viewed as a single robust integrated system, or should it be viewed as consisting of separate major database components and control mechanisms suitable for different purposes with not all systems necessarily integrated with others?
5. To what extent should a national spatial data infrastructure be centrally controlled? How might a spatial data infrastructure that has minimal central control be sustained? What are the reasonable alternative ways to implement federated spatial databases?
6. How might the behavior of individuals and organizations in networked environments affect the implementation of spatial data infrastructure strategies? (For instance, on the current Internet there exist many "volunteered services" such as software maintenance of Archie, Gopher, and WAIS [Wide Area Information Server], maintenance of list servers and news lists, and dedication of hardware for accomplishing such services.)

B. Specific Issues

1. For each potential infrastructure framework, what would be the range of likely implementation strategies? Can staged implementation plans be designed with key milestones and strategies for moving from each stage to the next?
2. What are the optimal roles that various institutions should play in developing a national spatial data infrastructure of practical use for a wide range of applications?

3. What are the key fundamental elements of a useful spatial data infrastructure at the global, national, state, and local levels?
4. How does the current infrastructure work, and what are the elements already in place, at the global, national, state, and local levels?
5. Is there a need for core data that would be supplied and maintained by a national general-usage spatial data infrastructure? If so, what data should be considered "core"? What are the alternate mechanisms that might be employed in making such a decision?
6. Is it possible to derive a minimum list of spatial functional procedures that, when combined with the substantive information peculiar to any specific application, will provide the decision support information essential to the needs of most users of that substantive information? If so, can this set of functional procedures be extrapolated to meet the newer areas of spatial decision support expected to emerge in the next ten years?
7. What is the likely profile of uses and users of the spatial data infrastructure of ten years hence?

Spatial information has a wider range of potential users and a longer life span than most other kinds of information. Its exchange, however, is still too often confined to a single organization or to even smaller units. The demand for efficient, equitable, and timely access to spatial data by the user community will continue to grow.

As stated at the outset of this book, sharing of geographic information is important because the more it is shared, the more it is used, and the greater becomes society's ability to evaluate and address the wide range of pressing problems to which such information may be applied. It is our hope that the issues raised and the research directions suggested in this book will spur interest on the part of academics and practitioners to engage in those activities that will provide increased understanding of problems and approaches in successfully sharing geographic information.

REFERENCE

Onsrud, H. J., and G. Rushton. 1992. Initiative 9. Report for the specialist meeting: institutions sharing geographic information. Technical Paper 92–5. Santa Barbara, CA: National Center for Geographic Information and Analysis.

Glossary

A

AACR2	*Anglo-American Cataloging Rules,* 2d ed.
AASHTO	American Association of State Highway and Transportation Officials
ACG	Address Coding Guide
AM	automated mapping
AM/FM	automated mapping/facilities management
AMS	Address Management System
APSRS	Aerial Photographic Survey Record System
ARJIS	Automated Regional Justice Information System
ASPRS	American Society for Photogrammetry and Remote Sensing
ATIS	Advanced Traveler Information System
ATMS	Advanced Traffic Management System
AVL	Automated Vehicle Location
AVM/C	Advanced Vehicle Monitoring and Communication System

B

BG	block group
BNA	block numbering area
BRA	Boston Redevelopment Authority

C

CAD	computer-aided design or computer aided dispatch
CADD	computer-aided drafting and design
CAGIS	Cincinnati Area Geographic Information System
CCC	collaboration to cooperation to coordination
CD	compact disc
CES	Committee on Earth Sciences
CETA	Comprehensive Employment and Training Act
CHIPS	Comprehensive Highway Information and Planning System
CLISC	County-wide Land Information Systems Committee
CLUIS	Centralized Land Use Information System
CMCC	computer mapping coordinating committee

CNDDB	California Natural Diversity Data Base
CUE	Correction; Update; Extension

D

DaneLINC	Dane County Land Information Consortium
DBMS	database management system
DCW	Digital Chart of the World
DEM	digital elevation model
DHHS	Department of Health and Human Services
DIME	Dual Independent Map Encoding
DLG	Digital Line Graph
DOQ	digital orthophoto quadrangle
DOT	department of transportation
DXF	digital exchange format

E

EDC	Environmental Data Center
EOEA	(Massachusetts) Executive Office of Environmental Affairs
EOS	Earth Observing System
ESRI	Environmental Systems Research Institute

F

FCCSET	Federal Coordinating Council for Science, Engineering, and Technology
FEMA	Federal Emergency Management Agency
FGDC	Federal Geographic Data Committee
FHLBB	Federal Home Loan Bank Board
FHWA	Federal Highway Administration
FICCDC	Federal Interagency Coordinating Committee on Digital Cartography
FIPS	Federal Information Processing System
FM	facilities management
FOIA	Freedom of Information Act

G

GAO	General Accounting Office
GBF	Geographic Base File
GICC	(North Carolina) Geographic Information Coordinating Council
GIS	Geographic Information System(s)
GIS-T	GIS for transportation
GMDNCC	(Florida) Growth Management Data Network Coordinating Council

GNIS	Geographic Name Information System
GPS	Global Positioning System

H

HUD	U.S. Department of Housing and Urban Development

I

IADB	Inter-Americas Development Bank
ICPSR	Inter-University Consortium for Political and Social Research
IHIS	Integrated Highway Information System
IHIS	Intelligent Highway Information System
IMAGIS	Indianapolis Metropolitan Area Geographic Information System
IOR	inter-organizational relations
ISTEA	Intermodal Surface Transportation Efficiency Act
IVHS	Intelligent Vehicle Highway System
IT	information technology
IVHS	Intelligent Vehicle Highway System

L

LANDNET	Land Network
Landsat	formerly ERTS—Earth Resources Technology Satellite program; renamed Landsat by NASA in 1975
LC	Library of Congress
LIS	land information system

M

MA	metropolitan area
MARC	Machine-Readable Cataloging
MIS	management information system
MOU	Memorandum of Understanding
MPA	Massachusetts Port Authority
MSC	Mapping Science Committee

N

NACIS	North American Cartographic Information Society
NAS	National Academy of Science
NASA	National Aeronautical and Space Administration
NCES	National Center for Education Statistics
NCGIA	National Center for Geographic Information and Analysis
NCHRP	National Cooperative Highway Research Program
NFS	network file systems
NGS	National Geodetic Survey
NHS	National Highway System
NIST	National Institute of Standards and Technology

NOAA	National Oceanic and Atmospheric Administration
NRC	National Research Council
NREN	National Research and Education Network
NSDI	National spatial data infrastructure
NSF	National Science Foundation

O

OB	organizational behavior
OCLC	on-line computer library center
OMB	Office of Management and Budget
OT	organizational theory

Q

QA	quality assurance
QC	quality control

R

RLIN	research libraries information network
ROM	read-only memory
RUIS	Regional Urban Information System

S

SACS	Street Address Conversion System
SCSD	street centerline spatial database
SDDB	school district data book
SDDPC	San Diego Data Processing Corporation
SDG&E	San Diego Gas & Electric
SDTS	Spatial Data Transfer Standard
SMAC	State Mapping Advisory Committee
SQL	structured query language

T

TAC	technical advisory committee
TIGER	Topologically Integrated Geographic Encoding and Referencing
TMS	transportation management system
TNC	The Nature Conservancy
TVA	Tennessee Valley Authority

U

UA	urbanized area
UCAR	University Corporation for Atmospheric Research
UIS	urban information system
UNAVCO	University Navstar Consortium
URISA	Urban and Regional Information Systems Association
USDA	United States Department of Agriculture
USGS	United States Geological Survey
USPS	United States Postal Service
UST	underground storage tank

W

WAIS	Wide Area Information Server
WLIA	Wisconsin Land Information Association
WLIB	Wisconsin Land Information Board
WLIP	Wisconsin Land Information Program

Contributors

Roemer M. Alfelor
The Urban Institute
2100 M Street, NW
Washington, DC 20037

Bijan Azad
196 Green Street
Jamaica Plain, MA 02130

William J. Bamberger
Regional Urban Information System
5975 Santa Fe Street, MS 51
San Diego, CA 92109

John D. Bossler
Center for Mapping
The Ohio State University
Columbus, OH 43212

Hugh W. Calkins
NCGIA/Dept. of Geography
Box 610023
SUNY—Buffalo
Buffalo, NY 14261-0023

Heather Campbell
Department of Town and
Regional Planning
University of Sheffield
Sheffield S10 2TN
United Kingdom

Donald F. Cooke
Geographic Data
Technology, Inc.
13 Dartmouth College Highway
Lyme, NH 03768-9713

William J. Craig
Urban and Regional Affairs
University of Minnesota
330 Humphrey Center
Minneapolis, MN 55455

Jack Dangermond
Environmental Systems Research
Institute (ESRI)
380 New York Street
Redlands, CA 92373

Kenneth J. Dueker
Center for Urban Studies
Portland State University
P.O. Box 751
Portland, OR 97207

Earl F. Epstein
School of Natural Resources
The Ohio State University
320-E Kottman Hall
Columbus, OH 43210-1085

John Evans
Computer Resources, Rm. 9-547
Massachusetts Institute
of Technology
77 Massachusetts Avenue
Cambridge, MA 02139

Joseph Ferreira, Jr.
Computer Resource Laboratories
Room 9-516
Massachusetts Institute
of Technology
77 Massachusetts Avenue
Cambridge, MA 02139

Steven Frank
Department of Surveying
New Mexico State University
Department 3SUR
205 Goddard Hall
Las Cruces, New Mexico 88003

Doyle Frederick
Federal Geographic Data Committee
U.S. Geological Survey
590 National Center
Reston, VA 22092

Michael F. Goodchild
NCGIA/Department of Geography
3510 Phelps Hall
University of California
Santa Barbara, CA 93106-4060

Michael J. Kevany
PlanGraphics
1300 Spring Street
Silver Spring, MD 20910

John Leslie King
Department of Information
and Computer Science
University of California
Irvine, CA 92717

Ian Masser
Department of Town and
Regional Planning
University of Sheffield
Sheffield S10 2TN
United Kingdom

Patrick McGlamery
Homer Babbidge Library
University of Connecticut
Storrs, CT 06268

Paul H. Meredith
Department of Management
and Quantitative Analysis
University of Southwestern
Louisiana
Lafayette, LA 70504-3570

Nancy J. Obermeyer
Department of Geography
and Geology
Indiana State University
Terre Haute, IN 47809

Harlan J. Onsrud
NCGIA
University of Maine
5711 Boardman Hall
Orono, ME 04469-5711

Jeffrey K. Pinto
Penn State University at Erie
School of Business
Station Road
Erie, Pennsylvania 16563

Bruce Rocheleau
Department of Political Science
Northern Illinois University
DeKalb, IL 60115-2854

Gerard Rushton
Department of Geography
The University of Iowa
Iowa City, IA 52242-1316

Jonathan Sperling
Geography Division
U.S. Bureau of the Census
Washington, DC 20233

David Stage
State of Florida
Growth Management Council
Executive Office of the Governor
Tallahassee, FL 32399-0001

Richard P. Taupier
Office of Geographic Information
and Analysis
University of Massachusetts
at Amherst
Blaisdell House
Box 30820
Amherst, MA 01003

Nancy Tosta
U.S. Geological Survey
National Mapping Division
590 National Center
Reston, VA 22092

Stephen J. Ventura
Institute for Environmental Studies
and Department of Soil Science
University of Wisconsin—Madison
439 King Hall
Madison, WI 53706

Ric Vrana
Center for Urban Studies
Portland State University
P.O. Box 751
Portland, OR 97207

Richard Weatherbe
NCGIA/Dept. of Geography
Box 610023
SUNY—Buffalo
Buffalo, NY 14261-0023

Lyna L. Wiggins
Center for Urban Policy Research
Rutgers University
P.O. Box 489
Piscataway, NJ 08855-0489